Statistics: Concepts and Applications for Science

David C. LeBlanc

Ball State University

2013 Edition

DLK Publishing

PO Box 434   Edgewater, FL 32132
(386) 424-1472   www.DLKPublishing.com

Published by DLK Publishing
PO Box 434 Edgewater, FL 32132
United States of America
Printed in USA

ISBN: 978-0-9885144-1-6

Contents

Preface for Instructors

Intended Audience for this Book

This book was written for college students majoring in the sciences. However, many examples and homework problems are based on studies reported in the popular media (newspapers and TV) and may be of interest to a more general audience. I have used working drafts of this book for several years in an introductory biostatistics course that includes both undergraduate and graduate students in life and health sciences. This text may also be appropriate for introductory statistics courses in medical schools.

This book does not require that students have mathematical skills beyond high school algebra. I find it is more important that students have general experience in representing concepts with numbers and translating word problems into calculations. Most students majoring in science develop these skills in their physics, chemistry, and biology courses.

Philosophy of this Book

This book is designed to show students how statistical reasoning and procedures are used in the practice of science. For 20 years, I have been a practicing scientist in a field that makes extensive use of statistical analyses. I have published in the top journal in my field, and have also served as one of its associate editors. Based on these experiences, I wrote this book to introduce science students to statistical concepts they must understand to be successful practicing scientists. To allow adequate time to cover the applications of statistics in science, I have reduced the coverage of the mathematical aspects of statistics to the bare essentials. Scientists rarely need to mathematically derive statistical procedures. Rather, a practicing scientist must be able to formulate valid research designs, choose and implement appropriate graphical and statistical analyses, correctly interpret results from these procedures, and explain their results in terms that are understandable to a general audience. This is a tall order for a single-semester course, and anything that does not contribute toward these objectives has been omitted from this book.

The pedagogical philosophy of this book is based on the "Teaching for Understanding" paradigm described by a number of authors in *Educational Leadership* Volume 51, Issue 5. The core precepts of this approach to teaching are:

1. Focus on a small number of critical concepts and issues.
2. Explore these concepts in a wide variety of contexts and applications.

3. Clearly communicate to the students what they must do to demonstrate that they understand these core concepts and issues. Ideally, students should be exposed to examples that illustrate different levels of understanding, from novice to journeyman to expert.

4. Give students many opportunities to practice "performances of understanding," including both risk-free formative assessment (guidance and suggestions for improvement) and summative assessment (to assign grades).

In this context, "understanding" can be defined as the ability to correctly explain and apply critical concepts and issues in novel contexts. Following the "Authentic Assessment" paradigm, I have developed performances of understanding that require the student to apply what they have learned in situations that are similar to the real-world practice of science. Many of these assignments and assessments require that the student explain the results of their analyses in context, using terms that are understandable to the general public. Many teachers have experienced "I never really understood this stuff until I had to teach it." I believe that if a student can explain the application of core statistics concepts in the context of a scientific study without resorting to memorized jargon, they are well on their way to understanding statistics.

Because a relatively small number of core concepts are the foundation of virtually all statistical applications used by scientists, I believe that an introductory statistics course provides an ideal context in which to apply the Teaching for Understanding paradigm. A single core concept is the foundation of all statistical analyses: the sampling distribution of a sample statistic. All applications of statistics in the practice of science are based on this concept, including confidence intervals, tests of significance, p-values, and power. I focus on developing student understanding of sampling distributions in the first half of this book. The approach is heuristic and conceptual, using empirical simulations instead of mathematical derivations. In the second half of the book, I repeatedly revisit core concepts and applications for a variety of statistical procedures, reinforcing understanding through repetition in novel contexts. For each procedure, I describe the connections between the sampling distribution and study design (including bias and precision), confidence intervals, p-values, and power. By focusing on this short list of essential concepts, students are provided ample time to practice the application of these concepts in a wide variety of contexts. Through repetition and practice, students learn how to correctly apply, interpret, and explain statistical reasoning and procedures in novel research situations. This is a widely recognized indicator that students understand what they are doing rather than simply implementing plug-and-chug numerical recipes and reciting semimagical incantations about "statistical significance" and "confidence" learned by rote memorization.

Organization and Content

The first seven chapters in Section 1 of this text introduce the fundamental concepts that must be mastered if the students are to understand how and why statistical analysis of data is a necessary aspect of most scientific research. The presentation of these fundamental concepts is organized as follows:

- Chapter 1: All scientific knowledge is based on observations of the real world (data). The methods used to obtain these data have a major influence on the usefulness of the data for describing the real world.

- Chapter 2: It is usually not possible to interpret the raw data values produced by scientific research. The raw data must be summarized to facilitate understanding. There are a variety of graphical and statistical procedures that can be used to summarize data. Which of these procedures is most appropriate will depend on the nature of the data and the intended use of the summary.

- Chapters 3–5: The results of scientific research are always influenced to some extent by random variation. This introduces uncertainty into conclusions based on these results. Probability is a measure of uncertainty that enables us to understand random variation and account for this in our estimates and conclusions.

- Chapters 6–7: Data from scientific research always represent "uncertain evidence" about the true nature of the real world. A primary role of statistical analysis is to assign probabilities to observed outcomes of scientific studies if specific hypotheses about the nature of the real world are true. These probability statements enable scientists to draw informed conclusions about the nature of the real world based on data from scientific studies, while also recognizing the degree of uncertainty associated with their conclusions. This same concept of "uncertain evidence" also requires that scientists evaluate the quality of their estimates of real world characteristics that are based on data. All such probability-based assessments of conclusions and estimates are based on a single core concept: the sampling distribution of a statistic.

The chapters in Section 2 of this book describe how the fundamental concepts described in Section 1 are applied in a variety of scientific research situations that require a variety of statistical analyses, including:

- Chapters 8–10: One- and two-sample tests for proportions, means and medians, and variances.

- Chapter 11: Tests for comparing more than two means or medians, including one-way analysis of variance, multiple comparisons tests, planned contrasts, Kruskal-Wallis test, and the Bonferroni procedure.

- Chapter 12: χ^2 tests for goodness of fit and homogeneity of proportions.

- Chapter 13: Tests for association between quantitative variables, including correlation and regression.

- Chapter 14: χ^2 test of independence (association) between categorical variables.

Each statistical procedure is presented in a standardized format, including Applications, Assumptions, Hypotheses, Test Statistics, Sampling Distribution, p-Value/Conclusions, and Confidence Interval. Core concepts are fully explained in the context of each procedure. While this approach is repetitious, it provides the student with opportunities to see commonalities among a wide diversity of statistical procedures and to practice the application of core concepts in novel contexts. This repetition also allows the instructor to pick and choose among the many statistical procedures described in Section 2 without skipping important concepts.

Ancillaries

This book is intended to be used in conjunction with an accompanying student workbook that contains homework problems, exercises, and study problems that have fully explained answer keys. I firmly believe that students learn statistics better by DOING statistics, rather than by simply listening to a professor talk about statistics. The workbook is designed to provide for students ample opportunity to practice skills described in the text. The workbook is designed so that the students write their answers on the same page as the question. This juxtaposition of questions and answers allows the student to better use their homework problems to learn from their mistakes and study for examinations. In my own introductory biostatistics course, much class time is focused on going over homework problems, identifying student difficulties, and resolving them. Students get immediate feedback on what they do and do not understand. Corrections written on the workbook pages provide "red flags" that can help students better focus their study efforts. The homework questions require that the student regularly apply important fundamental concepts and practice critical skills. Learning to use statistical procedures and reasoning in the practice of science is difficult for most students. To succeed, students need lots of practice and feedback. The workbook is designed to facilitate this process.

The workbook is produced as perforated, 3-hole punched pages so that students can submit their homework to the instructor for evaluation, and then reassemble their work into an organized portfolio. I use homework problems as formative assessments, giving students credit for completing the assignment, but not penalizing their errors. This is consistent with the philosophy that students should be given many opportunities to practice "performances of understanding." I use the more involved Exercises at the end of each chapter as summative assessments, to assign grades based on the students ability to perform and explain statistical analyses.

Statistics: Concepts and Applications for Science is supported by a website at **http://bioscience.jbpub.com/statistics/**. The website will provide resources to support both the student and the instructor, including electronic versions of data files used in homework problems and exercises that allow the student to copy and paste the data into most statistical software products, thus avoiding tedious data entry while still allowing for realistic data analysis experiences. The secure instructor website area includes extensive materials that make the instructor's job easier, including:

- Fully elaborated answer keys to all workbook problems and exercises, in large-print, Adobe Acrobat PDF format to facilitate both printing to overhead transparencies and direct presentation in classrooms with computer projection systems.

- A test bank of questions for each chapter, including take-home problems with multiple versions that allow for "authentic assessment" of the students' ability to actually perform and interpret statistical analyses.

- Pedagogical guides that describe how recent innovations in "best practices" in teaching can be applied to the teaching of statistics to science students, including Teaching for Understanding and Authentic assessment.

Some Words About Software

In the twenty-first century, scientists have ready access to powerful computers and do not need to create graphics or perform statistical calculations by hand. Hence, I wrote this book under the assumption that students will have access to computer software that produces graphics and performs a wide range of statistical analyses. However, there are many different products on the market that perform these functions, and it is not practical to write a separate book to display examples for the use of each product. The examples of graphics and statistical print-outs are purposefully simple and general, with a minimum of features specific to any one software product. I have two reasons for taking this approach: (1) Most graphs and statistical print-outs are sufficiently similar across various software products that it really doesn't matter which one is used, and (2) when students go out into the real world, they will have to learn to adapt to the software provided by their employers. The student can begin to learn this skill by adapting the general examples provided in this book to the specific statistics software they use for their statistics course.

I fully understand that most instructors of statistics courses will have their students learn to use only one statistics software product. The students will need detailed instructions for performing data management, graphical analyses, and statistical procedures for whatever software they use. An additional feature of the website for this book will be a set of tutorials for using various statistics software products to perform the homework problems and computer exercises in the student workbook. These tutorials will be made available for downloading by students and instructors via a website maintained by the publisher, Jones and Bartlett. Initially, tutorials for Data Desk (part of ActivStats), SYSTAT, and Minitab will be provided. Tutorials for other software products may be provided later, in response to instructor demand.

Preface for Students

The discipline of "statistics" can be described as the art and science of using quantitative information (data) to gain understanding and to make informed decisions. Statistics includes such activities as evaluating the quality of data, designing procedures for producing data, using graphs and numerical summaries to condense large volumes of raw data into forms that facilitate understanding, and protocols for using data as evidence to test competing hypotheses and theories about how the world works. Throughout most of history, as humans have practiced the process we call "science," the primary obstacle to advancement has been insufficient information. However, with recent technological advances in data acquisition, large national research projects, and worldwide computerized information distribution via the Internet, the modern practicing scientist often faces the problem of TOO MUCH information. This huge volume of information presents new challenges of data retrieval, organization, analysis, integration, and presentation to the ultimate users of scientific knowledge. In an earlier era, an illiterate person suffered from a lack of access to the wealth of knowledge contained in books. In the Information Age, an individual who is statistically illiterate will suffer from an inability to use the wealth of available information to better understand his or her world.

While many students come to a statistics course believing it will be irrelevant and boring, I believe that a good statistics course is the single most important course a science student can take. Statistics is POWER! The average college student is an uncritical consumer of scientific knowledge. Like a typical first-time car buyer, they know little about the product they are "buying" when reading a scientific paper or book, and they are willing to accept the word of the seller (author) with regard to the quality of the product (information). While the vast majority of scientific publications use a rigorous review process to ensure quality in their publications, most scientists know of published papers that should never have appeared in print. An understanding of statistical concepts is an important part of becoming a discriminating consumer of the scientific literature. An even more important aspect of the power that statistics brings to students is the important role that statistical thinking plays in DOING science. An understanding of statistical concepts is critical to collecting samples, doing experiments, and analyzing data. Hence, students who want to make the transition from "consumer" of knowledge to "producer" of knowledge must understand the fundamental concepts of statistics. While many college science courses deal implicitly with the logic of the scientific method, a good applied statistics course explicitly describes the logical thought processes involved in using data to draw conclusions about the world around us.

The Role of Statistics In the Scientific Method (Why You Need To Take a Statistics Course, Even If You Don't Want To)

While there is no single method by which all scientists develop new knowledge, one process that is often called "THE" scientific method has the following steps: Observation → Question → Hypothesis (Informed guess at the answer) → Prediction (expectation of future events IF the hypothesis is correct) → Critical Test (to determine if the prediction actually comes true) → Re-evaluation of the hypothesis in light of the outcome of the critical test. See Figure A for a more detailed diagram of this process. A scientific hypothesis MUST be able to produce a prediction that is "testable" by a critical test based on observation and(or) experimentation. If the prediction is not supported by the data from the critical test, the hypothesis is "falsified." That is, the hypothesis must be partially or totally incorrect because in did not explain (predict) the observations from the test. In this case, the hypothesis must either be modified (in light of the observations from the critical test) or completely rejected. If the predicted outcome of the critical test does occur, this is taken as evidence that *supports* the hypothesis. However, this outcome does not *prove* the hypothesis is true. There may be other explanations that could be equally supported by available information, but you simply haven't yet thought of them. In science there is no certain PROOF, only hypotheses and theories that are well supported by the outcomes of many studies.

Many scientific questions that develop from our observations of the universe tend to encompass a number of interacting entities and processes. Often, to address these general scientific questions we must formulate a number of more specific questions for each of these components of the phenomenon of interest. We must also recognize that there are many possible hypotheses, and that none, one, or more than one of these hypotheses may be partially or completely correct. Formulating specific questions, hypotheses, and critical tests involves not only knowledge and experience, but also imagination and inspiration (complete with the little lightbulb above the head). For example, if we wanted to know why the population of an endangered species is declining, we might propose the following hypotheses: (1) The mortality rate in this population is increasing, (2) the birth rate in this population is decreasing, (3) the migration rate into or out from this population is changing, and that these changes are caused by (4) habitat degradation (e.g. pollution, climate change), (5) increased hunting, (6) competition from another species, or (7) disease outbreak. Each of these hypotheses is sufficiently specific so that we could design a study to collect the data necessary for a critical test. Based on each hypothesis, we could state a prediction that could be tested by observation or experimentation. Our prediction would state the types of observations or measurements that should be made and, as precisely as possible, what we expect to observe from these data if the hypothesis is correct. Only after we have tested these, and perhaps other hypotheses, could we arrive at an informed, "best available answer" about why the population is declining.

The discipline of statistics provides important tools for implementing the scientific method, including: (1) methods for obtaining observations and measurements that will provide a "fair" (unbiased) test of a hypothesis, (2) methods for summarizing large amounts of quantitative information so that the investigator can better understand the outcome of the critical test, and (3) methods for evaluating data as evidence for or against a specified hypothesis. While the scientific method does not always require statistical analysis, the concepts and methods of statistics are critical tools in many areas of scientific inquiry.

FIGURE A **The Role of Statistics in the Scientific Method.** This diagram summarizes what is called the Hypothetico-Deductive Method, as described by Carl Popper. The beginning of this process, in the upper-left corner of the diagram, occurs when an observation results in a question. A common scenario involves an observation that is not consistent with some prior concept of how the world operates. Sometimes these prior concepts about how the world works are formal statements from a theory or paradigm, based on accumulated prior observation and experience. The inconsistency between new observations and prior understanding results in an "informed question." Potential answers to this question are framed as hypotheses (educated guesses). Based on many considerations (e.g., nature of the question, predisposition and training of the scientist, availability of equipment, financial support, time/effort required), the scientist chooses the format for a critical test (e.g., field study or lab experiment) and the variables that will be measured. Once these choices have been made, the scientific hypotheses are restated as statistical hypotheses that predict specific outcomes of the critical test in terms of specific values that will be computed based on data obtained from the test. Statistical theory then guides the investigator in making choices about how to obtain data for the critical test. Statistical analyses are used to evaluate the strength of the evidence supporting conclusions regarding whether or not a predicted outcome actually occurred. Based on this evaluation, conclusions are drawn regarding the validity of the scientific hypotheses, and ultimately, the scientific theory or paradigm. The use of statistical understanding and analysis in the practice of science occurs in those steps to the right and below the diagonal line in the diagram.

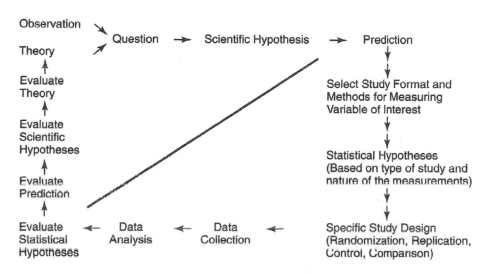

Acknowledgments

I was a Ph.D. research scientist when a colleague showed me *Introduction to the Practice of Statistics* by David Moore and George McCabe. After reading their description of sampling distributions I had an epiphany along the lines of, "So THAT'S what all this statistics stuff means!" That moment of blinding insight was the beginning of a journey that has resulted in this book. I wondered why I had not learned this concept in the many statistics courses I had taken during my college career. Looking at my old textbooks, I discovered that this critical concept was either missing, or buried in an obscure description in a single chapter. Like many recent converts, I needed to share my newfound insight with anyone who would listen. I thank Moore and McCabe for giving me this powerful insight, and the authors of all my old statistics textbooks for showing me the job that needed to be done.

I owe a great debt to the students who have taken my biostatistics course during the years I worked on this book. Their questions and editorial feedback contributed greatly to improving the clarity and organization of this text.

I thank the administration of Ball State University for recognizing that scholarship comes in many forms, and is not measured only in grants received and papers published. The institutional commitment to high quality teaching provided an environment that has allowed me to focus my efforts on this long-term project to improve statistics education for science students.

Writing a textbook is a time-consuming exercise that requires total commitment. Without the loving support of my wife Cheryl, my life would have fallen apart while I was engrossed in this effort. I also greatly appreciate the understanding of my children, Nathan and Maya, who continued to love Daddy even when he wasn't around as much as they would have liked.

SECTION 1
Fundamental Concepts

1 What Is "Data" and Where Does It Come From?

THE GOAL OF SCIENCE The goal of science is to understand the true nature of our universe. However, even the narrowest, most specific scientific question usually pertains to a group of individuals or objects that is too large to be studied in its entirety (called the *population of interest*). Even though scientists may be interested in understanding these large groups, they usually have time and resources to actually measure only a small subset of individuals from the population of interest (called a *sample*). Based on the knowledge gained from measurements on a sample, scientists make estimates about various characteristics of the larger population (such as the average blood lead concentration or the proportion of individuals in the population infected with a disease).

Because most science is based on information gained from samples taken from larger populations, it is important that those samples be representative. That is, we want our samples to be similar to the entire population. Information gained from a truly representative sample provides us with the best possible knowledge about the larger population that can be gained without measuring the whole. This chapter presents the fundamental concepts and procedures for obtaining representative data about populations that can be used to make accurate estimates and provide a sound basis for evaluating hypotheses about the population. This chapter can make you aware of important issues and thus give you an introduction to a much larger body of knowledge.

The process of "doing science" is often comprised of three parts: (1) collecting data so that the sample information is a useful representation of the real world, (2) summarizing data to make it easier to understand and use for describing the real world, and (3) using data to critically evaluate evidence for or against a specific hypothesis or theory. This chapter focuses on the first, and arguably most important, step in this process. Whether or not the data collected for a study are representative of some part of the real world is determined in large part by the methods used to obtain the data. No matter how much effort is expended obtaining data, it will be of little or no use unless the sample is representative of some larger population. How you obtain your data determines, and sometimes limits, what you can do with it. It is always a good idea to consult with an experienced scientist or applied statistician before undertaking a new scientific research project. Modest efforts spent to ensure that your data collection methods are appropriate to your needs and objectives can save you much grief and wasted effort later. ■

CHAPTER GOAL AND OBJECTIVES

In this chapter you will learn the fundamental procedures for producing data and the underlying concepts that determine appropriate study design.

By the end of the chapter, you will be able to:

1. Describe what constitutes data and the concepts of cases and variables.
2. Explain the relationship between sample statistics and population parameters, based on the concepts of random sampling variation, bias, precision, and accuracy.
3. Describe how to implement completely random and systematic random sampling to obtain sample data to describe a population, including using the processes of randomization and replication.
4. Describe how to implement a completely randomized experimental design, including the components of random assignment, replication, control, and comparison.
5. Describe the distinction between controlled vs. natural experiments and the advantages and disadvantages of these two types of study for testing hypotheses.
6. Identify the problem of pseudo-replication, based on descriptions of the population of interest and the sample unit (replicate), and describe appropriate ways to address this problem of pseudo-replication.

1.1
What Is Data?

Data can be defined as a set of observations and/or measurements made on some part of the universe to address a particular information need or question. Mere collections of disjointed, unrelated observations and measurements obtained for no particular purpose do not constitute data. Data usually have a specific structure, with observations/measurements for one or more variables made on a number of individuals from the population of interest. Data are usually organized in the manner described in Example 1.1. The term **variable** refers to any characteristic of an object or group of objects that can be represented with a number and that has more than one possible value. (For instance, height and weight are variables in Example 1.1.) The columns of **values** in Example 1.1 contain the actual numbers that each variable takes for the individuals measured. Each **case** of the data is marked in the lefthand column, showing in that row the measurements made on a single individual. *Note:* The word "data" is a plural term, indicating multiple observations or measurements. Each observation or measurement is a "datum." When using the word "data" as the subject in a sentence, the verb should be plural (e.g., the data were obtained from a representative sample).

■ EXAMPLE 1.1

The structure of data. The following is part of a collection of numerical and categorical information that could have been obtained for a study of obesity in university students. Each row defines a case of all measurements/observations made on a single individual. Each column contains the measurements/observations for a single variable.

		Variables			
Case	Name	Gender	Weight (lbs)	Height (ft)	% Body Fat
1	Tom	M	154	5.7	15
2	Dick	M	284	5.8	40
3	Harry	M	198	6.2	20
4	Jill	F	135	5.4	18
5	Betty	F	101	5.2	9
6	Sue	F	220	5.6	25

■

Many types of variables can be included in a collection of data. The nature of the variables helps determine the nature of statistical analyses that can be done using the data.

Quantitative variables have values that can be meaningfully manipulated with arithmetic operations. Some examples include temperature, % body fat, or the number of leaves on a plant. *Ratio-scale variables* are quantitative variables measured on a scale that has a constant increment between successive values and that has a true zero value (e.g., concentration and mass). All arithmetic operations can be validly applied to data values for this type of variable. *Ordinal- or rank-scale variables* have values that represent only the ranked order of the objects or individuals with regards to a variable. For example, the student with the highest grade point average is designated as rank 1 in the class, the student with second highest grade is designated as 2, the third highest grade is 3, etc. However, the actual differences between individuals of adjacent ranks can differ. For example, the students ranked 1 to 3 could have GPAs of 3.95, 3.94, 3.8. This characteristic of rank-type variables limits the arithmetic operations that can be validly applied. Quantitative variables can also be discrete or continuous. *Discrete variables* are often based on counting something, and can take only specific values (e.g., number of flowers on a plant, number of females in a population). *Continuous variables* are those that can take an infinite number of possible values, limited only by the number of decimal places to which the value can be precisely measured (e.g., blood alcohol concentration, pH, weight).

Categorical (qualitative) variables have values that indicate the individual belongs to a particular class or category, but these values cannot be manipulated with arithmetic operations (e.g., gender, blood type, color). Although categorical variables are not inherently represented by numbers, they are often analyzed in terms of the count or proportion of individuals in a group that fall into a specified category.

1.2
Sample Statistics and Population Parameters

The goal of scientific study designs is to obtain *valid data* for the variable(s) of interest, from individuals that constitute a *representative sample* from some larger population of interest. The entire group of objects or individuals about which information is desired is called the **population of interest** (or simply **population**). A **parameter** is a number that describes some characteristic of a specific population (e.g., the average weight of Antarctic blue whales, the proportion of HIV infected persons in the United States, the "intelligence" of babies born to crack-addicted mothers). Note that in each of these examples a variable of interest and a particular population of individuals to which the parameter value applies have been specified. The investigator must specify the variable(s) and population of interest as the first step in designing a scientific study to address a specific question. To know the value of a population parameter, *all* individuals in the specified population must be measured for the variable of interest. Although the investigator really wants to know the value of population parameter, this is usually not possible.

Most scientific understanding is based on data from samples of individuals selected from various populations of interest. A sample is a sub-set of individuals from the population of interest that is measured/observed to obtain data. A **sample unit** is an individual unit that comprises the population of interest and the sample, and each sample unit is measured for the variable(s) of interest. For example, if the population is comprised of people, the sample unit is often an individual person. Data values for the variable of interest, obtained from sample units, are used to compute a **sample statistic**, which is a number that describes the *sample* with regard to the variable of interest (e.g., the average height of a sample of 25 students). A fundamental purpose of scientific study design is to use sample statistics to estimate the values of population parameters. However, the value of a sample statistic rarely is exactly equal to the value of the population parameter for the variable of interest, even when the sample data are obtained by a valid study design. Sample statistics can be more or less accurate, but they should never be thought of as representing the one "true" value of the population parameter.

To better understand why sample statistics are only estimates of population parameters, consider the following thought experiment. Suppose each student in your statistics class was given the assignment to measure the height of individuals in a sample of 25 students attending your university and to use these data to compute an estimate of the average height for all students at the university. Further suppose that each of the students in your class used exactly the same procedures to obtain valid data for this variable and population of interest. There is only one true value for the parameter "average height of students at your university." Would you expect every student in your class to come back with exactly the same average height value, computed from their different samples of 25 students? Would you expect all these estimates derived from different samples of 25 students to be exactly equal to the true average height for all students at your university? You would probably answer no to both questions. Each sample of 25 students would include a different group of individuals, and so would produce different values for the sample statistic "average height." Some samples might include a tall basketball player, others not. Because each sample was obtained using the same valid procedure, all the different values for average height computed from these samples are equally valid estimates of the one true value for average height of students at your university. However, it is not possible to say that any one of these values for the sample statistic is correct (equal to the true parameter value).

Variation in the values of a sample statistic computed from different, independent samples taken from the same population is called **random sampling variation**. Random sampling variation is a consequence of the randomness of the process by which individuals are selected from the population to create a sample. This randomness is unavoidable as long as scientists use samples to understand larger populations. However, if the data are obtained by appropriate methods, the amount of sampling variation, and the consequent uncertainty associated with our estimates of population parameters, can be minimized. Under these conditions, we can be confident that the value of a sample statistic is a close approximation of the true value of the population parameter. The bottom line is that all knowledge based on information obtained from samples is more or less uncertain and approximate. This fundamental fact of life in the practice of science is the reason why statistics is employed as an integral part of scientific research. A primary purpose of statistical analysis is to allow scientists to estimate population parameters, evaluate hypotheses, and make decisions based on the uncertain information provided by sample data.

The relationship between sample statistics and population parameters is described based on the concepts of bias, precision, and accuracy. Good scientific study design provides sample statistics that are unbiased, precise, and accurate estimates of population parameters.

Bias

Bias is any systematic deviation of the values of sample statistics away from the population parameter value. When a biased study design is implemented, sample statistics computed from repeated, independent samples from the same population have a tendency to be greater than (or less than) the true population parameter value. When an unbiased study design is implemented, half the sample statistic values from multiple independent samples from the same population will fall above the true parameter value, and half will fall below. Hence, the average of these multiple, independently computed values for a sample statistic would be very close to the true parameter value.

Whether or not a study design is biased is determined in part by the method used to select individuals from the population. Suppose that statistics students collecting data to estimate average height of students at your university did *not* randomly choose who would be included in their sample. Rather, members of fraternities and sororities measured their friends (men tend to be taller than women), athletes measured fellow athletes (who might be larger than average), and so on. Estimates derived from such samples would *not* be representative of the larger population of interest, and therefore would provide biased estimates of the population parameter (average height of all students).

A second aspect of study design that determines whether or not sample statistics are biased estimates of population parameters is the choice of method used to measure the variable of interest. The investigator must first choose a measurable aspect of the subjects that best represents the variable of interest. This choice should be based on the concept of **measurement validity** (the idea that a measurement made on the study subjects accurately quantifies the phenomena or variable of interest). For example, if "size" is the variable of interest, do you measure length, volume, or mass? The answer to this question will often depend on the nature of the question or information needed. How best to measure an abstract variable such as "intelligence" presents an even greater challenge. In this case, you must first carefully define what you mean by intelligence and then devise a method of measurement that is equally applicable to all subjects in the population of interest. For example, using a test such as the Scholastic Aptitude Test (SAT) to quantify intelligence has been shown to be biased because it measures only a part of what constitutes intelligence. Determining the most appropriate way to measure a variable of interest is often based on widely accepted "standard methods" published in scientific literature and the state-of-science for measurement technology.

A third source of bias occurs when the implementation of an otherwise appropriate measurement method is *consistently* done incorrectly. For example, suppose that the students assigned the task of estimating the average height of students at their university were given a measuring tape with metric units on one side and English units on the other. Further suppose that some students used the metric side of the tape but recorded the data as English units (e.g., the height of a 2-meter-tall person was recorded as 2 yards). Obviously, their sample averages would be biased estimates of the average height of students at your university. Many measurement procedures in science are very complex and require training to ensure that the data values are not biased due to consistent measurement error.

Precision

The concept of **precision** refers to the amount of variation among the values of a sample statistic derived from repeated, independent samples from the same population. If a study design is repeatedly implemented on the same population and the values of a sample statistic derived from these repeated samples are very similar, the study design produces *precise* estimates of the population parameter. On the other hand, if the values of the sample statistic from these repeated samples vary widely, the study design provides *imprecise* estimates.

One source of variation among the values of a sample statistic derived from repeated samples from the same population is **random sampling variation**. This variation occurs because the repeated samples included different subsets of individuals who vary with regard to the variable of interest. Often, random sampling variation occurs because some samples, by chance, included unusual individuals while other samples did not. The effect of a few unusual observations on the value of the sample statistic depends on the **sample size** (how many individuals are in the sample). If the sample size is small, one unusual value can have a large effect on the value of the sample statistic. However, if the sample size is large, the effect of one unusual value is diluted (its effect diminished) by the many normal data values and has less influence on the value of the sample statistic. Hence, the primary means for reducing random sampling variation is to compute sample statistics based on a large sample size.

A second source of random variation is **measurement error**. This term refers to deviation of the value of a variable measured on an individual away from the true value for that individual. A key component of learning any measurement procedure is developing the ability to correctly perform the measurement exactly the same way every time you do it.

> **NOTE** Scientists rarely take repeated, independent samples from the same population. Typically, a scientist will take a single sample from the population and use the sample statistic from that one sample to estimate the population parameter. However, to understand the concept of precision you need to think, "What if I did my sampling study repeatedly?" (More about this in later chapters.) Because random sampling variation and measurement error cannot be completely eliminated, the value of a sample statistic will vary randomly among samples taken from the same population. Hence, you should never think that any sample statistic is exactly equal to the population parameter. However, by understanding the causes of this random variation, the investigator can design study methods to minimize this random variation and obtain better estimates of population parameters. ■

Accuracy

Accuracy refers to how close the value of a sample statistic comes to the true population parameter value. Accurate statistics must be both unbiased and precise. The value of a statistic based on data from an unbiased study is equally likely to be greater than or less than the true population parameter value. Hence, the average of values for a statistic derived by repeatedly performing an unbiased study on the same population will be close to the value of the population parameter. The values of a precise statistic derived by repeatedly sampling the same population tends to fall within a very narrow range. Taken together, the values of a precise sample statistic derived from an unbiased study design are very likely to fall close to the true parameter value. *Accuracy and precision are not the same thing,* even though these terms are often used interchangeably by persons who are not trained in statistics. Sample statistics can be *precise but not accurate* if an inappropriate study design or flawed measurement procedure is applied consistently to a large sample. For example, if the value of a parameter is 100, and 10 independent samples produce the following statistics: 98, 99, 100, 97, 98, 98, 97, 99, 99, 98, this would indicate the measurements are inaccurate (all but one value is below the true value of 100), but with good precision (values are very close together).

In most real-world situations, it is not possible to determine the accuracy of a sample statistic by direct comparison to the true population parameter value. The parameter value is rarely known, and if it were known, you wouldn't need to perform a sampling study to estimate it. Rather, the accuracy of sample statistics is evaluated based on the description of methods used to obtain the sample from the population and to measure the variable of interest. If all individuals in the population of interest have an equal chance to be selected for the sample *and* if the measurement procedure provides valid data for the variable of interest, the statistic will be an unbiased estimate of the parameter. If the sample size is sufficiently large, the statistic should be precise. If all these criteria are met, the statistic should be an accurate estimate of the parameter.

To summarize, sample statistics will be accurate estimates of population parameters if the study design and measurement procedure *minimize bias and maximize precision*.

- *Bias is minimized by:* (1) using an appropriate procedure for selecting study units/subjects so that the samples are representative of the larger population of interest, (2) selecting a measurement procedure that adequately quantifies the variable of interest, and (3) making sure that personnel are adequately trained to correctly implement the measurement procedure.

- *Precision is maximized by:* (1) collecting data on samples with a large number of subjects, and (2) training personnel to make measurements using a consistent methodology.

You should know not only how to correctly obtain data to make accurate estimates of population parameters, but also how to identify when someone else is presenting you with biased estimates. Television, radio, Internet, and print media bombard us with statistics purported to represent one population or another. Many of these statistics are derived from scientific study designs that meet the criteria for producing accurate estimates of population parameters, as described above. However, there are two types of so-called data commonly presented in popular media that virtually never provide accurate statistics.

Anecdotal data are obtained by the haphazard collection of individual cases that come to our attention because they are unusual. For example, John Doe had terminal cancer. He ate nothing but prunes and his cancer went into remission. Therefore, prunes are a cure for cancer. Because anecdotal data are not representative of larger populations, using such data for decision-making results in erroneous conclusions and inappropriate actions. This type of data is most commonly used by politicians, talk-show hosts, and the advertising industry, whose primary objective is not necessarily to determine the truth of a matter.

A less obvious source of biased statistics is the **voluntary (call-in) survey**, which leads to *voluntary response bias*. Even reputable news organizations use this very dubious method and report the results on national television. The general procedure is to pose a question and ask people to respond via the telephone or Internet. For example, television news programs routinely ask their viewers to register their opinions on controversial social issues such as gun control, abortion, or the death penalty, and then report the percentage of respondents in favor or opposition. The fundamental problem with the results of such call-in surveys is that the people who respond are typically not representative of the entire population. People who have strong negative opinions are often more likely to expend the effort to respond than people who have positive opinions. Television news organizations know this and usually caution viewers that the statistics they are reporting from such surveys are "not scientific" (meaning that they are biased). I often wonder why such reputable organizations spend expensive air time to report a statistic and then admit that it is biased. On the other hand, there is little mystery as to why some radio talk show hosts report statistics from call-in surveys without admitting the inherent bias in such data. Be wary of people who try to support their personal or political convictions with statistics.

> **NOTE** Because of ethical considerations, virtually all research involving human subjects is voluntary. The voluntary response bias described above is peculiar to opinion poll studies. Studies of weight-loss treatments, new drug trials, exercise regimes, and similar studies are also done using human volunteers. However, these experimental studies usually include methods to reduce the bias associated with volunteer samples. (More about this bias will be described later in this chapter.) ■

1.3
Sampling Study Design

General Principles

A **study design** is a description of the methods that the investigator will be using to acquire data. Before you begin a scientific study, you should spend time formulating your study design and have it reviewed by an experienced scientist. The specific characteristics of a study design depend on both the requirements for statistics to be accurate estimates of population parameters and the specific objectives of the study. Those objectives might typically be to estimate the value of a population parameter or to test a hypothesis about some aspect of how the population of interest works. Your choice of methods for collecting data is governed, in part, by which of these objectives you want to pursue. The types of study design described in Sections 1.3 to 1.5 of this chapter include sampling studies, controlled experiments, and natural experiments.

A **sampling study design** is generally used when the study objective is to estimate the value of a population parameter. A sampling study design describes the specific methods that will be used to select a representative sample from the population of interest (including specifying the sample size) and describes exactly how the variable of interest will be measured. Good sampling study designs stipulate sample size and measurement procedures to obtain the required level of precision for sample statistics while minimizing cost and effort. For example, to estimate the average wing span of a newly discovered species of beetle, a sampling study design might state the following: Traps will be used to collect a sample of 200 beetles, and the wing span of each will be determined by measuring the length of one fully extended wing and doubling this value. By measuring only one wing the researcher's effort per beetle is reduced, allowing time to measure a larger sample size of beetles. Given that wings must be equal size for stable flight, this time-saving protocol would probably not result in any significant bias or loss of information.

Because sampling studies use data from a sample to estimate some parameter of a larger population, the methods used to select the sample of individuals from the population of interest determine whether or not the sample statistic is an accurate estimate. The principles for obtaining a representative sample that will provide an unbiased and precise estimate of a population parameter are: (1) All individuals in the population should have an equal chance to be included in the sample, and (2) the sample should include a sufficient number of individuals in the sample to represent the range of variation that is present within the population. **Randomization** is a process for selecting individuals who will be included in a sample based on some random mechanism (dice, random numbers table, random numbers generator in a computer program). Whatever random mechanism is used, each individual in the population must have an equal opportunity to be selected for the sample. *The purpose of randomization is to minimize bias.* The values of a sample statistic computed from repeated, random samples from the same population are equally likely to fall above or below the true value of the population parameter. **Replication** refers to the number of individuals that are included in a sample. *The purpose of replication is to control sampling variability and increase precision* (Example 1.2). Increasing the number of individuals in a sample reduces variability in the values of sample statistics computed from repeated samples because variation among sample units "cancels out" within samples. Also, the chance inclusion of unusual individuals is diluted in a large sample, so that the unusual data value has little effect on the value of the sample statistic (Example 1.3). This begs the question, "How large a sample size is enough?" In general, the sample size should be as large as possible, given constraints of time and money available for the study. In later chapters I will describe methods to estimate the sample size needed to attain a desired level of precision.

■ **EXAMPLE 1.2**

Random sampling variation vs. sample size for sample proportions. Suppose the parameter of interest is the proportion of the student body at a particular university who are women. Further suppose that the true value for this parameter is 0.5 (or 50%). The following table presents the results of repeatedly sampling the student body by taking 1,000 independent samples of the indicated sample size.

Sample Size	Middle of 1,000 Sample Proportion Values		Range of 1,000 Values	
	Mean (%)	Median (%)	Minimum Value (%)	Maximum Value (%)
25	49.6	52	20	80
100	49.9	50	32	67
500	49.9	50	40	57

The sample statistic is the proportion expressed as the percentage of students in each sample who are women. If the sampling procedure is randomized and unbiased, the mean and median of the 1,000 sample proportion values should be very close to the true population proportion value (50%). (The mean and median are two measures of the "middle" of a set of numbers.) The range (minimum to maximum) of the 1,000 values for the sample proportion computed from the 1,000 independent repeated samples reflects random sampling variation. Note that for all the different sample sizes (ranging from 25 to 500), the "middle" (mean and median) of the 1,000 sample proportion values is very close to the true value of 50%. This indicates the samples are unbiased. As the sample size increases from 25 to 100 to 500, the range of values for the sample statistic "proportion of women" obtained from the 1,000 independent samples becomes smaller. This demonstrates that sample statistics computed from large samples exhibit less random sampling variation than samples computed from small samples. When a sample size of 25 students was used, the "middle" of the 1,000 values for the sample statistic was close to 50% (unbiased), but the range of values went from 20% to 80% (very imprecise). It is much easier to imagine that, by chance, a random sample of 25 students would include only 5 women than it is to imagine a random sample of 500 students would include only 100 women. This is the nature of the relationship between sample size and random sampling variation.   ■

■ EXAMPLE 1.3

Random sampling variation vs. sample size for sample averages. Suppose a random sample of 10 marathon runners has an average blood concentration of low density lipids (LDL or "bad" cholesterol) of 100 mg/dl. Further suppose that a runner was added to this sample who had a genetic predisposition for high LDL levels and a blood concentration of 400 mg/dl. The average blood cholesterol level for this sample of 11 runners would now be 127 mg/dl. Adding this one unusual individual to a sample of 10 changed the value of the sample average 27%. However, if this one unusual individual was added to a sample of 100 runners with an average LDL concentration of 100 mg/dl, the sample average for the 101 runners would change from 100 to 103 mg/dl (only a 3% change). Values of sample statistics computed from large samples are less affected by a few unusual individuals than the values of statistics computed from small samples.   ■

Types of Sampling Study Designs

Completely random sampling

Conceptually, this is the simplest procedure for selecting a representative sample from a population. For a sample to be completely random, all members of the population must have an equal chance to be selected for the sample. This requires that the investigator be able to identify each individual in the population of interest, a task that can be difficult for large populations. This identification is often accomplished in one of two ways:

1. *Random selection from a list.* To select a **random sample** of 100 students from the student body at a university with 20,000 students, you might obtain a list of all students and assign each student a five-digit random number from a random numbers table (e.g., Table 1 in the Appendix). Select a random starting point in the table and assign sequential random numbers (down a column or across a row) to each name in the list. Next, sort the list of names by ordering their corresponding random numbers from smallest to largest. This has the effect of putting the list of names in random order. Select the first 100 names on this randomly sorted list to obtain a completely random sample.

   For large databases, like the 20,000 name list of students, this process is best done using a computer database program. These programs typically have a function that generates random numbers. Use this function to produce a random number for each name in the list, then use the Sort tool to sort both the column of names and the column of random numbers based on the value of the random number.

2. *Random location of sample points.* In field sampling studies, it is not possible to list all individuals in the population of trees, flowers, or birds that is under study. An alternative to the random selection from a list is to identify random locations within the study area and then select the nearest individual at each location for the sample. Suppose you wanted to randomly locate 10 points within a woods to collect data regarding the size and species of trees. You might

start with a map of the woods and overlay a transparent sheet with a *x-y* grid on it. You would then use a random numbers table to select pairs of random numbers that represent *x*- and *y*-coordinates on the grid. Values from the random numbers table that fall outside the boundary of the woods are simply ignored. The points on *x-y* grid are used to randomly locate positions for sampling on the map of the woods. The map scale, geographic reference points in the woods (e.g., fence corners), and a compass can be used to locate the actual sample points in the woods in accordance with the random points on the map.

NOTE Haphazard selection or location of sample units is not the same as random sampling. Examples of haphazard selection include: (1) scanning the phone directory to "randomly" pick names and (2) walking around in the woods aimlessly to pick "random" locations. These selection processes are controlled by both conscious and unconscious processes. Hence, it is not possible for you to know whether or not the selection is truly random (all individuals or locations equally likely to be selected). Students in my ecology class who haphazardly locate study areas in the local woods always seem to "randomly" miss sampling in large areas with difficult shrubbery or dense poison ivy. I wonder if the source of this apparent bias in their sampling is conscious or unconscious. ■

Randomized systematic sampling

This type of selection is commonly used in field sampling when it is very difficult to travel to random points or when a relatively small number of sample points will be used to describe a relatively large area. With randomized systematic samples, a random starting point is chosen, and then sampling locations are located at fixed distance intervals or travel-time intervals proceeding away from this random point. This can significantly reduce the cost and effort of locating sampling points relative to completely random sampling. Also, this ensures that sample points are evenly distributed through the study area. In contrast, completely random sampling often results in some areas being heavily sampled while other areas are not sampled at all (see Figure 1.1). One problem with randomized systematic sampling is that bias can occur if a spatial pattern in the population coincides with the systematic sample grid. For example, in many mid-western states the county roads are organized on a square grid of north-south and east-west roads that are spaced one mile apart. If a systematic sampling design for estimating plant cover stipulated the sample points should be located at one mile intervals along lines that run north-south or east-west, and the first randomly located point was on a road, the resulting sample would indicate that there is no vegetative cover (the whole area is covered by asphalt). This would obviously be a biased estimate. Such problems are theoretically possible, but they are generally rare and avoidable if the investigator is paying attention. Randomized systematic sampling is commonly used by field biologists because it is easier to implement (therefore providing a larger sample size for the same effort) and because it produces even sampling from the area (population) of interest.

FIGURE 1.1   Random vs. Systematic Samples.   Part A shows the location of sample points in a study area located using an *x-y* grid and a random number table. Part B shows sample points located using a systematic grid. Note that the random location of points has left some areas undersampled and some areas oversampled.

Random Sampling

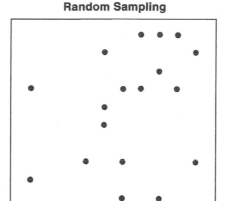

Part A

Systematic Sampling

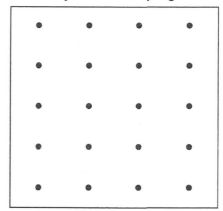

Part B

Advanced Topics

Stratified Sampling Design

Sampling study designs more complex than completely random or randomized systematic sampling are often used to make the logistics of sampling easier or to provide better information about complex or heterogeneous populations with less effort or cost. The brief description of stratified sampling below is intended only to introduce you to this subject area. You can often save yourself much effort and aggravation by consulting a statistician or a sampling design text to determine if one of the more complex sampling designs would be more efficient for your study.

Some populations are quite heterogeneous, often with distinctly different classes of individuals (e.g., males and females). In such cases, taking a simple random sample and computing an average may constitute "combining apples and oranges." **Stratified random sampling** involves identifying the various subpopulations (called **strata**; singular, **stratum**) and taking separate random samples from each. Usually, the number of individuals sampled from each stratum is proportional to its relative size within the overall population. (For example, if African-Americans constitute 13% of the U.S. population, then 13% of the overall number of individuals in a stratified sample should be African-Americans.) Separate sample statistic values are computed for each stratum. Computation of overall population descriptive statistics combines separate stratum statistics in a way that takes into account the relative size of each stratum.

■ **EXAMPLE**   The prevalence of HIV-infected individuals differs among various strata in the population of the United States (heterosexual, homosexual, Caucasian, African-American, Asian-American, IV drug users, hemophiliacs, etc.). A sampling study to determine the overall prevalence of HIV infection in a population could sample individuals from each of these strata in proportion to the percentage of the

U.S. population that falls into each category. A stratified sampling design would provide additional information about differences in HIV infection rate among these different categories of people, as well as a basis for computing more precise estimates of overall prevalence of HIV infection in the entire population. ■

EXAMPLE   A study of density (number per unit area) of breeding pairs of a bird species conducted in an area of mixed broadleaf and conifer forest might define strata by forest type. For many bird species, conifer trees provide quite different habitat quality (food availability, cover, nest sites) than hardwood trees. Hence, investigators might have reason to believe that density of breeding pairs is substantially different in these two forest types. If 70% of the study area was covered by broadleaf trees and 30% by conifer trees, the investigators would randomly locate 70% of their sampling points within the area covered by broadleaf trees and 30% in the area occupied by conifer forest. Average density of breeding pairs would be computed for each area separately. The investigator could then use a formula specific to stratified sampling to combine the statistics computed for each area to estimate the total number of breeding pairs in the study area. ■

1.4 Experimental Study Design

An **experiment** is defined as the deliberate imposition of a treatment by the investigator on a sample of subjects to evaluate the response of the subjects to the treatment. The primary purpose of experimental study designs is to determine *cause-effect relationships* between a "treatment" variable and a "response" variable. The **treatment (explanatory) variable** measures the condition(s) that the investigator imposes on the study subjects (e.g., temperature, nutrient concentration, drug dose). The **response variable** measures some characteristic of the study subjects that is hypothesized to change as a result of the treatment (e.g., pulse rate, growth, mortality, blood pressure). The treatment is the presumed "cause" and the response is the presumed "effect." Experiments are often done in controlled settings to minimize the possibility that nontreatment variables might influence the response of study subjects. The investigator randomly assigns individuals to the different treatment groups so that the groups are similar before the treatment is imposed. The objective of experimental design is to establish conditions such that there are only two possible explanations for why groups that received different treatments are different at the end of the experiment: (1) the treatment caused the difference or (2) the difference is due only to random sampling variation. This enables the most unambiguous determination of the treatment effect. If nontreatment factors are allowed to influence some study subjects, or if the treatment groups were different before the treatment was imposed, it is impossible to determine if the treatment was the "cause" of observed differences at the end of the experiment.

EXAMPLE   To determine the effect of ozone (an air pollutant) on soybean yield, investigators grew randomly selected groups of plants in two greenhouses. One greenhouse has ozone removed from the air by passing it through a charcoal filter; the other gets unfiltered air that contains ambient amounts of ozone. The difference in average bean yield per plant between the two groups is a measure of the effect of ozone, but only if all other conditions in the two greenhouses are held the same and if the soybean plants in the two chambers were similar at the beginning. ■

Basic Principles of Experimental Design

The term "**control**" in *controlled experiments* refers to restricting the variation of all extraneous (nontreatment) variables so that differences between the treatment groups can be clearly interpreted to be a result of the treatment. Specifically, all conditions other than the treatment are controlled such that all experimental subjects are subjected to exactly the same conditions, except for receiving different treatments. This often means conducting the study under the artificial conditions of a lab setting. However, experiments can be conducted in field settings if the investigator has knowledge that conditions are uniform across the area where the experiment will be conducted.

Good experimental design involves *comparison between groups* given different treatments. Such comparisons allow the investigator to distinguish responses to the treatment as distinct from responses to other factors that influence subjects or other unavoidable changes in the study subjects over time (e.g., maturation or learning). This comparison often includes a group of study subjects that received no treatment but are otherwise treated the same as the treatment groups. This group of subjects is often called the **control group** (not to be confused with experimental control of extraneous variables). However, it is not necessary that all experiments have a control group.

An essential component of good experimental design is the concept of **equivalent treatment groups**. This means that prior to the imposition of the treatment the various treatment groups were similar. Hence, any differences among the treatment groups that develop during the course of the experiment are more clearly attributable to something that happened during the experiment (e.g., the treatment). Randomization of assignment of study subjects to treatment groups is used to ensure that groups are equivalent before the treatment is imposed. Replication involves including many subjects in each treatment group. All individuals are different from each other. To make sure that treatment groups are similar *on average*, we must include many subjects to "average-out" individual differences. Adequate replication minimizes the possibility that differences between groups are due to random chance differences between individuals assigned to groups.

Components of Experimental Design

Some steps to consider in designing an experiment include the following:

1. *Define the basic experimental units (replicates).* The **experimental unit** is the individual that comprises the treatment groups and that is measured for the variable of interest. The experimental unit is also the basic unit of the population of interest. When experiments are done on animals (including humans) or plants, the organism is typically considered to be the experimental unit. Thus the definition of the experimental unit is based primarily on the nature of the question and how it defines the population of interest. For example,

   a. Does lowering blood pressure reduce the likelihood of a heart attack?
      Population: Patients with high blood pressure   Replicate: Patient

   b. Do SAT coaching seminars increase student test performance?
      Population: People who take the SAT test   Replicate: Student

   c. Do conflict resolution programs reduce physical aggression in schools?
      Population of interest: Schools in the U.S.   Replicate: School

   d. Does fertilizer increase growth of corn?
      Population of interest: All corn plants   Replicate: An individual plant

e. Does fertilizer increase corn yield per acre?

Population of interest: All corn fields in the U.S.   Replicate: A field or plot

When determining the experimental unit, the concept of "Independence" is an important consideration. That is, the value for the response variable measured on one experimental unit should not be influenced by the value measured on other units. For example, an investigator doing an experiment to assess the effects of fertilizer on plant growth could put several plants in each pot, rather than one plant per pot, to deal with a space limitation in a greenhouse. However, plants within a single pot compete with each other for light, so that the growth of one plant is not independent of the growth of its pot-mates. Because measurements on the multiple plants in each pot are not independent, the individual plants are not true replicates. In this case, the pot itself is the true, independent, replicate. Each plant in a pot would be measured, but these values might be averaged to get a single, independent growth measurement for each pot. I will deal more with this issue in a later section.

2. *Select subjects or units to be used in the experiment from a larger population of interest.* Although an investigator is interested in the response to a treatment of specific subjects or units, ultimately this information should tell us something about how a larger population of interest would respond to similar treatment. Hence, the response of the experimental subjects should be representative of the larger group of interest. Ideally, the experimental subjects would be chosen by random selection from the population of interest. Of course, this is not possible for studies on human subjects, and there are also limitations on completely random selection in many other situations. For example, many biological experiments are performed on genetically uniform "lab rats" to minimize random variation among individuals for the purpose of more clearly measuring the effect of a treatment on some aspect of physiological function. The investigator is not interested in curing disease in lab rats, but uses lab rats in an experiment to address other experimental concerns. Even though the response of lab rats to a new drug treatment does not provide sufficient data to determine if the treatment will be effective on humans, experiments with rats are often an important part of assessing drugs intended for human consumption. Nonetheless, consideration should be given to defining the larger population of interest, and the process of selecting study subjects should, whenever possible, work toward obtaining a sample of subjects that is representative of that population.

3. *Determine the specific levels of the treatment variable.* Treatment variables cause different levels of response depending on the level of treatment. For example, different temperatures, light intensity, or amounts of a drug, fertilizer, or pesticide might cause effects of different size. Keep in mind the following practical considerations for choosing experimental treatment levels:

a. Responses to treatments are often nonlinear. (See Figure 1.2.) Some treatment levels are too low to cause any response (see section A in the top graph within Figure 1.2). Within a higher range of treatment levels, a one-unit change in the treatment variable causes a 2.5 unit change in the response variable (section B). At very high treatment levels, the subject cannot respond any further to increase in the treatment (section C) or could even die (section D). As shown in the lower graphs of Figure 1.2, a larger number of treatment levels in an experiment provides a more accurate estimate of the true shape of the response curve.

FIGURE 1.2 Choice of treatment levels influences conclusions about response curves. The figure in the top panel displays the true (and unknown) response curve of the population of interest to the different levels of the treatment variable. The six panels below display the estimated response curves that might be obtained with different combinations of treatment levels in an experiment. The dots in these six panels represent the average response for a group of subjects exposed to one of the treatment levels, 0 to 6. Experiments that include only two treatment levels can produce only linear response curves (the first two of the six panels). The slope of that line depends on which treatment levels were chosen. An experiment must include at least three treatment levels to show nonlinear responses. As more treatment levels (and groups) are included in the experiment, the response curve obtained can display greater complexity and detail.

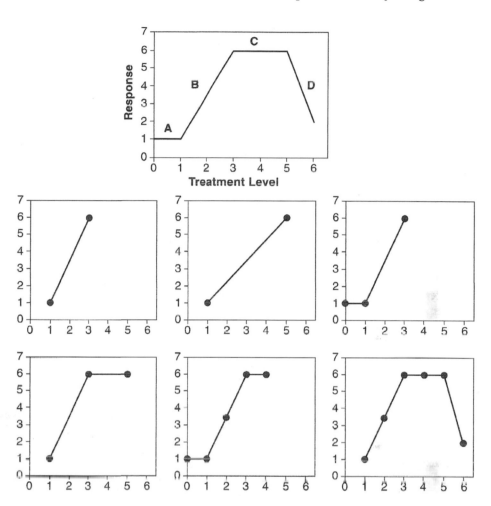

b. Increasing the number of treatment levels can reduce replication per group. Although including many different treatment levels in an experiment provides a more detailed picture of the true response curve, a greater number of treatment levels also requires that a greater number of subjects be included in the study. More subjects mean more time, effort, resources, and money. However, if the number of study subjects in the experiment is limited by logistical or economic constraints, increasing the number of treatment levels (and groups) means decreasing the number of individuals in each group (replication). This increases random sampling variation among groups and increases the risk that groups will not be equivalent prior to the imposition of the treatment.

c. *The linkage between experimental results and real-world applications can be enhanced by setting at least one treatment level close to "ambient" (values typical of real world settings).*

■ **EXAMPLE** An infamous experiment was done to determine if the artificial sweetener saccharin was carcinogenic. The concern was that daily ingestion of small amounts saccharin in diet soft drinks and foods over many years might increase risk of cancer. However, the investigators could not perform an experiment that involved low doses consumed over extended time periods. Instead, the investigators fed ultra-high doses of saccharin to lab rats over relatively short time periods to attain the same total lifetime dose. The equivalent dose for a human would have required consumption of pounds of saccharin per day. Rats that were fed ultra-high doses of saccharin did develop cancer, but it was subsequently shown that this data was useless for assessing the cancer effects of long-term exposure to low doses of the sweetener.   ■

4. *Assign individuals (experimental units) to treatment groups.* A fundamental requirement in good experimental design is that groups of individuals that will receive different levels of the treatments should be equivalent before the treatment is imposed. In practical terms, prior to the treatment, the value of whatever sample statistic you will use to compare the group responses after treatment should be approximately the same across the groups. Procedures used to attain equivalent treatment groups have two components: (1) random assignment and (2) replication.

   a. You can use the following techniques to achieve random assignment to treatment groups.

      * *Using a random numbers table.* If the experimental subjects have names, make a list of these names. Consult a table of random digits and assign sequential **random numbers** from the table to subject names on the list. If the experimental subjects do not have names, then simply assign each subject an identification number that is taken from a random number table.

      * *Using a computer.* Use a random-number generator function to create a variable that contains a column of random numbers; then assign these numbers in the sequence printed by the computer to the study subjects.

      * Re-order the list of experimental subjects so that their random numbers are in rank order from smallest to largest. In a spreadsheet, use the Sort tool to sort the two columns of names and random numbers by using the random numbers as the sort variable.

      * Assign each subject to a treatment group based on the subject's position in the now randomly sorted list. Usually, there are N objects to be randomly assigned to K different treatment groups, with N as some multiple of K so that each of the K treatment groups has the same number (n) of units. The first n units in the ranked list $(n = N/K)$ are assigned to treatment group 1, the second block of n subjects to group 2, and so on.

   ■ **EXAMPLE**   If $N = 40$ subjects were to be randomly assigned to $K = 4$ groups, the first $n = 10$ (40/4) subjects in the list sorted by random numbers would be assigned to group 1, the second 10 subjects on the list to group 2, and so on.■

   b. Replication to attain equivalent treatment groups.

      * We know that sample statistics computed for two independently selected groups of individuals from the same population will rarely be exactly the same value because of random variation among individuals and the fact that the two different groups contain different individuals. However, this kind of random variation can be minimized by computing statistics based on measurements on a large sample of individuals in each group. Hence, com-

bining random assignment with large sample sizes in each group is one way to attain the goal that treatment groups be equivalent prior to treatment.

• Reduce the amount of variation among the study subjects. When logistical or economic constraints preclude doing experiments with large numbers of subjects, the goal of equivalent treatment groups is often attained by reducing the amount of individual variation among the study subjects. This is a major justification for doing experiments on genetically uniform lab rats. In experiments in which humans are study subjects, investigators can reduce individual variation by including subjects from homogeneous subpopulations defined by (say) gender, age, or racial group. Of course, this limits how the results of the experiment can be validly generalized to a larger population of interest. For example, a large study on whether or not a daily low dose of aspirin reduces the risk of heart attack was implemented on male Caucasian physicians. The results indicated that the aspirin treatment did reduce risk of heart attack. However, the results of this study can apply only to well-educated Caucasian males in upper socio-economic classes. Extrapolation of these results to women or men of other races or socio-economic groups is problematic, as individuals from these groups were not included in the study. (A similar study of the effect of daily low aspirin doses on heart attack risk in women is ongoing.)

Types of Experimental Design

There are many different experimental designs that provide options for the investigator who must perform experiments to address a variety of questions under a range of conditions and limitations. Several important considerations determine which experimental design is appropriate for a given investigation.

1. How many different treatment variables will be included in the experiment?
2. If more than one variable is included, will the effects of one variable interact with or be independent of the effects of the others?
3. Are there extraneous (nontreatment) variables the investigator cannot eliminate from the experimental situation that might affect responses of the study subjects to the treatment variables?
4. Are there constraints that limit the investigator's ability to randomly assign study subjects to treatment groups?

Depending on these considerations, the investigator can choose from a variety of experimental designs that will best provide the answers to these questions. Again, it is not within the scope of this introductory textbook to cover this complex subject in its entirety. The following is intended only as an introduction to experimental design. As questions become more complex to reflect the natural complexities of the real world, so too the experimental designs become more complex. Before you begin an experiment, you should always consult an experienced scientist, statistician, or text on experimental design.

Completely randomized experimental design (one treatment variable, two or more levels).

A sample of study subjects are randomly assigned to two or more treatments groups. All subjects in a given group are given the same treatment, while different groups receive different levels of the treatment. The response of the individual study subjects is observed/measured and a value for the sample statistic of interest is computed for each group. The values of the group sample statistics are compared among the groups to assess the treatment effect. This is the simplest of experimental designs, and can be used only under the following conditions: (1) for the study of the effect of a single treatment variable, (2) when completely random assignment is possible, and (3) when all extraneous factors can be controlled. Example 1.4 diagrams a completely randomized experimental design.

■ **EXAMPLE 1.4**

Diagramming a completely randomized experimental design.

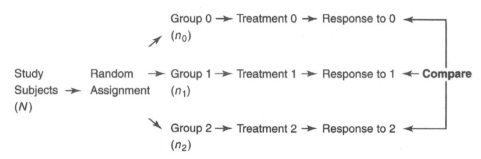

When you diagram an experiment, replace the words Treatment 0, 1, 2 with the actual treatment levels (e.g., No aspirin, 350 mg aspirin, 1,000 mg aspirin), and replace the words Response to 0, 1, 2 with the actual response variable observed or measured on each study subject (e.g., Level of pain, Heart attack: Yes/No). The value N is the total number of subjects used for the experiment, and the values n_0, n_1, and n_2 are the number of subjects in each of the treatment groups. It is important to clearly indicate where random assignment of subjects is done; this placement is a major criterion for defining the type of experimental design and for determining which statistical procedures are most appropriate for analyzing the data produced by the experiment.   ■

Advanced Topics

Before-After Experimental Design

In a Before-After design, each study subject is measured for the response variable *before* any treatment is imposed; then the subjects are given the treatment and remeasured for the response variable. The difference between the Before and After response measurements reflects the treatment effect. If, on average, the differences from a sample of subjects are consistently above (or below) zero, this would constitute evidence that the treatment had an effect. This type of design is typically used when the effect of the treatment is expected to be small relative to variation among the study subjects. For example, a random sample of students usually shows SAT scores that vary widely (from 300 to 800). A workshop that teaches students skills for taking the SATs could improve each students score by 10 to 20 points. However, because this individual improvement is small compared to the large differences among individuals, a study of the efficacy of the workshop that used a completely randomized design would probably fail to detect this improvement in scores.

If there is more than one treatment, Before-After designs may suffer from *carry-over effects*, which include any effect of one treatment on the subjects that influences their responses to subsequent treatments.

⚛ **EXAMPLE**   Suppose you wanted to compare three different SAT preparation workshops to determine which, if any, increased SAT scores the most. In a Before-After design, each study subject (student) would take a test beforehand, attend a workshop, take an SAT exam, then attend another workshop and take another SAT exam, and so on. There are at least two possible carry-over effects. First, the students might learn from past experiences (workshops and exams) and get progressively higher scores on the SAT exams due to this learning. If student scores were higher after the third workshop than after the first, you would not be able to determine if this was because the third workshop was better or if the higher score occurred because of carry-over learning. Alternatively, if all three workshops and tests were done on the same day, test scores might be lower after the third workshop due to fatigue. The most common way to eliminate the influence of carry-over effects on the outcome of a study is to randomize the sequence in which the subjects are exposed to the multiple treatments.   ⚝

Matched-Pairs Experimental Design

In a Matched-Pairs design, pairs of study subjects are matched based on their similarity to each other with regard to the response variable. For example, a study of the efficacy of a weight-reduction drug might pair individuals of similar weight. The treatment is randomly assigned to one member of the pair, while the other is used as a control. After imposing the treatment, each member of the pair is measured for the response variable again. The difference in the response measurements computed between the two individuals in a matched pair is used as the measure of the treatment effect. If, on average, the difference values from replicated matched-pairs are consistently above (or below) zero, this would constitute evidence that the treatment had an effect. By basing the assessment on comparisons of similar study subjects, the treatment effect can be more clearly discerned in spite of substantial random variation among subjects. Unlike the Before-After design, there is no carry-over effect in a Matched-Pair design.

Advanced Topics

Randomized Block Experimental Design

A **randomized block experimental design** is used when the response to a treatment is influenced by an extraneous variable that cannot be controlled or eliminated. For example, the response of females to a drug may differ from that of males. In this situation, the study subjects are first divided into *blocks* based on the characteristic that makes them different (e.g., gender or race). Obviously, this is not a random assignment. Within each block, individuals are randomly assigned to treatment groups. All treatment levels are applied within each block. Comparisons to determine the treatment effect are performed *within* blocks, not between blocks. For example, in a drug study that included both sexes, comparisons would be made among groups of women that received different treatments, and comparisons would be made among different groups of men, but male treatment groups would *not* be compared to female groups.

■ **EXAMPLE**   A randomized block design for an experiment to determine the effectiveness on men and women of a drug to reduce blood pressure.

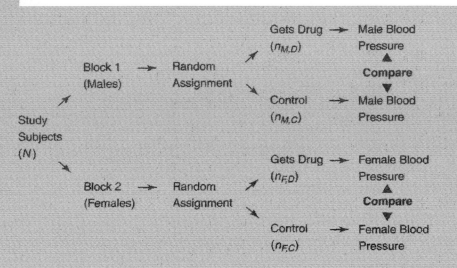

Factorial Experimental Design

A **factorial experimental design** is used when the investigator wants to determine the response of the study subjects to two or more treatment variables and has reason to believe that the effect of one variable will interact with the effects of other variables. For example, the response of crop plants to fertilization often depends on the availability of water. Plant growth is increased by fertilization (increased nutrient availability), but only if water availability is sufficient. In a factorial design, there are multiple treatment variables, each with multiple treatment levels. A full factorial design creates treatment groups for all possible combinations of the different treatment variables and levels. Comparisons of the response variable are made between all groups, representing the various combinations of treatments.

■ **EXAMPLE**   A factorial experimental design to assess the interactive effects of fertilization and irrigation on crop plant yield. The diagram that follows is for a simple 2 × 2 factorial design (two treatment variables × two treatment levels for each variable). The diagram lists treatment groups for all combinations of fertilization (F) and irrigation (I) required for a full factorial design. Treatment levels are indicated by + (treatment applied) and − (treatment not applied). Effects are measured in bushels per acre.

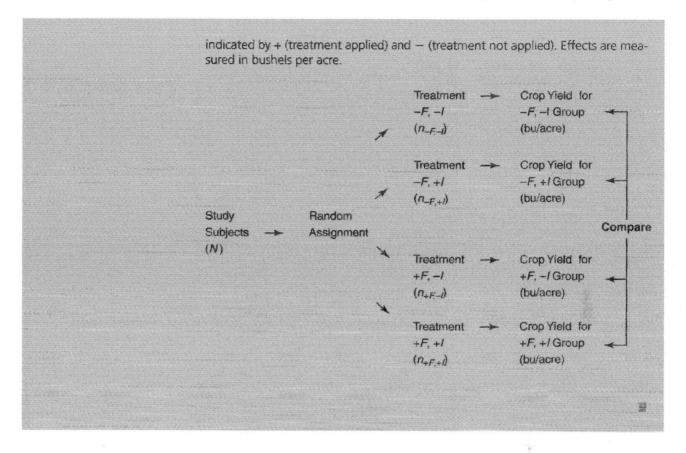

Potential Problems with Cause-Effect Inferences from Experiments

The purpose of experimental study design is to provide a realistic test of the effect of a treatment on individuals in a population of interest. The treatment effect is measured by differences between experimental groups that received different treatments. However, an investigator might reach an erroneous conclusion regarding the effect of the treatment on the population of interest due to any of several types of problems.

1. *Confounding factors.* Some factors other than the treatment that were not controlled by the investigator might actually be the cause of differences between treatment groups. These factors are considered **confounding**.

   Example: A scientist performing a study of fertilizer effects on corn yield plants all the treatment (fertilized) plots in one part of a field, and all the control (no fertilizer) plots in another part of the field. If the soil characteristics differ between these two areas (e.g., water drainage, natural nutrient availability, or other factors), the investigator would not be able to determine if differences in corn yield between treatment and control plots were due to the fertilizer treatment or due to differences in soil fertility. That is, the fertilizer treatment is confounded with any differences that might exist in the soil or the environment between the field areas. To avoid this problem, the investigator should randomly intersperse treatment and control plots across all areas of the field.   ▣

2. *Poor measurement validity:* Measurement *validity* refers to how well the measurements of treatment and response variables actually quantify the phenomenon of interest. Measurements that are poor representations of the phenomenon of interest provide misleading information about the true nature of the universe, and so lead to inappropriate conclusions.

   Example: In a drug-dose response experiment, the amount of drug consumed may be a less valid measure of "dose" than the concentration of drug in the blood. Drug substances that are not assimilated into the blood stream cannot be expected to cause a physiological response. If a drug delivery system (e.g., pill) inhibits assimilation, an otherwise effective drug might appear in an experiment to produce no response. This could lead to the erroneous conclusion that the drug is not effective, when in fact it is the delivery system that is ineffective.   ▣

   Example: In a study of plant growth responses to pollutants or fertilization, using plant height or above-ground mass as a measure of response may be invalid. Many studies have documented that both of these types of treatment can reduce root mass while above-ground mass is unaffected or even increases.   ▣

3. *Improper assignment of subjects to treatment groups.* Another potential problem is that the groups are different *before* the treatment. This problem is generally associated with lack of randomization or inadequate replication or both. If subjects are improperly assigned to treatment groups, the experiment identifies between-group differences as a treatment effect when the treatment actually had no effect on the study subjects.

4. *Nonrepresentative subjects included in the experiment.* Suppose subjects who are not representative of the larger population of interest are included in the experiment. The response or nonresponse of the experimental subjects to the treatment does not provide an accurate assessment of how members of a larger population might respond to similar treatment. This problem is common when investigators try to control for extraneous factors by including subjects only from a homogeneous group. Results of such experiments should not be extrapolated to other groups that were not included among the study subjects.

5. *Investigator bias.* If the investigators know which subjects were in the treatment and control groups, they are more likely to see, or not see, a treatment effect due to their preconceived expectations about how the experiment should turn out. The procedures for obtaining some kinds of data involve a subjective judgment, as opposed to a strictly quantitative measurement. Even basically honest investigators may be more likely to perceive an effect if they know that the study subject received the treatment, and less likely if they know the subject was in the control group. This problem is usually avoided by using a **blind experimental design**, meaning that the investigator does not know which subjects received the various treatments at the time that data are collected. This information is kept secret, often by using random numbers for subject identification.

6. *Placebo effect.* Investigators are not the only persons who can be influenced by knowledge of who did or did not receive the treatment. People who are told they are being given a drug to alleviate some ill sometimes report that they feel better, regardless of whether or not they actually received the drug. Hence, results of experiments that use people as study subjects may indicate the treatment produces an effect, when in fact the treatment has no effect. To address this problem, the subjects in an experiment must not be able to know whether or not they are receiving the treatment. A **placebo** is a treatment that looks identical to the actual treatment (pill, injection, or other treatment), but does not actually contain the active component of the treatment. Individuals in all experimental groups are given a so-called treatment, but only some groups actually receive the active ingredient. A real treatment effect is identified only if the group(s) that receive the actual treatment report a different response than the group that received the placebo.

7. *Lack of realism.* The response of study subjects to treatments under highly controlled laboratory conditions may not be representative of their responses under natural environmental conditions. *All experiment studies suffer from the following dilemma:* The more realistic the treatment and experimental conditions, the less control the investigator may have over confounding factors. The more control the investigator exerts over confounding factors, the less realistic the experimental conditions and the less likely the study subjects will respond as they might in their natural environment.

1.5 Natural Experiments

Natural experiments involve comparing samples obtained from two or more populations in their natural environment. These populations have previously been exposed to different conditions (i.e. natural treatments). Other names for this type of study include *quasi-experiments* (psychology literature), *comparative studies* (ecology literature), and *case-controlled studies* (medical literature). The natural experiment is a hybrid of sampling and experimental studies and as such can provide information on subject responses to treatments under conditions that are more realistic than those of controlled experiments in laboratory settings. However, this type of study differs from true experimental studies in two important ways: (1) The investigator has limited or no control over which subjects received treatments and (2) the investigator has limited control over how the subjects have been influenced by extraneous, nontreatment factors. Hence, natural experiments often provide a less-than-clear basis to infer that differences observed between the treatment groups were caused by the natural treatments. However, there are a number of justifications for natural experiments, in spite of these drawbacks.

1. Responses of experimental subjects to the treatments are more likely to be realistic if the subjects are in their natural state.

2. Certain types of experimental subjects (e.g., trees, large carnivores) cannot be readily studied in controlled experiments. Natural experiments may provide the only means for addressing questions regarding such subjects.

3. Certain types of treatments cannot be imposed by the investigator due to ethical concerns. Natural experiments allow the study of subjects who have been exposed to such treatments during their life. For example, it is important to know how

humans are affected by exposure to radiation. However, it is completely unethical to perform a controlled experiment that involves purposefully exposing humans to potentially harmful levels of radiation. Instead, a natural experiment could be done to compare cancer rates and other responses among people who have and have not been exposed to high levels of radiation during the course of their life.

Two significant problems must be addressed before the results of natural experiments can be used to infer whether or not the natural treatments caused a response:

1. Because experimenters cannot randomly assign individuals to natural populations, *there is substantial risk of having nonequivalent groups*. The populations that were differentially affected by a natural treatment may have been different *before* the treatment occurred. Therefore, differences between the populations may not have been caused by the natural treatment variable.

2. Natural experiments cannot completely control various extraneous nontreatment variables that might also influence the response of individuals in a natural population. It is possible that *observed differences between populations may be due to a variable other than the treatment variable of the natural experiment.*

Both of these problems can be addressed by selecting study subjects based on criteria that make different natural treatment groups as similar as possible in other regards. However, these criteria may restrict the scope of the conclusions that can be drawn from the study.

■ **EXAMPLE**   Some suggest that having an abortion increases a woman's risk of later developing breast cancer. A controlled experiment to test this claim would be ethically abhorrent. Instead, investigators could study medical records of women who have or have not had an abortion and compare the proportion of women in each group who subsequently developed breast cancer. One problem with this study design is that the population of women who have an abortion may differ in other regards from the population of women who have not (e.g., socio-economic category, life style, ethnicity, and so on). These differences may also influence the women's risk of developing breast cancer, regardless of whether or not they had an abortion. To address this concern, the investigators could select for their study only women who are similar in all regards except whether or not they had an abortion (e.g., only women of European ancestry, who have never been exposed to radiation or carcinogenic pollutants, and whose weight, blood cholesterol levels, and other medical factors are within a specified range). This screening process would help to form equivalent treatment groups and eliminate potential confounding factors, but the final results of the study might apply only to the larger population of women who also meet these exact criteria. The study results might not apply to women of other ethnic backgrounds, women who have been exposed to carcinogenic agents, or women whose weight, blood cholesterol levels, and other medical factors are outside the range included in the study.   ▨

For all natural experiment studies, it is critically important to identify and measure many variables in addition to the presumed treatment and response variables. These additional measurements can be used to substantiate the claim that the populations being compared are equivalent in many regards, and that variables other than the treatment can be excluded as possible causes of any differences observed between the different groups. If general equivalence of populations and exclusion of nontreatment variables as potential causes can be documented, a stronger case can be made for a cause-effect linkage between the natural treatment and differences in the response.

Because no single study can eliminate all potential problems described above, scientific research generally requires supporting evidence from many studies (done by different scientists and involving different populations) before a hypothesis or conclusion is given widespread credibility.

1.6
Independence and Pseudo-Replication

A fundamental assumption of all statistical analyses is that the data values obtained from a sampling or experimental study represent independent observations on a representative sample from the population of interest. I have already described what constitutes a representative sample, based on randomization and replication (Section 1.2). Measurements or observations are **independent** if the value of each observation is in no way influenced by, or related to, the value of other observations in the sample. There are two general types of violation of the independence assumption: (1) pseudo-replication and (2) failure to implement truly random sampling or assignment. These are described below.

Pseudo-Replication

Pseudo-replication is the term applied to the error of treating multiple measurements made on each individual in the sample the same as single measurements made on multiple individuals. Multiple measurements are sometimes made on each individual to provide a more accurate measure of the individual's condition, or to look for change in an individual's condition over time. For example, if a doctor suspects that a patient may have high blood cholesterol, the doctor might request that several blood samples be taken over several days because blood cholesterol varies from day-to-day. The doctor would then average the data values from different days to obtain a more precise measurement of that patient's blood cholesterol. This would not be pseudo-replication because the multiple data values for this one individual were reduced to a single "best" data value. However, suppose an investigator was doing an experiment to determine if a new drug reduced blood cholesterol. The investigator obtained 24 volunteer study subjects and randomly assigned 12 subjects to each of a control and treatment group. The control group subjects are given a placebo pill and the treatment group is given a pill containing the new drug. Recognizing that blood cholesterol varies from day-to-day, at the end of the treatment period the investigator takes three blood samples from each patient on three successive days. Hence, for each experimental group the investigator has 12 subjects $\times$ 3 data values per subject = 36 data values. If the investigator treated these 36 data values as if they were independent observations, when he performed statistical analyses to determine if the drug reduced cholesterol, this would constitute pseudo-replication. The three repeated measurements on each patient are not independent observations. A person with high cholesterol on day 1 would likely still have high cholesterol on days 2 and 3. To avoid pseudo-replication, the investigator must first average the data values from the three blood tests done on each patient to obtain a single data value per patient. These average values would constitute independent observations. The average blood cholesterol values for the 12 subjects in the control group and the 12 subjects in the treatment group would provide for a valid statistical analysis to determine if the drug reduced blood cholesterol.

Another reason that pseudo-replication occurs is that the investigator has not correctly specified exactly what constitutes an individual (also called a *replicate*) in the population of interest. This is a much more subtle type of pseudo-replication and can be identified only by carefully specifying the population of interest. For example, if the student body at a university was your population of interest, the basic unit that constitutes an individual in this population would be a student. A single measurement on one student who is randomly chosen from the student body would represent a single independent observation. Although it may seem that identifying the basic unit that comprises the population of interest should be easy, this is not always the case. For example, an analysis of 156 papers in the ecological literature found that 27% of the papers incorrectly specified what constituted an individual in the population of interest (Hurlbert 1984).[1] Hurlbert and White (1993)[2] performed a similar literature review to determine if any progress had been made in reducing the occurrence of this error since the publication of Hurlbert (1984). They found pseudo-replication in 32% of papers describing studies of invertebrate zooplankton published from 1986 to 1990. All of these scientific papers had gone through rigorous review prior to publication, yet this problem had not been identified.

The following description demonstrates the subtle distinctions that separate a valid study from a study that suffers from this second form of pseudo-replication.

■ **EXAMPLE**   Suppose an investigator wanted to determine the effect of sewage pollution on fish communities in rivers that receive this pollution. The population of interest here is "river fish communities." Suppose that this investigator studied the fish community in the river that flows through his city, taking samples at multiple randomly chosen locations upstream (control) and downstream (treatment) of the sewage overflow pipe. Is this pseudo-replication? The answer depends on the specific wording of his study question and the conclusions he draws from his study. This study design would be valid if he phrased his study question as "What is the effect of sewage pollution on fish communities in the White River, Muncie, Indiana?" and his conclusion about the effect of sewage pollution explicitly stated that his results apply only to this local population of interest. On the other hand, suppose his study question was more general ("What is the effect of sewage pollution on fish communities in rivers of the Midwest United States?") and he wanted to draw a conclusion regarding this much larger population of interest. The study design described above would be invalid due to pseudo-replication. The population of interest for this study question is "fish communities in rivers of the Midwest U.S." His multiple random samples from the White River, Muncie, Indiana would constitute multiple observations of fish communities from an individual river (i.e., pseudo-replication). To assess the effect of sewage pollution for this larger population of interest, the investigator would need to randomly select a sample of rivers located in this geographic region. The investigator would obtain data on the fish communities at locations upstream and downstream of sewer pipes in each river randomly selected for this sample. This would be a very ambitious, expensive study, and would likely exceed the capacity of any one investigator. In real-world scientific research, most investigators would perform their study in one river close to home, and limit their study question and conclusion to this smaller population of interest. ▨

Small-scale studies performed in a single location or on a single individual of the population of interest are called **case studies** because they address the scientific question in the context of one specific case. In the example above, each river has its own characteristic flow volume and speed, bottom structure, watershed geology, and vegetation. Different combinations of these characteristics in different rivers may influence both the nature of the fish community and its response to sewage pollution.

To address the general question of sewage effects on fish communities of rivers in the Midwest U.S., a researcher must obtain data for multiple rivers in the region. Often the investigator would perform a literature search and find published results from a large number of small-scale case studies. By identifying commonalities among the many case studies the investigator could come to an informed conclusion about the general effects of sewage pollution on fish communities for the larger region.

Failure to Implement True Random Sampling or Assignment

Observations can fail to be independent not only because of pseudo-replication but because investigators do not implement truly random sampling or assignment. Two common causes of this lack of independence are *genetic relationships* among study subjects and *close physical proximity* of study subjects. Data obtained from organisms that are genetically related to each other (siblings, parents) often do not represent independent observations because related individuals share genetically determined characteristics. Hence, a sampling protocol that randomly selects families, then measures the variable(s) of interest on all members of each family would suffer from a lack of independence among the observations from each family.

When substantial effort is required to travel from one randomly located point to another, the investigator will sometimes obtain data from multiple individuals at each randomly chosen location while reducing the number of randomly chosen locations in the sample. Because multiple individuals at each random location are in close physical proximity, they share a common environment and may even interact (compete, cooperate). Hence, data values from multiple individuals at a single randomly chosen location may not be independent.

The bottom line: Sometimes it is appropriate to make multiple measurements on each study subject or to measure multiple individuals at each randomly chosen location. This is *not* a problem. However, treating data values obtained in this manner as if they were independent *is* a problem. If data are obtained in this manner, standard statistical procedures should not be applied to the raw data. Multiple measurements on single individuals should be averaged to obtain a single independent observation. Measurements on multiple individuals located around a single randomly located point must be analyzed using procedures specifically developed to handle data from this type of study design. Usually, multiple measurements on the same individual, genetically related individuals, and individuals in close physical proximity exhibit less variation than the variation observed among measurements from a truly random sample of individuals. These errors in study design and data analysis often result in underestimation of the amount of variability in the population, leading to biased sample statistics and erroneous conclusions.

More Examples of Pseudo-Replication

No replication, observations not independent.

The question of interest is: "Does the number of times a male frog "peeps" change as temperature changes?" To address this question, a male frog is placed into a controlled-temperature chamber and the number of peeps per minute are recorded as the temperature is varied. The frog is placed in the chamber for 30-minute intervals at each of the following temperatures (5°, 10°, 15°, 20°C). This procedure is repeated with the same frog on 10 different days, and the measurements for each day are treated as replicates.

This design illustrates pseudo-replication at a number of levels. First, a single frog does not constitute a representative sample from the population of all frogs. The true replicates for this questions would be "frogs," not multiple measurements on a single frog. Second, repeated measurements on the same individual are *not* independent in this design. If the frog is the "strong silent type" today, it will likely behave similarly tomorrow and all other days during the experiment. Hence, the value for peeps/minute today will be related to the values on all other days, and the repeated measurements will not be independent.

A valid study design for this question might involve the following: A random sample of 40 frogs is chosen from the local population, 10 frogs are randomly assigned to each of the 4 temperature treatments, and each frog is kept in the controlled-temperature chamber for 30 minutes and the number of peeps is recorded. Multiple observations might be done for each frog over several days to get a better estimate of each frog's peep rate, but these values should be averaged to a single value for each frog at each temperature. The effect of temperature on peep rate would be tested by comparing the average peep rate of the 10 frogs for each group between the four temperature treatments.

Replicate observations not independent (due to spatial proximity).

The question of interest is, "What is the effect of fertilization on corn plant growth?" To address this question, two plots are established (one fertilized, the other an untreated control plot), and 100 corn plants are grown in each plot. To determine the effect of fertilization, the average mass of the corn plants (root and shoot) is compared between the two plots.

This design is invalid because the observations of growth of adjacent corn plants within a plot are not independent. If one corn plant is genetically superior and grows particularly large, it will tend to shade its inferior neighbors and cause them to grow more slowly. In many cases, the genetically superior plants are better able to take advantage of fertilization and this accentuates the suppression of adjacent, inferior plants. If the focus of the question is to be on the response of individual plants, an appropriate experimental design would be to put the plants in separate pots so there is no competition for resources; here the individual plants would be the replicates. If the focus of the question is to be on the yield of corn per acre, an appropriate design would establish stands of corn in several fertilized and unfertilized plots of ground; the replicate data values would be corn yield per plot.

The Efficiency of Study Designs and Pseudo-Replication

Efficiency of a study design can be defined as the amount of useful information that is obtained from a specific amount of time/effort and(or) for a specified cost. Each measurement of a data value requires the time and effort of personnel, and direct costs for salary, supplies, and equipment. Indirect costs must also be considered; there are the costs incurred when the researcher's time is spent doing the study rather than being gainfully employed elsewhere (such as on graduate student research projects). A hallmark of good study design is that it obtains all the data necessary to address the question of interest at a minimal cost in time and resources.

By definition, pseudo-replication reduces the efficiency of study designs. Each measurement of a data value incurs a cost, regardless of whether the measurements are on independent replicates or repeated measurements on one replicate. Generally, there is far more variation among measurements from independent replicates than among repeated measurements on the same replicate. Hence, more information about the population is gained by making measurements on independent replicates. When it is necessary to make multiple measurements on each replicate, the number of measurements per replicate should be kept to the minimum necessary to obtain the desired measurement precision. This plan allows the investigator to maximize the number of independent replicates that can measured within the limitations of time and resources.

■ **EXAMPLE**   How does the abundance of various alpine plant species differ between the north, east, south, and west sides of Mt. Washington? To address this question, five square 2 meter by 2 meter plots are randomly located on each side of the mountain. Within each plot a systematic grid of 200 points is established, and the plant species present at each point is recorded. The abundance of each alpine plant species on each side of the mountain is quantified as the percentage of points where the species was found (out of 5 plots × 200 points per plot = 1,000 points on each side of the mountain). However, the observations at the 200 points within each plot are not independent, so the 1,000 points are not replicates. With so many points in each plot, adjacent points can fall on the same individual plant. In fact, the true replicate for this study is the "plot," and the points within the plots are repeated observations on the same sample unit. In an appropriate study design, the replicate measurement for abundance of an alpine plant species would be the percentage of points in a plot in which that species occurred. Since the question of interest is specific to this particular mountain, the five plots on each side of the mountain constitute replicated observations from this limited population of interest. Not only is the study design described above invalid due to pseudo-replication, but it is also very inefficient in terms of time and effort. So much effort is spent observing 200 points within each of the plots that the number of replicate plots that can be studied in the time available was limited to five on each side of the mountain. A better design might have only 50 or 100 points per plot, freeing time to establish 10 or 20 of plots on each side of the mountain. ■

■ **EXAMPLE**   What is the response of brain cells in the visual cortex to a cessation of stimuli resulting from damage to the eye? To address this question, a laboratory rat is blinded on one side (one eye is surgically removed). After a period of six months the rat is sacrificed, and thin sections are taken from the visual cortex. Neuron length is measured on 100 neurons from the part of the visual cortex linked to the damaged eye and compared to the length of 100 neurons in the part of the visual cortex linked to the undamaged eye.

Obviously, many measurements of neuron length from a single rat do not constitute independent replicate measurements. However, doing this experiment with many rats and measuring only one neuron from each side of the brain per rat would be very costly in time, effort, and resources. An appropriate study design might impose the treatment on 10 to 20 rats, on which 10 to 20 measurements of neuron length would be made. Again, the multiple neuron length measurements from each rat could be averaged to obtain a good estimate of neuron length for each side of the brain for each rat replicate. Sometimes the relative costs of repeated measurements on each replicate versus the costs of measuring multiple replicates is an important consideration in determining the optimal balance between repeated measures and the number of independent replicates. ■

CHAPTER SUMMARY

1. All scientific knowledge is based on observations and measurements made on small parts of the universe, called **samples**. However, scientists are primarily interested in understanding the nature of some larger part of the universe called **the population of interest**. Because scientific understanding about these larger populations is based on observations of only small subsets (samples), all scientific understanding is incomplete, uncertain, and approximate.

2. **Data** refers to a collection of observations and/or measurements for one or more variables, made on one or more individuals from the population of interest for the purpose of addressing a specific question.

3. Data obtained from individuals in a sample are used to compute **statistics**, which are numbers that describe characteristics of the sample. Sample statistics are used to estimate or infer something about the values of population **parameters**, which are numbers that describe characteristics of the entire population of interest. Because sample statistics are usually the only information available to gain understanding about population parameters, it is essential that samples used to compute statistics be representative of the populations from which they are taken.

4. **Bias** is the systematic deviation of sample statistics away from the true value of population parameters. The three most common reasons for bias are: (1) Individuals included in the sample are not representative of the larger population of interest. (2) Measurements do not adequately represent the variable of interest. (3) An otherwise appropriate measurement is consistently done incorrectly.

5. **Precision** is the amount of random variation exhibited by the values of a sample statistic computed from repeated, independent samples that have been obtained from the same population. There are two sources of this variation: (1) Different samples include different subsets of individuals, and there is random variation among the measurements for these individuals (called **random sampling variation**). (2) Inconsistent measurement error due to sloppy technique. If many individuals are included

in a sample, random variation among individuals will average out, and sample statistics computed from repeated samples will be less variable, and therefore, more precise.

6. Representative samples are obtained through two processes. If a researcher randomly samples from the population, all individuals will have an equal opportunity to be selected and bias is minimized (**randomization**). If a researcher obtains data for a large number of individuals in a sample (**replication**), various sources of random variation are averaged out and the values of sample statistics will exhibit less random variation (be more precise). If a researcher combines randomization and replication, the value of a sample statistic can be assumed to be close to the value of the true population parameter (i.e., the statistic will be an **accurate** estimate of the population parameter).

7. The purpose of **sampling study designs** is to estimate some characteristic (parameter) for a population of interest based on statistics computed from sample data. **Completely random sampling, randomized systematic sampling**, and **stratified sampling** are but three of a variety of sampling study designs that can be used to obtain representative samples that provide a valid basis for estimating population parameters. Systematic sampling is used to increase logistical efficiency and to ensure "even coverage" of the population of interest. Stratified sampling is used when the population is known to be comprised of heterogeneous groups.

8. The primary purpose of **experimental study designs** is to test hypotheses about the response of study subjects to some treatment. In true experiments, the investigator must impose the treatment on the subjects. Good experimental design includes four features:

   a. **Random assignment** of study subjects to different treatment groups.

   b. **Replication**. By including many study subjects in each group and by using random assignment, a researcher can define **equivalent treatment groups** that will be similar prior to the imposition of the treatments.

c. **Control** of nontreatment (extraneous or confounding) factors that might influence the response of the study subjects to the treatment.

d. **Comparison** of groups that receive different treatments, but are otherwise similar, to determine if the treatment causes an effect.

The combination of these four features allows the investigator to attribute any differences observed among the experimental groups to the differences in the treatments applied and not to preexisting or nontreatment factors.

9. There are many different types of experimental study design. The choice of design appropriate to any particular situation is based on: (a) The number of different treatment variables to be included, (b) whether or not multiple treatment variables are expected to interact in their effect on the study subjects, (c) the existence of extraneous factors that cannot be excluded from the experiment, and (d) constraints on the investigator's ability to randomly assign individuals to all combinations of treatment variables.

10. **Natural experiments** compare representative samples obtained from two or more populations in their natural environment. These populations have previously been exposed to different conditions (the "natural treatments"). A natural experiment is sometimes necessary when the nature of the subjects or the treatment preclude controlled experiments.

11. A fundamental assumption of all statistical analyses is that data values represent independent observations from the population of interest. This assumption can be violated if:

a. Sampling of each individual in the sample is not truly random, such that some individuals in the sample are located in close spatial proximity to each other or are genetically related.

b. Multiple measurements made on individuals from the population of interest are treated the same as single measurements from different, randomly chosen individuals (**pseudo-replication**).

END NOTES

1 Hurlbert, S. H. 1984. Pseudoreplication and the design of ecological field experiments. *Ecological Monographs 54*: 187–211.

2 Hurlbert, S. H. and White, M. D. 1993. Experiments with freshwater invertebrate zooplanktivores: Quality of statistical analysis. *Bulletin of Marine Sciences 53*: 128–153.

GO!   *Enhance your understanding of the material presented in this chapter by doing the practice problems and exercises found on pages 1–1—1–36 in the workbook companion to this textbook.*

2 Exploratory Data Analysis: Using Graphs and Statistics to Understand Data

INTRODUCTION Often, students will immediately try to use their collected data in complex statistical analyses to "get the answer" to their question without spending even a modest amount of effort to look at the data values for unanticipated patterns or oddities. At best, this highly focused emphasis on statistical tests to address specific questions misses opportunities to gain new insights. At worst, the student fails to see that the nature of the data is inappropriate for the statistical test he or she has chosen, resulting in misguided and mistaken conclusions. Experienced investigators always "look before they leap." All statistical analyses, from the simplest to the most complex, are bound by the GIGO principle: Garbage In, Garbage Out. If you apply a statistical analysis to data that is not appropriate for your analysis, your apparently scientific and precise results will, in actuality, be nonsense.

The goal of **exploratory data analysis** (**EDA**) is to enable the investigator to assimilate and understand the information embodied within the data, even when the amount of data is great or when the data are complex. This goal can be attained by summarizing the data values in ways that accentuate **patterns** (systematic variation in data values) and that distinguish patterns from **noise** (random variation or deviations from the pattern). Patterns are the "bread and butter" of science. Patterns allow us to classify similar objects into groups, so that we don't have to study and describe multitudes of individuals. *Associations among variables* are a type of pattern that allows us to make informed predictions of future events, the true goal in science.

We usually begin a study expecting to find a specific pattern or association in the data. We design data collection procedures and data analyses to test for this expected pattern. We may not find the pattern we expected. However, a goal of EDA is to look for evidence of *any* pattern without preconceived notions as to its nature. The purpose really is to explore. Sometimes you will be reassured and able to proceed to more complex statistical tests with confidence. Sometimes you will be unpleasantly surprised and forced to change your study plan to accommodate unanticipated difficulties in your data. Sometimes you will catch a measurement or data entry error. Occasionally, you will experience the wonder of a completely novel insight that will lead you down unexplored paths of inquiry. All of this could be yours for the relatively small cost of time and effort required to simply look at your data.

The primary tools of EDA are graphics and summary statistics that convert a confusing multitude of numbers into simple pictures and a few descriptive numbers that are easily assimilated and understood. Some **graphics** that we will discuss might be familiar to you (histograms, scatter plots), while others might be new (stem-leaf plots, box-plots). Some graphics are easy to produce by hand, while others are best done using computer software. Some graphics retain virtually all the information in the raw data, while others summarize data in the interest of reducing complexity and enhancing patterns. **Summary statistics** are the ultimate summarization (e.g., a million data values can be summarized to a single average value). However, summary statistics lose much of the detailed information in the raw data, and can obscure a wide variety of problems and patterns. Hence, the usual plan in exploratory data analysis is to first look at the data graphically and then to choose the most appropriate summary statistics to describe the data. ■

CHAPTER GOAL AND OBJECTIVES

In this chapter you will understand the importance of fully exploring the features of your data before proceeding with statistical summarizations and statistical tests. You will also learn the basic graphical and statistical options to enhance your understanding of the data.

By the end of this chapter, you will be able to:

1. Produce graphical presentations for the distribution of data values for a single variable (histograms, stem-

leaf plots) and interpret them. You will be able to interpret (a) the characteristics of the data distribution (center, spread, shape, gaps, outliers) and (b) the frequency and relative frequency of data values.

2. Compute and interpret descriptive statistics representing the center (mean, median, mode) and spread (variance, standard deviation, interquartile range) of data distributions. You will be able to describe which of these statistics are sensitive or resistant to outliers.

3. Produce and interpret graphical presentations that compare the distributions of data values for a single variable between multiple samples from different populations (back-to-back stem-leaf plots and side-by-side histograms and boxplots).

4. Produce and interpret graphical presentations for the association between two variables (scatterplots and time plots).

2.1
Data
Distributions

A data distribution is a summarization of the data values for a variable that shows the range of values (smallest to largest), and how common or rare each specific value is in the sample. There are two ways to describe how common the specific values are in a data distribution. **Frequency** is the number of times that a specific data value occurs in the sample data. But frequency can be interpreted only if you know the total number of values in the sample. For example, a data value that occurs 3 times in a sample of 10 values means something entirely different from when a data value occurs 3 times in a sample of 1,000. **Relative frequency** is the proportion or percentage of data values in a sample that are a specific value. For example, if a data value occurs 3 times out of a sample of 10 values, the relative frequency of that data value is 0.3 or 30%. If a data value occurs 3 times in a sample of 1,000 values, its relative frequency is 0.003 or 0.3%.

Describing the data distribution for each of the variables is often the first step in summarizing the data. For example, 1,000 data values for a single variable might be summarized to a list of 20 distinct values that actually occur in the data and to 20 frequency or relative frequency values that tell you how many times each of the 20 distinct values occur. That is, 1,000 numbers could be completely described by $20 + 20 = 40$ numbers. It is much easier for the human mind to assimilate the information in 40 numbers than that same information in 1,000 numbers.

Characteristics of Distributions

There are five characteristics of distributions that the investigator should always determine when they practice exploratory data analysis. These include:

1. *Center:* The center is the middle of the range of values, where the most frequently occurring data values are often found.

2. *Spread:* The spread describes amount of variability in the values away from the center. Distributions with little spread have values of the variable that exhibit little variability; distributions with large spread have values that exhibit great variability.

3. *Shape:* This characteristic includes the number of "peaks" in the distribution and whether the distribution is symmetric or asymmetric. Multiple peaks might indicate that the population of interest is composed of two or more different classes of individuals (e.g., males and females). Many statistical tests require the data distribution to be symmetric, with a single peak.

4. *Gaps:* Gaps are segments within the range from the minimum to maximum data values for which no data values occur.

5. *Outliers:* Data values that are very different from all the other values are called **outliers**. They may indicate measurement error, that individuals were sampled from outside the population of interest, or novel situations that represent "exceptions to the rule" that may challenge previous theoretical expectations. The first two cases constitute mistakes, so that it might be appropriate to drop the outlier from the data. In the latter case (a challenge to accepted theory), outliers can be important for defining the limitations of current theory or, in rare cases, even bring down an existing theory.

2.2

Graphical Displays of Data Distributions

Histograms and Bar Charts

Histograms are graphics commonly used to display data distributions for quantitative variables, and **bar charts** display distributions for categorical variables. Even though these graphics can be produced by hand, they are usually done by computer graphics programs. In both types of graphics, the range of values of the variable is listed along one axis of an *x-y* graph. Above each value is a bar, and the length of the bar indicates the frequency or relative frequency of that value, as indicated by the scale on the other axis of the graph. See Figures 2.1 and 2.2 for examples of a histogram and bar chart, respectively.

> **NOTE** The concepts of center and shape apply only to quantitative variables, not to categorical variables as displayed in bar charts. For example, the order of tree genus names in Figure 2.2 is entirely arbitrary. Hence, the center and shape of this bar chart would vary, depending on the order chosen for the genus names. The concept of variability applies to categorical variables such as genus only in the sense that some categorical variables have more distinct classes than others.

Histograms for continuous quantitative variables

To make a histogram for values of a continuous quantitative variable, the data values must first be assigned to **classes**. Few data values have *exactly the same value* when the variable of interest is a continuous quantitative variable. A histogram of

FIGURE 2.1   Example of a histogram with a relative frequency scale. In this histogram, there are many more small trees than there are large trees. The distribution of values is very asymmetric, with a center around 20–30 cm and a range from 11–100 cm.

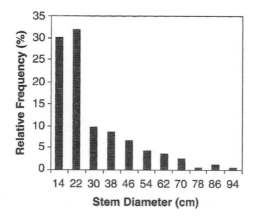

FIGURE 2.2 Example of a bar chart with a frequency scale. In this bar chart, there are many more maples in Christy Woods than other species. There are five classes for the categorical variable genus represented.

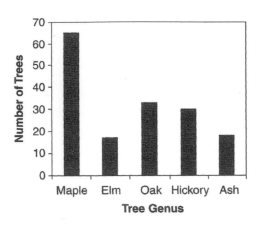

such values would list many values on the X-axis, and each value would occur only once or twice. Classes are defined so that all possible values of the variable fall into one and only one class. That is, classes should be mutually exclusive and all encompassing. The width of the intervals that define classes (called **bin width**) is usually the same for all classes, and the value assigned to each class is usually the average of the upper and lower limit values of the class interval. (See Example 2.1.)

■ **EXAMPLE 2.1**

Putting values for a continuous variable into classes. Suppose we wanted to make a histogram of the following data for the variable "weight." The first step is to convert the continuous quantitative variable "weight" to the discrete quantitative variable "weight class."

Data Values

Weight	43	48	60	64	66	67	69	71	73	76	77	77	78	79	79	81	83	88	94	97
Weight Class	45	45	65	65	65	65	65	75	75	75	75	75	75	75	75	85	85	85	95	95

Data Organized to Produce a Histogram

Class Intervals Weight Values (lbs)	Bin Width	Midpoint Values (Listed on X-axis)	Frequency	Relative Frequency
40 to 49.9	10	45	2	0.1
50 to 50.9	10	55	0	0.0
60 to 60.9	10	65	5	0.25
70 to 70.9	10	75	8	0.40
80 to 80.9	10	85	3	0.15
90 to 99.9	10	95	2	0.1

There is usually no obviously "correct" bin width; selection of the best bin width is often done by trial-and-error. If the intervals are too narrow, the histogram will have many relatively short bars, and random variation in the data values will produce many peaks and gaps. If the intervals are too wide, the histogram will have few, tall bars, and any gaps, multiple peaks, and outliers in the data will be obscured. Somewhere in the middle of these extremes is a bin width that will produce a histogram that best displays the center, spread, shape, gaps, and outliers of the distribution.

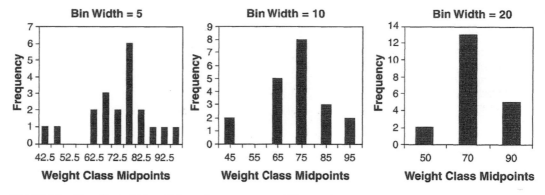

FIGURE 2.3 The effect of bin width on the appearance of histograms. These three histograms are all based on the same weight data presented in Example 2.1. Note how changing the bin width alters the appearance of the data distribution.

Usually, this optimal interval width is found only after looking at multiple histograms with different interval widths (see Figure 2.3).

When to use histograms.

Histograms that use an appropriate bin width can display all the basic characteristics of data distributions (center, spread, shape, gaps, outliers). Histograms are particularly useful when there are many (> 100) data values. Histograms have the advantage of having a "professional" appearance and of being familiar to many people. Hence, histograms are a good choice if you want to display a data distribution as part of a presentation. However, producing a good histogram by hand is tedious work, so a computer graphics program is generally required. Also, histograms are larger than some other types of graphical presentations of distributions, which may be a disadvantage if data for many variables or many different populations must be displayed.

Stem-Leaf Plots

The **stem-leaf plot** is a graphical display of data distributions that uses the data values themselves to make a picture of the distribution. To produce a stem-leaf plot, three steps are followed.

1. Each numeric data value is divided into two parts: (a) the **stem** (all the digits in the data value except the rightmost digit) and (b) the **leaf** (the rightmost digit of the data value).

2. The stem values are listed in a column. All stem values between the minimum and maximum stem values in the data distribution must be listed, regardless of whether or not they occur in the data. For example, the minimum data value in Example 2.2 is 291 and the maximum is 409. The stem values are the two leftmost digits of each data value. Therefore, the column of stem values should list all integers between 29 and 40.

3. The leaf values are listed in a row adjacent to their respective stem values. Within each row, the leaves are listed in numerical order, with the smallest leaf value next to the stem and the largest leaf value at the other end of the row (see Example 2.2).

Condensed Stem-Leaf Plot.

A **condensed stem-leaf plot**, formed by dropping the ones digit and using the tens digit as the leaf, is used in a special circumstance—when the range between the smallest and largest data values is large and there are relatively few data values. In this situation, the standard stem-leaf plot will have many stem rows, but with few leaves in any one row (and often many empty rows). To produce a more useful stem-leaf plot, the investigator can truncate (not round) the less significant digits (ones and tens digits), then separate stem and leaf from the remaining digits to create the plot.

■ **EXAMPLE 2.2**

Creating a stem-leaf plot. To create a stem-leaf plot of the data values listed below: (1) Find the smallest and largest data values, in this case 291 and 409. (2) Make a column of all whole numbers between stem values 29 (the stem for 291) and 40 (the stem for 409). (3) Write the leaf digit for each data value on the row with the appropriate stem value (for example, for the data value 320, the leaf is 0 and the stem is 32. Write the 0 on the row with the stem 32). (4) Reorder the leaf digits on each row such that the smallest leaf value is next to the column of stem values and the largest leaf is at the end of the row.

Data Values:

291, 303, 309, 310, 311, 316, 320, 320, 325, 332, 335, 337, 337, 338, 341, 342, 345, 350, 355, 363, 409

Stem | Leaf
29 | 1
30 | 39
31 | 016
32 | 005
33 | 25778
34 | 125
35 | 05
36 | 3
37 |
38 |
39 |
40 | 9

Condensed Stem-Leaf Plot

2 |9
3 |0011122233333444556
4 |0

Description of the data distribution: The data values range from 291 to 409. With the exception of the outlier value 409, the distribution is symmetric and centered around the value 330. There is a gap in the data distribution, with no data values for the stems corresponding to values 370 through 399.

The *condensed stem-leaf plot*, shows that the center of the data is in the 300s, but fails to show the gap from 360–400 and does not adequately display the 409 value as an outlier. For the condensed plot in Example 2.2, the ones digit was truncated, the tens digit was used as the leaf, and the hundreds digit became the stem. In this case, the condensed plot displays much less information about the distribution of data values than the uncondensed plot. Sometimes you will not be able to determine beforehand whether or not the data values should be truncated to make the stem-leaf plot (much like the problem of determining the optimal bin width for histograms). In these cases you should create the graphics both ways so that you can determine which way provides the clearest display of the distribution characteristics.

When to use stem-leaf plots.

Stem-leaf plots are most appropriate for quantitative variables with fewer than 100 data values; the rows of "leaves" become too long with larger data sets. This problem of larger data sets can be addressed by splitting each stem row so that leaf digits 0 to 4 are on one row and leaf digits 5 to 9 are on a second row with the same stem value (see Example 2.6). This graphic is very easy to create by hand and retains all the detailed information in the data; the original data values can be determined from the plot (unless digits were truncated). The stem-leaf graphic is an excellent tool for an initial quick-check exploratory data analysis. Also, a stem-leaf plot can be an efficient intermediate step in the determination of some summary statistics (described in Section 2.3). However, the appearance of stem-leaf plots is less "professional" looking than that of histograms, and the display of all those digits can be distracting to the viewer. Because more people are familiar with histograms than stem-leaf plots, histograms are more commonly used for presentations and in the popular media.

2.3
Summary Statistics to Describe Data Distributions

A common aim in modern science is to standardize the measurement of characteristics, including those of data distributions. This objective is usually attained by the development of standard ways to quantitate the characteristic of interest. In this context, "quantitate" refers to the numerical representation or measurement of an abstract concept. Although the five characteristics of a data distribution (center, spread, shape, gaps, outliers) can be shown graphically in histograms or stem-leaf plots (see Figures 2.1, 2.2, and Example 2.2), these graphical displays are somewhat subjective—different people looking at the same graphic often have somewhat different interpretations.

Summary statistics are numerical values that describe the characteristics of data distributions. As we shall see, there is often more than one way to quantitate concepts such as the center or spread of a data distribution, but the interpretation of a quantity is less subjective than that of a graphical display. This section will describe how various summary statistics are computed, and under what conditions each is a more or less appropriate way to describe the characteristics of data distributions. These summary statistics concisely convey the main pattern of the data in just a few numerical values. Another benefit of summary statistics is that they can be used in statistical tests; they provide for more objective and precise assessments of differences or associations than visual comparisons of graphics.

Summary Statistics for the Center or Location of a Data Distribution

Summary statistics can represent the **center** of a data distribution (the "middle" value of the data) and the **location** of data values along a numerical scale or graph axis. Often, but not always, the most common data values are near this middle value. There are three commonly used summary statistics for describing the center:

- The **mode** is the most frequent data value in a data distribution (i.e., the data value that corresponds to the highest bar or "peak" in a histogram). The value of the mode is of interest mainly when the data values are for discrete quantitative variables and there are relatively few distinct values in the data. The mode is rarely, if ever, used for statistical analyses.

- The **mean** (given the symbol $\bar{x}$ pronounced "*x*-bar") is the sum of the data values divided by the number of values in the sample (represented by the symbol n). This is also called the *average*, and is computed as:

$$\bar{x} = \sum_{i=1}^{n} x_i/n$$

The symbol Σ indicates summation of the individual data values (x_i) from the first ($i = 1$) to the last ($i = n$). This notation will be shortened to $\Sigma x/n$ hereafter.

The mean is the most widely used descriptor for the center of data distributions, and many statistical tests are based on it. However, if the sample size is small, a single outlier data value from an unusual individual in the sample can cause the value of the mean to change substantially. In such cases the value of the mean is not really in the center of the vast majority of the other data values (see Example 2.5).

- The **median** (given the symbol M) is the middle value of a ranked list of data values so that 50% of data values are less than the median and 50% are greater. The median is determined by following these steps:

1. Arrange the data values in a ranked list from smallest to largest (most efficiently done by constructing a stem-leaf plot).

2. If there is an *odd* number of values, the median is middle observation in the list.

3. If there is an *even* number of values, the median is computed as the average of the two middle observations in the list (see Example 2.3). *Note:* There are other, more complex methods for computing the median under various circumstances. Most scientists rely on statistics computer programs to compute the median.

When outliers in a small sample cause the value of the mean to deviate from the center of the bulk of the data values, the median should be used to describe the center of the data distribution. Even a large outlier has little or no effect on the value of the median.

Percentiles.

The *p*th **percentile** of a data distribution is the value of the variable such that p% of the values in the sample are less than that value. For example, the median is the 50th percentile. Two commonly reported percentiles are the 25th and 75th percentiles, also called the first and third *quartiles*, respectively. Twenty-five percent of the data values are less than the 25th percentile, and 75% are less than the 75th percentile.

Computing the quartiles (25th and 75th percentiles).

The 25th percentile is determined by finding the middle (median) of the data values *that are less than the median*, and the 75th percentile is the value that is in the middle of the data values *that are greater than the median*. *Important:* Do not count the median value when determining the quartiles, if the median is actually a data value (see Example 2.3).

■ EXAMPLE 2.3

Using a stem-leaf plot to determine the mean, median, and quartiles. (Data values are from Example 2.2.) *Note:* The leaves in each row must be in numerical order and the data values cannot be truncated if the plot is to be used to compute the median and quartiles.

There are 22 data values. Therefore, the median is the average of the middle two (11th and 12th) data values (the values 332 and 335). There are 11 data values that are less than the median of 333.5. The 25th percentile is the middle value of these 11 numbers. There are 11 data values greater than the median of 333.5. The 75th percentile is the middle value of these 11 numbers. (Although the data values 332 and 335 were used to compute the median by averaging them, neither of them is the median value. Hence, these data values *are* counted when determining the quartiles.)

```
29|1
30|39
31|016      25th percentile = 316
32|0059
33|2578     The median = (332 + 335)/2 = 333.5
34|1257     75th percentile = 345
35|05
36|3        The mean of this distribution is computed as the sum
37|         of the data values divided by the number of data
38|         values ( = 7318 / 22 = 332.6). Note that the mean is close to
39|         the median 333.5, indicating the outlier 409 had only a
40|9        modest effect on the value of the mean.
```

Summary Statistics for the Spread of a Data Distribution

Summary statistics for the **spread** of a data distribution quantify the amount of *variability* in the data values. This variability can be quantified in a number of ways, but a summary statistic that is based on a large number of data values contains better information than a statistic that is based on only a small subset of those data values. However, summary statistics for spread that are based on all data values may be overly affected by unusual outlier values.

- The **range** is the difference between the largest and smallest data values; it is computed as:

  Range = Maximum Data Value − Minimum Data Value

Although the range indicates the full spread of the data values, it is determined entirely by the two most extreme data values and does not reflect the variation among the remaining values. This measure of spread is rarely used.

- The **variance** (S^2) is the average of the squared differences between individual data values and the mean of the distribution. In this context, spread is interpreted as how far the data values are from their center. The simplest quantitative expression of this interpretation would be the average of the differences between data values and their mean. However, since the mean is in the middle of the data values, half these differences would be positive and half negative. Hence, the mean of these differences would always be zero. To eliminate the negative difference values for $(x_i - \bar{x})$, these differences are squared and then averaged.

<table>
<tr><td align="center">**Variance:**
Conceptual Formula</td><td align="center">**Variance:**
Computational Formula</td></tr>
<tr><td align="center">$S^2 = \dfrac{\Sigma(x_i - \bar{x})^2}{n - 1}$</td><td align="center">$S^2 = \dfrac{\Sigma x_i^2 - (\Sigma x_i)^2/n}{n - 1}$</td></tr>
</table>

Usually a conceptual formula is shown for ease of comprehension, and an equivalent formula is used for computation. Neither formula is "user-friendly," especially for large data sets. However, most scientific calculators have functions to compute the mean and variance for a set of data values. From the conceptual formula, it is easy to see that values for the variance are always positive (due to the "squared differences"), and that the variance increases when the data values are more spread out from the center ($\bar{x}$).

While variance is the most commonly used measure of spread in the calculations for statistical tests, it is less commonly used to describe the spread of the data distribution. Typically, scientists report both a summary statistic for center and a statistic for spread. The mean is reported in the same units as the original data values (meters, grams, etc.), but the variance is in squared units (e.g., m^2, g^2) due to the squaring of differences $(x_i - \bar{x})^2$.

- The **standard deviation** (S, sometimes abbreviated as SD) is the square root of variance ($\sqrt{S^2} = S$). This descriptive statistic for the spread of the data distribution is in the same units as the mean and the original data values, and is more commonly reported than the variance. To compute the standard deviation, first calculate the variance, then find the square root of that value.

- The **interquartile range** (IQR) is the difference between the 25th and 75th percentiles; it expresses the range of the middle 50% of the data values. For example, the IQR for the data values in Example 2.3 is $345 - 316 = 29$. If the median is used as the measure of center, the IQR is generally used as the measure of spread because both are based on the ranking of data values rather than the data values themselves.

Example 2.4 displays the calculations for various summary statistics that describe the center and spread of a data distribution.

■ **EXAMPLE 2.4**　　　　**Calculations of summary statistics.**

Data Values (x)	x^2
8	64
12	144
15	225
21	441
23	529
26	676
27	729
35	1225
38	1444
40	1600
43	1849
46	2116
49	2401
52	2704
54	2916
57	3249
63	3969
66	4356
74	5476
77	5929
Σx = 826	Σx^2 = 42,042

Stem-Leaf Plot

```
0 | 8
1 | 25
2 | 1367
3 | 58
4 | 0369
5 | 247
6 | 36
7 | 47
```

25th percentile = (23 + 26)/2　　　= 24.5

Median　　　= (40 + 43)/2　　　= 41.5

75th percentile = (54 + 57)/2　　　= 55.5

IQR　　　= 55.5 – 24.5　　　= 31

Sample Size (n)　　　= 20

Mean ($\bar{x}$)　　= $\Sigma x / n$ = 826/20　　= 41.3

Variance (S^2)　　= $\{\Sigma x^2 - [(\Sigma x)^2/n]\}/(n-1)$
　　　= $\{42042 - [(826)^2/20]\}/(20-1)$　= 417.3

Standard Deviation (S) = $\sqrt{417.3}$　　　= 20.43

Resistant (Robust) vs. Sensitive Measures of Center and Spread

A **sensitive statistic** is one whose value can be greatly altered by even a single unusual data value (outlier). Because all data values are used to compute the mean, variance, and standard deviation, these summary statistics are sensitive to outliers. When a outlier is present, the mean may not fall in the center of the majority of the data values, and the variance and standard deviation will overstate the amount of variation in the data values. When outliers are present, sensitive statistics may not be the best descriptors the distribution.

> **NOTE** Sensitive statistics become less sensitive if sample size is increased. As the number of observations becomes larger, the influence of any single outlier is reduced. Hence, when the sample size is large, the mean and standard deviation may be used. ■

A **resistant (robust) statistic** is one whose value is relatively unaffected by the presence of outliers in the data. The median and IQR are determined by the relative ranking of data values and data values near the middle of the data. These summary statistics are minimally affected by outliers at the extreme ends of the data distribution. See Example 2.5 for a comparison of sensitive vs. resistant summary statistics for data distributions that do and do not include outlier values.

■ EXAMPLE 2.5

Effect of an outlier on resistant vs. sensitive summary statistics.

Small Sample Size No Outlier	Small Sample Size Outlier Present	Larger Sample Size Outlier Present
Data Values	**Data Values**	**Data Values**
1 2 3 4 5 6 7 8 9 10	1 2 3 4 5 6 7 8 9 **100**	1 1 2 2 3 3 4 4 5 5 6 6 7 7 8 8 9 9 10 **100**
Summary Statistics	**Summary Statistics**	**Summary Statistics**
$\bar{x}$ = 5.5	$\bar{x}$ = 14.5	$\bar{x}$ = 10
Median = 5.5	Median = 5.5	Median = 5.5
S = 3.03	S = 30.2	S = 21.4
IQR = 5	IQR = 5	IQR = 5

We can make to following observations on the effect of an outlier on summary statistics for these three samples.

- Because all data values are involved in the computation of $\bar{x}$ and S, these measures of center and spread are much changed when the largest data value is changed from a 10 to 100, creating an outlier.

- Because the median and IQR are determined mainly by the rank order of data values (*not* by the values themselves), changing the maximum data value 10 to 100 had no effect on these measures of center and spread.

- When the sample size is small ($n = 10$), changing the largest data value from 10 to 100 substantially changes the value of $\bar{x}$ and S. When the sample size is doubled to 20 data values, the impact of the outlier on the mean and SD is reduced.   ▩

Describing the Shape of a Distribution

While summary statistics that describe the center and spread of data distributions are the most commonly reported, the overall **shape** of a data distribution is also important. Many statistical procedures are appropriate only if the distribution is **Normal** (bell-shaped). Hence, descriptions of the shape of distributions are often expressed in terms of how the shape deviates from this symmetric, single-mode bell shape. In this chapter we will describe the shape of data distributions based on the visual interpretation of graphics because quantitative descriptions are difficult to compute by hand and are rarely used. The following four terms are used to describe the shape of a data distribution.

- **Symmetry.** If the two halves of a distribution to the either side of the center are mirror-images, the distribution is called *symmetric*. An example of a symmetric distributions is the bell-shaped Normal distribution (See the top panel in Figure 2.4).

- **Skewness.** If most of the data values are clumped at one end of the range of values and there are few data values across the rest of the range, the distribution is called *skewed*. If most values are at the low end of the range and the upper end of a distribution is a drawn-out "tail" of values that have low frequency, the distribution is called *positively skewed*. By *tail* we mean the outer end of the distri-

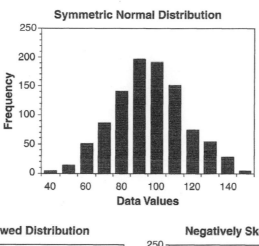

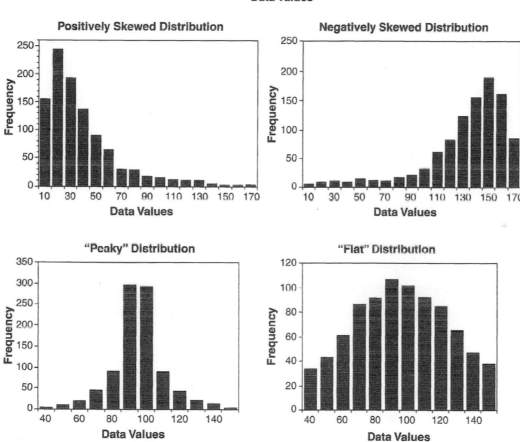

FIGURE 2.4   **Examples of different distribution shapes.**

FIGURE 2.5   Example of a multimodal distribution.

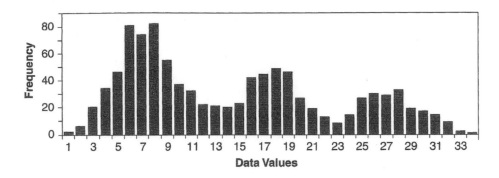

bution that has few values. If most data values are clumped at the upper end of the range of values and the lower end of the distribution has a long tail, the distribution is called *negatively skewed*. By definition, a skewed distribution is not symmetric. (See the middle two panels in Figure 2.4 for examples.)

- **Kurtosis.** This term compares the shape of a distribution *relative to the bell-shaped Normal distribution*. A "peaky" (leptokurtic) distribution has more data values near the center and fewer values in the tails of the distribution than does a Normal distribution. A "flat" (platykurtic) distribution has a smaller proportion of data values near the center and more data values in the tails of the distribution than a Normal distribution. (Compare the bottom two panels in Figure 2.4 to the bell-shaped distribution in the top panel.)

- **Unimodal vs. Multimodal.** A **mode** is a peak in a data distribution corresponding to data values that occur frequently. A *unimodal* distribution has only one peak, a multimodal distribution has two or more peaks, often indicating that the sample included individuals from two or more distinct subpopulations. (See Figure 2.5.)

Box-and-Whisker Plots (Boxplots)

Box-and-whisker plots display what is called the *five-number summary* of a data distribution, which includes the maximum data value, the 75th percentile, the median, the 25th percentiles, and the minimum data value (see Figures 2.6 and 2.7). The "box" part of the boxplot is shown as a vertical rectangle whose horizontal lines indicate the three values of the 25th percentile (bottom of the box), median (center of the box), and 75th percentile (top of the box). Extending from the box are the two "whiskers," indicating the smallest and largest values (see Figure 2.7).

Modified boxplots also display individual data values that are outliers; these outliers can be identified by the following procedure:

How To Identify Outlier Values for a Modified Boxplot

1. Compute the 25th and 75th percentiles and the interquartile range (IQR).
2. Any data value less than [25th percentile − 1.5(IQR)] is an outlier. That is, subtract 1.5 times the value of the IQR from the value of the 25th percentile. If the data value is less than this number, it is an outlier.
3. Any data value greater than [75th percentile + 1.5(IQR)] is an outlier. That is, add 1.5 times the value of the IQR to the value of the 75th percentile. If the data value is more than this number, it is an outlier.

If an outlier is present, it is indicated as a point, and the end of the whisker for that side of the distribution becomes the smallest (or largest) data value that is not an outlier (see Figure 2.6). Some computer graphics software distinguish regular outliers (defined above in Steps 1–3) from extreme outliers, which are defined as 3(IQR) beyond the quartiles, rather than just 1.5(IQR). Different plot symbols are used for each. (See Figure 2.10 for an example.)

FIGURE 2.6. Interpreting a modified boxplot. This modified boxplot displays the distribution of the data values used to make a stem-leaf plot in Example 2.3. The data value 409 is an outlier because it is greater than 75th percentile + 1.5(IQR) = 345 + 1.5(29) = 388.5. Therefore, the upper whisker in the modified boxplot ends at 363, the largest data value that is not an outlier.

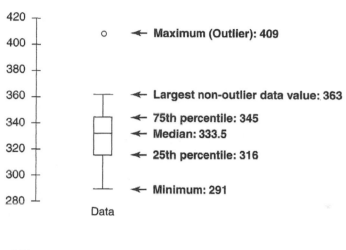

FIGURE 2.7 Simple boxplots for data distributions of different shapes. Symmetric Normal distributions have whiskers of equal length and the median line is in the middle of the box. Symmetric "peaky" distributions have a short box and long whiskers, while flat distributions have a long box and short whiskers. Skewed distributions have one whisker that is longer than the other and the median line is not in the middle of the box. These boxplots are based on data similar to that displayed using histograms in Figure 2.4.

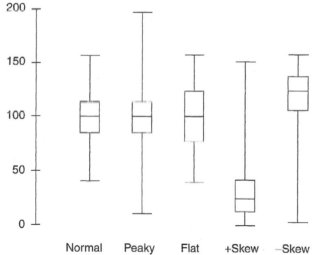

Use of boxplots.

Boxplots are relatively easy to draw by hand, once the five-number summary is determined; they are equally useful for both large and small numbers of data values. Boxplots can display the center (median), spread (range, IQR), shape (see Figure 2.7), and outliers of a data distribution. However, boxplots do not display gaps or multiple modes in the data (see Figure 2.8). Hence, boxplots present less information about the distribution of a variable than do histograms or stem-leaf plots. However, boxplots are smaller than histograms and stem-leaf plots, and provide for clearer comparisons of data distributions for many groups (see Section 2.4). Finally, box-plots are unfamiliar to most people, and so require more explanation if used in a presentation to the public.

FIGURE 2.8 Comparison of boxplot and histogram graphics derived from the same data. Note that the boxplot does not display the multimodal nature of the distribution.

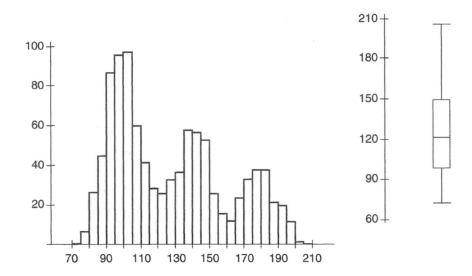

2.4
Graphics for Comparing Multiple Distributions

A common exercise in data analysis is to compare the data distribution for a particular variable between two or more groups. For example, in a study of hypertension (high blood pressure) in people over age 65, we might want to compare the data distributions for men and women (who tend to have fewer heart attacks than men). In an experimental study, we might want to compare the blood pressure data values from a treatment group that received a new drug to the distribution of values from a control group who did not receive the drug. Statistical tests compare the values of summary statistics computed from samples taken from various populations to determine if there is strong evidence that the populations differ with regard to some parameter. However, much of the detailed information embodied in the data values is lost in the calculations of statistical tests. Graphical comparisons of data distributions provide for a more complete comparison and are often more easily assimilated and understood than statistical tests. Also, graphical comparisons of data distributions are often used in public presentations of study results (journal articles or oral presentations at scientific conferences).

A major trade-off in selecting graphic formats for comparing distributions is between presenting much detailed information versus the clarity of the comparisons. Graphs that include much detailed information are often cluttered, and variation among individual data values (noise) may obscure the pattern you wish to display. If the purpose of a graph is to clearly display the pattern of differences among groups, the graphic format should be one that reduces the amount of detailed information presented. For example, variations on the basic histogram and stem-leaf graphics can be used to compare distributions (see Example 2.6 and Figure 2.9). However, both these graphics contain much detailed information, and are not effective for comparing more than two or three distributions. In contrast, side-by-side boxplots present the main features of the data distribution while greatly reducing the amount of detailed information. This graphic is the most effective for comparing many distributions in a single graph.

By further elaborating on the basic graphics of histograms, stem-and-leaf plots, and boxplots, we can get even more effective comparisons.

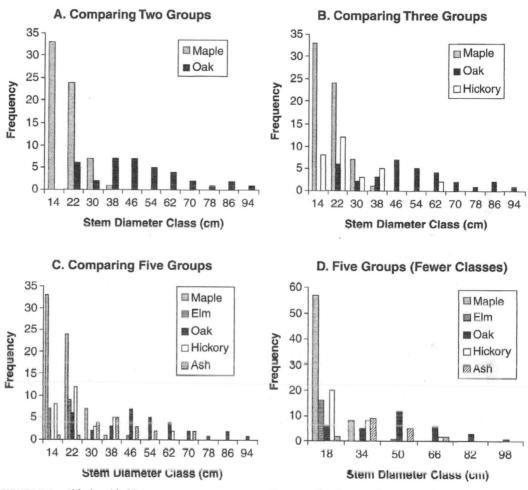

FIGURE 2.9   Side-by-side histograms to compare stem diameter distributions for 2, 3, and 5 groups of tree species.   Note that as the number of compared groups increases, it becomes progressively more difficult to make comparisons among the distributions. This problem can be alleviated somewhat if the number of diameter classes on the X-axis is reduced by increasing the bin width (e.g., compare graph C and graph D). However, increasing bin width can reduce the amount of information in the histogram, especially regarding multiple peaks and gaps in the distributions.

Side-by-Side Histograms

We can combine histograms for a single variable measured on multiple samples (groups) to facilitate comparisons of distributions among the different groups. A separate histogram bar for each group is presented above each class of the variable listed on the X-axis. **Side-by-side histograms** can be quite effective for comparing data distributions among two to three groups. However, if four or more distributions are compared, the side-by-side histogram becomes very difficult to interpret and another graphic format should be used. See Figure 2.9 for examples of side-by-side histograms.

Back-to-Back Stem-Leaf Plots

Back-to-back stem-leaf plots are an easily produced graphic for comparing *two* data distributions. They are produced just like the simple stem-leaf plot, except that the leaf values for two groups of data values are listed on opposite sides of a single column of stem values. On both sides of the stem values, the leaf digits are listed in order, with the smallest leaf values next to the stem value and the largest leaf values at the end of the row (see Example 2.6). Obviously, this format can be used to compare only two distributions. Also, if you are creating a graphic for a public presentation, side-by-side histograms provide a more professional appearance and are more familiar to most people.

■ **EXAMPLE 2.6**

Back-to-Back Stem-Leaf Plot. The data plotted below are Math SAT scores from a random sample of 100 students at a university. This plot was constructed to compare scores between male and female students. The original data values spanned a range from 321 to 800, so a stem-leaf plot using the ones digits for leaf values would have 48 stem values for displaying 100 data values. Because it would be of little use to make a plot showing many stem values, each with just a few leaf values, the ones digit of the original data values was dropped. For example, the lowest data value (321) was converted to 32. The resulting condensed stem-leaf plots displayed relatively few rows (stems 3 to 8), with many values per row. Therefore, the row for each stem value was split into two rows. The first row for each stem value lists leaf values from 0 to 4, the second lists leaf values from 5 to 9. The actual range of SAT data values included on each row is listed to the right of the back-to-back stem-leaf plot to assist you in interpreting the plot. For example, on the first row of the plot, there are two male SAT scores (one in the 320s and one in the 340s) and one female SAT score (in the 330s).

Male		Female	
Leaf	Stem Values [in hundreds]	Leaf	SAT Values
42	3	3	300–349
985	3	7899	350–399
432100	4	00112334	400–449
998777665	4	56678899	450–499
4443332221	5	01122233334	500–549
998655	5	556689	550–599
43321111	6	1223	600–649
755	6	567	650–699
40	7	2	700–749
8	7	557	750–799
	8	0	800–849

NOTE Simply comparing the two stem-leaf plots does not clearly identify any obvious differences between the distributions of SAT scores for male vs. female students. The approximate locations of the median and 25th and 75th quartiles of each distribution are indicated in boldface type. When you compare these summary statistics for the two distributions, male students in this sample have slightly higher math SAT scores than female students. ■

Side-by-Side Boxplots

To compare the center, spread, shape for many groups of data values you can use a **side-by-side boxplot**. This graphic also emphasizes clarity of the comparison by reducing detailed information about individual data values in each group the five-number summary of maximum, 75th percentile, median, 25th percentile, and minimum. For an example of this trade-off between detailed information and clarity of comparisons, compare the five-group side-by-side histogram in Figure 2.9 with the five-group side-by-side boxplots in Figure 2.10.

FIGURE 2.10   Side-by-side boxplots for the tree diameter distributions presented in Figure 2.9C and D.   Outliers (1.5 × IQR criterion) are indicated by an open circle plot symbol. Extreme outliers (3 × IQR criterion) are indicated by an asterisk plot symbol. Boxplots lose much of the detailed information in the data, but display differences of center, spread, and shape among multiple data distributions more clearly than do histograms.

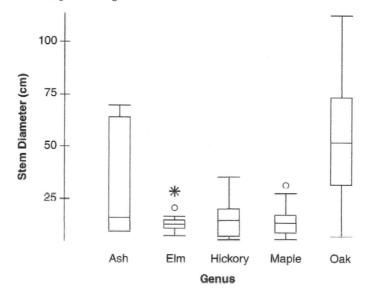

To summarize, graphics for comparing distributions include: (1) side-by-side histograms, (2) back-to-back stem-leaf plots, and (3) side-by-side boxplots. Each type of graphic has advantages and disadvantages, depending on the number of groups to be compared and the amount of information about the distributions contained in the graphic, as listed below:

Graphic	Number of Groups That Can Be Compared	Information Displayed by Graphic
Back-to-Back Stem-Leaf Plot	Only 2	Maximum amount of information, easy to produce, but unfamiliar to many people.
Side-by-Side Histogram	2–3	Emphasis on detail, easily understood by most people, "professional" appearance, but difficult when making specific comparisons.
Side-by-Side Boxplots	2 or more	Emphasis on clarity of comparisons, but sacrifices some detail (gaps, multiple modes) and is unfamiliar to many people. Best graphic to compare distributions for more than three 3 groups.

2.5

Graphics to Display Associations Between Variables

Although studying differences among the data distributions for a single variable measured on multiple groups (the topic of Section 2.4) is a common exercise in science, equally common is the study of associations between two variables. The simplest study design to collect data to study an association would involve selecting a representative group of sample or experimental subjects and making measurements for both variables on each subject. That is, the data are comprised of *paired measurements* of two variables. We say there is an **association** between the two variables if, as the value of one variable changes, the value of the other variable also changes in a systematic manner. For example, as the values of one variable increase, the matched values of the second variable also tend to increase.

Examples: There is an association between a person's height and weight, in that taller people tend to weigh more than shorter people. In organisms that do not regulate their body temperature (plants, invertebrates, fish, amphibians, and reptiles), as environmental temperature increases, so does the organism's metabolic rate. When analytical chemists use a colorimetric test, they use an association between the color density of a solution (as measured by absorption of light of a specific wavelength) and the concentration of the substance in standard solutions. In toxicology experiments for new chemicals, organisms like *Daphnia* (water flea) are exposed to several concentrations of the chemical. If the chemical is toxic, increasing the concentration will be associated with decreasing growth and reproduction. ▪

The graphic that is most commonly used to study associations between variables is the **scatterplot**, which shows the association between two variables as a set of points in an *x-y* coordinate system.

Constructing a Scatterplot

1. Values for two variables, X and Y, are measured on each sample or experimental subject. If there is reason to believe that the association between the two variables is the result of a cause-effect linkage, X is the "cause" variable and Y is the "effect" variable.

2. The pairs of values are treated as ordered pairs (x, y) and plotted on an *x-y* coordinate system (X on the horizontal axis and Y on the vertical). See Figure 2.11.

FIGURE 2.11   Example of the construction of a scatterplot.

Data Values

X	Y
1	1
2	5
3	2
4	4
5	3
6	8
7	6
8	7
9	10
10	9

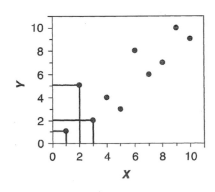

Interpreting Scatterplots: Describing the Nature of an Association

- If the array of plotted points indicates a pattern, such that large values of the X variable tend to be paired with large values of the Y variable, and vice versa, this indicates a **positive association** between the variables (Figure 2.12).

- If the array of points indicates that when X is large Y is small, and vice versa, this indicates a **negative association** between the variables.

- If the array of points has no clear single axis (i.e., no pattern), or if the central axis of the array of points appears to be a horizontal or vertical line, there is **no association** between X and Y.

- If the array of points in the scatterplot is oblong and clustered about a central, linear axis, the association is **linear** (i.e., a unit change in X is associated with a constant amount of change in Y, regardless of the value of X).

FIGURE 2.12 Example scatterplots for various types of association between two variables.

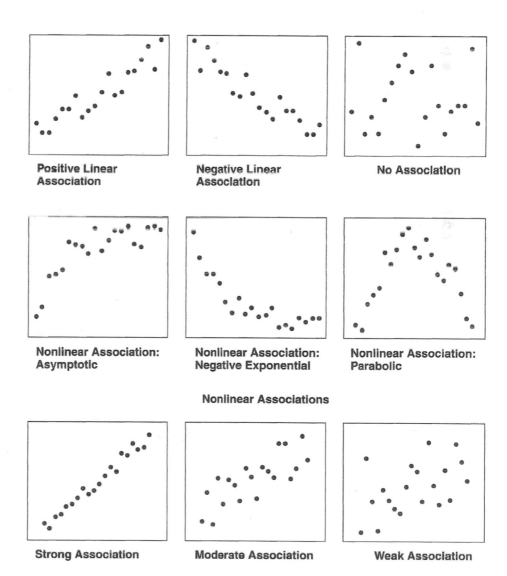

Positive Linear Association

Negative Linear Association

No Association

Nonlinear Association: Asymptotic

Nonlinear Association: Negative Exponential

Nonlinear Association: Parabolic

Nonlinear Associations

Strong Association

Moderate Association

Weak Association

- If the array of points in the scatterplot is clustered around an axis that curves, the association is **nonlinear** (i.e., the amount of change in Y for a unit change in X depends on the X values involved; the slope of the line is not constant).

Interpreting Scatterplots: Assessing the "Strength" of Associations

While the shape of the cloud of points in a scatterplot tells us much about the nature of the association between two variables (positive, negative, linear, nonlinear), the amount of "scatter" in the cloud of points indicates the strength of the association. The concept of "strength of association" is most easily explained if we put it in the context of a true cause-effect relationship. A strong association is one where the cause variable is the only factor that controls the values of the response variable. For example, when chemists perform a colorimetric test, the concentration of the substance in the solution is the only factor that influences the color density (as measured by absorption). A scatterplot for a strong linear association would have all the points representing paired values of the two variables fall close to a straight line with a non-zero slope (see bottom left panel of Figure 2.12). In contrast, a weak association is one where many "cause" variables influence the values of the response variable. For example, a person's weight is influenced not only by their height, but also by variables like bone density and amount of body fat. Because people of the same height can weight different amounts, the association between weight and height is weak. A scatterplot for a weak linear association will have a dispersed "cloud" of points, oriented about a central linear axis (see bottom right panel of Figure 2.12). The greater the scatter of points away from the central linear axis, the weaker the association. When interpreting scatterplots, the central linear axis is the "Pattern," and the scatter of points about the linear axis is the "Noise." The closer the points fall to the central linear axis, the greater the pattern in the data relative to noise.

Time Plots to Display Associations among Variables

Time series data are multiple measurements on the same subject(s) at various points in time. A **time plot** has the time scale on the X-axis, and the data values are on the Y-axis. Each point is an ordered pair (x, y), with the time (year, day, hour, minute, or second) as the x-value and the measured data value as the y-value. These ordered pairs of values could be plotted as points, producing a graphic that looks like a scatterplot. It is common to connect the points in a time plot with a line, which makes it easier to determine the temporal sequence of data values. Sometimes, the data values are plotted against time simply to determine if the values change over time. For example, a time series of height values measured on a child during the time interval between birth and maturity would display the height growth curve of that child. In the description of scatterplots, I indicated earlier that the X-axis variable is often the "cause" variable in a cause-effect relationship. In time plots, "time" is *not* a cause variable. Many variables change over time, but these changes are caused by factors that happen to occur over time (e.g., growth of a child occurs over a period of years but is caused by cell division and tissue maturation, not simply by time).

One kind of evidence that is used to identify an association between two variables is **coincident variation over time**. That is, if we measure variables Y and Z over time, the change in Y values over time *coincides* with the change in Z values. If there is a positive association among Y and Z, when Y increases, so does Z (e.g., Figure 2.13). If there is a negative association, whenever Y increases, Z decreases at the same time (e.g., Figure 2.14). In Figures 2.13 and 2.14, scatterplots and time plots display the association between the width of annual growth rings for white oak trees in southern

FIGURE 2.13   The time plot in part A shows the coincident variation of oak annual ring width and June rainfall that indicates a positive association. The solid line represents annual ring width and the dashed line represents June rainfall. In years when June rainfall is high, annual ring width is also larger, and vice versa. This is evidence for a positive association, as indicated by the scatterplot in part B.

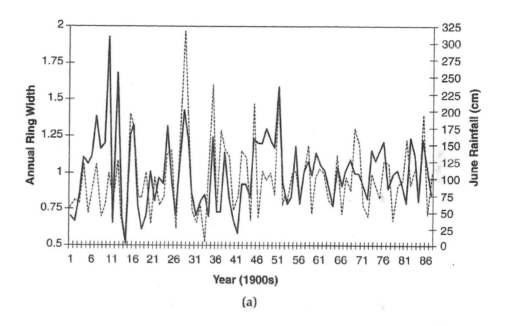

(a)

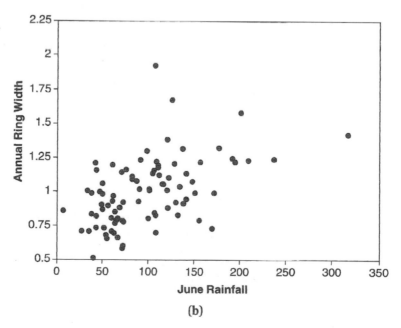

(b)

Illinois and June rainfall and June temperature. The growth of oak trees in this location is often limited by summer droughts (long periods of low rainfall and high temperature). Hence, when rainfall is abundant and temperatures cool, the growth of trees is enhanced, producing a wide annual ring of wood. When rainfall is low and temperatures is high, the growth of the trees is inhibited and a narrow annual growth ring is produced. The scatterplots display the nature and strength of these associations. The time plots allow us to see the coincident year-to-year variation and provide detailed information that cannot be obtained from the scatterplot.

FIGURE 2.14   The time plot in part A shows the coincident variation of oak annual ring width and June temperature that indicates a negative association.   The solid line in part A displays ring width; the dashed line displays June temperature. In years when June temperature is high, annual ring width is small. This indicates a negative association, as displayed by the scatterplot in part B.

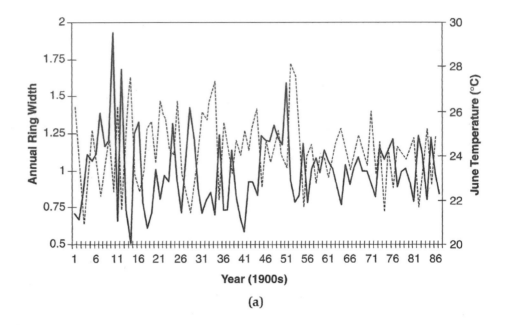

(a)

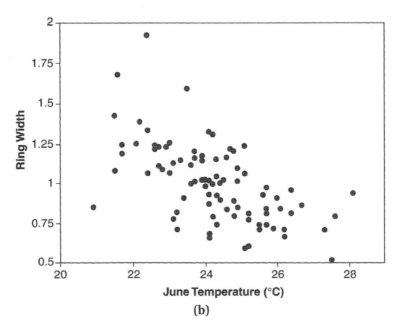

(b)

CHAPTER SUMMARY

1. Exploratory data analysis is an essential tool Always use EDA
   a. To identify unexpected patterns or unusual data values
   b. To identify unusual values that might be due to error (measurement or typographical)
   c. To determine if the assumptions for statistical tests are fulfilled
2. **Data distributions** display the pattern of variation in values of a variable from a sample All data distributions have the following two parts (a) a list or range of all values in the sample data and (b) for each distinct value, the number (frequency) or proportion/percent (relative frequency) of times the value occurs in the data
3. **Graphical displays** of data distributions help us to understand our data
   a. **Histograms** and **stem-leaf plots** display the details of the data distribution (including center, spread, shape, gaps, and outliers)
   b. **Boxplots** reduce the details of the data down to a **five-number summary** (maximum, 75th percentile, median, 25th percentile, and minimum), but do not display gaps or multiple modes
4. **Summary statistics** are quantitative descriptors of the center, spread, and shape of data distributions, but do not describe characteristics such as gaps and outliers
   a. The **mean** and **standard deviation** are quantitative expressions of the center and spread of a data distribution, respectively These summary statistics are *sensitive* to unusual values (**outliers**) That is, their values are much affected by the presence of outliers, especially if sample size is small These statistics should not be used in these cases

   b. The **median** and **interquartile range** (IQR) are also quantitative expressions of the center and spread of a data distribution, respectively These summary statistics are *resistant* to outliers (not much affected) and are the most appropriate for describing data distributions when sample size is small and outliers are present
5. In the choice of graphics and summary statistics the researcher must balance a desire for detailed information against the need for clarity of comparisons and patterns
   a. Histograms and stem-leaf plots retain all or most of the detailed information contained in the raw data values However, these graphics are less useful for comparing data distributions because the detailed information makes it difficult to see patterns and differences
   b. Side-by-side boxplots sacrifice much detailed information but are the best graphic for comparing distributions of data values for multiple groups
   c. If comparisons need to be made among many groups or variables, boxplots and/or summary statistics are generally used in presentations and reports to emphasize the main points of comparison
6. **Scatterplots** are used to display the association between two variables and are used to determine (a) the *nature of the association* (positive, negative, linear, nonlinear) and (b) the *strength of the association* (based on the clustering of the points about their axis)
7. **Timeplots** display *temporally coincident* variation that may indicate an association

GO!

Enhance your understanding of the material presented in this chapter by doing the practice problems and exercises found on pages 2–1—2–21 in the workbook companion to this textbook.

3 Probability: How to Deal with Randomness

INTRODUCTION The application of statistics in the process of science can be divided into three parts: (1) obtaining data (experiment and sampling design), (2) summarizing and describing data (exploratory data analysis, descriptive statistics), and (3) using data from samples and experiments to make estimates and test competing hypotheses about the universe (inferential statistics). Chapters 1 and 2 dealt with obtaining and summarizing data. The remaining chapters in this book deal with concepts and methods for estimation and hypothesis testing.

A fundamental fact of life in science is that *all estimates and conclusions derived from sample data are uncertain*. Different samples randomly taken from the same population will produce different estimates of the population parameter value. This **random sampling variation** is always present, even in the best-designed studies. Because of random sampling variation, we can never say with certainty that the value of a sample statistic exactly equals the true value of a population parameter. Hence, any conclusion we draw based on the value of a sample statistic might be incorrect. Scientists must quantify just how uncertain their estimates and conclusions are to convince others of the validity of their judgments.

In many aspects of our daily life we deal with uncertainty by using the concept of probability as a measure of how likely or unlikely is a particular outcome or event. **Probability** can be defined as the relative frequency of an event. That is, if you observed a very large number of outcomes from a random phenomenon, the proportion of outcomes that meet the description of a specific event is an estimate of the probability of that event. For example, when a meteorologist says there is an 80% probability of rain today, this means that it rained on 80% of days when similar conditions prevailed in the past. Given this information, you decide to bring an umbrella when you go outside. If the forecast had indicated there was only a 10% chance of rain, you might not have bothered. The statement of probability gives you a quantitative measure of how likely it is that it will rain. You use this information to make an informed decision (bring the umbrella or not). Sometimes you bring the umbrella, but don't need it. Sometimes you decide to leave the umbrella behind, and later get wet. And sometimes, you bring the umbrella and use it to stay dry when the anticipated rain comes. When you make a decision based on uncertain evidence, sometimes your decision is correct, other times it is not.

A primary use of statistics is to persuade others that the estimates or conclusions you derive from sample data have a high probability of being correct, in spite of the uncertainty due to random sampling variation. Scientists evaluate the validity of hypotheses by determining the probability of getting the observed value of a sample statistic if the parameter value proposed by a hypothesis is true. The fundamental purpose of most statistical tests is to provide this quantitative estimate of probability. If the probability of getting the observed value is small, this is taken as evidence that the hypothesis is false and the scientist makes the decision to reject the hypothesis. Sometimes this decision will be wrong, but the vast majority of the time it will be correct. This is how scientists deal with uncertainty. The probability-based approach to using data as evidence for testing competing hypotheses requires that scientists have at least a basic understanding of probability. ▪

CHAPTER GOALS AND OBJECTIVES

In this chapter you will learn the basic concepts of probability and how to apply them to assign probabilities to the outcomes of simple random phenomena.

By the end of this chapter, you will be able to:

1. Define random phenomena and how they are described using probability distributions.

a. Distinguish random vs. haphazard phenomena.

b. Describe the sample space of possible outcomes.

c. Use a probability distribution to determine which outcomes in the sample space are more or less likely.

2. Use the rules of probability to:

a. Identify invalid probability statements and probability distributions.

b. Assign probabilities to single events, the union of disjoint events (probability of event A *or* event B), and the intersection of independent events (probability of event A *and* B) for simple random phenomena.

3. Describe how simple random phenomena are quantitatively represented using random variables.

4. Describe how empirical probabilities and theoretical probabilities are derived, and explain the relationship between these probabilities, as stated in the Law of Large Numbers.

3.1

Describing Random Phenomena Using Probability Distributions

A **random phenomenon** has individual outcomes that are not predictable, but the probabilities associated with the possible outcomes are well-defined. This is different from a **haphazard phenomenon,** for which the probabilities associated with the various possible outcomes are unknown. For example, when you roll a standard six-sided die, it is not possible to predict the outcome of any single toss. However, if the die is "fair," we do know that in the long run each of the six sides has an equal probability of being face up ($1/6 = 0.167$). Suppose you did not have a die, but claimed that you could "randomly" choose numbers from 1 to 6. In this case, not only is it impossible to know the outcome of any one number you might choose, it is not possible to know the probability associated with each value from 1 to 6. We cannot know if you have a "favorite number" that is more likely than others to be chosen. In fact, we have no idea about the process by which you would choose numbers. No one has the ability to produce truly random sequences of numbers off the top of their head. When people try to be "random," they are usually being haphazard.

The **probability distribution** of outcomes for a random phenomenon is comprised of two parts: (1) a listing of all possible outcomes for a random phenomenon (called the **sample space**) and (2) the probabilities associated with each outcome. For example, if the random phenomenon is flipping a fair coin once, the sample space is {heads, tails}, and each of these two outcomes has a probability of 0.5. If the random phenomenon is rolling a single fair six-sided die, the sample space is {1, 2, 3, 4, 5, 6 }, and each outcome has a probability of $1/6 = 0.167$.

When we study random phenomena, we are usually interested in determining the probability associated with some specific event. An **event** is defined as a combination of outcomes from a random phenomenon that meet a specific criterion. For example, rolling a six-sided die is a random phenomenon with a sample space of six possible outcomes {1, 2, 3, 4, 5, 6}. One event might be "An odd number of dots appears." The sample space of outcomes that fall within the definition of this event is {1, 3, 5}. Another event could be "The number of dots is less than 3." Outcomes that meet the definition of this event are {1, 2}. Given knowledge or assumptions about a random phenomenon (e.g., the coin or die is fair), the rules of probability can be applied to determine probabilities associated with events defined by outcomes of that phenomenon.

3.2

Rules for Probability and Probability Distributions

Basic Rules of Probability

- *The value of a probability must fall within the range from 0 to 1.* Sometimes statements about probability are made in terms of percents; in that case valid probability values must be between 0 and 100%. If a coach is asked by a reporter to estimate the probability his team will win in the upcoming game, and he replies "110%", he is engaging in exaggeration by using an invalid probability statement, and likewise, if the coach is asked the likelihood of his team losing the game and he replies, "Less than zero."

- *The sum of the probabilities associated with all possible outcomes in the sample space of a random phenomenon is always 1.0.* Note that the examples of probability distributions for flipping a coin or rolling a die in the previous section meet both Rules 1 and 2. Any probability distribution that does not meet these two rules is invalid. When you are asked to determine probability distributions for random phenomena in later chapters, you should always apply these two rules to determine if you have made any errors of logic or calculation.

- *For any event A in the sample space, the probability that A does not occur (written as A^c) is 1 minus the probability that A does occur.* This is called the **complement rule**:

$$P[A^c] = 1 - P[A]$$

For example, in a single coin toss, if A is the event {Getting a head} then A^c is the event {Not a head} or "getting a tail." A^c is the **complement** of A. This is a logical consequence of Rule 2 that the sum of probabilities for all events in a sample space must equal 1.0. *Note:* Throughout this text, the notation $P[\ ]$ is used to represent probability of whatever is within the square brackets.

■ **EXAMPLE** The probability of getting 1 when you roll a fair die is 0.167. The complement of this event is "Not getting 1," which includes individual outcomes {2}, {3}, {4}, {5}, {6}. We usually write them as the sample space {2, 3, 4, 5, 6}. The probability of the event "Not getting 1" can be computed as $1 - 0.167 = 0.833$. This rule is used often as a labor-saving device. For example, to determine P [Not 1], you could sum the probabilities associated with the set of outcomes {2, 3, 4, 5, 6}. Alternatively, you could use the complement rule: P [Not 1] = $1 - P$ [1]. ■

Probability of the Union of Events

The **union of events** is the combination of their outcomes. For example, suppose that a fair six-sided die is cast, event A is "Getting a 1 or 2" and event B is "Getting 6". The union of these two events, indicated as **A or B**, is the combination of all outcomes {1, 2, 6}. This union of events is said to have occurred if any one of the outcomes in the combination occurs.

Disjoint events.

When we want to determine the probability associated with the union of two events, it is important to determine if the events are disjoint. **Disjoint events** are events that do not have any outcomes in common. Events A "Getting 1 or 2" and event B

"Getting 6" in the example above are disjoint. Suppose I rolled a fair six-sided die and defined event C as "Getting a number of dots evenly divisible by 2" and event D as "Getting a number of dots evenly divisible by 3." The possible outcomes for event C are $\{2, 4, 6\}$ and the outcomes for event D are $\{3, 6\}$. Because events C and D share a common outcome $\{6\}$, they are *not* disjoint.

Suppose we randomly selected an individual from the U.S. population and event A is that the person be male and event B is that the same person is Republican. Are these events disjoint? To answer this question you must determine whether or not it is possible for a single person to be both male and Republican. Because a single person can be both male and Republican, these two events are *not* disjoint.

The **simple addition rule of probability** is used to determine the probability associated with the union of two events. *If events A and B are disjoint*, the probability associated with the union of these events is obtained by summing the probabilities associated with each event:

Simple Addition Rule of Probability

$$P[A \text{ or } B \text{ or } C] = P[A] + P[B] + P[C]$$

In this case, this is the probability that any one of events A or B or C will occur. (See Example 3.1.) This simple addition rule does not apply to events that are not disjoint. (The general addition rule must be used in such cases, and is described later in the chapter.)

■ **EXAMPLE 3.1**

Finding the probability of the union of disjoint events. Suppose a fair die is rolled and event A is "Getting a number < 2" that has outcome $\{1\}$, and event B is "Getting a number > 4," that has outcomes $\{5, 6\}$. These two events are disjoint, and union of these two events is $\{1, 5, 6\}$. The probability of this union of events can be computed as:

$$P[< 2] = 1/6$$
$$P[> 4] = 2/6$$
$$P[< 2 \text{ or } > 4] = 1/6 + 2/6 = 3/6$$

■

Nondisjoint events.

Suppose a fair die is rolled and Event A is "Getting a number divisible by 2" that has outcomes $\{2, 4, 6\}$, and event B is "Getting a number divisible by 3" that has outcomes $\{3, 6\}$. The union of these events is $\{2, 3, 4, 6\}$. Because all outcomes of rolling a fair die are equally likely, with a probability of $1/6 = 0.167$, the probability associated with this union of events is $4/6 = 0.667$. If the simple addition rule described above were applied to compute the union of these events:

Event A: $P[\text{Evenly divisible by 2}] = 3/6$

Event B: $P[\text{Evenly divisible by 3}] = 2/6$

$P[A \text{ or } B] = P[A] + P[B] = 3/6 + 2/6 = 5/6$ Not the true probability

Note that the probability computed using the simple addition rule (5/6) is different from the true probability (4/6). This discrepancy is due to the "double-counting" of the outcome $\{6\}$ that is part of the descriptions for both these nondisjoint events. We will see how to compute the true probability for the union of nondisjoint events later in the chapter.

Probability of the Intersection of Events

The **intersection** of two or more events is the event that *all* events will occur. For example, if event A is {A person selected is a male} and event B is {That person is a Republican}, the intersection of these events, indicated as **A and B**, would occur if a randomly chosen individual from the U.S. population was both male *and* Republican. For two events to have an intersection, it must be possible for a single observation of the random phenomenon to meet the criteria for both events. If events A and B are disjoint, they have no intersection and it is impossible for them to occur together, i.e., P [A and B] = 0.

When determining the probability associated with the **intersection of two events, P [A and B]**, it is important to determine if the two events are independent. Events A and B are **independent** if the probability P [B] is in no way related to, or affected by, whether or not event A has occurred, and vice versa. For example, flipping a fair coin twice produces two events (tosses) of a random phenomenon. Whether or not the outcome of the first toss is {heads} is in no way related to whether or not heads will occur on the second toss. Hence, these two events are independent.

The **simple multiplication rule of probability** can be used to determine the probability associated with the intersection of two events. *If events A and B are independent*, the probability that both events will occur can be computed by multiplying the probabilities associated with each of the two events:

Simple Multiplication Rule of Probability

$$P \text{ [A and B]} = P \text{ [A]} \times P \text{ [B]}$$

(See Example 3.2.) The converse of this statement is also true. If P [A and B] = P [A] × P [B], then A and B are independent events.

■ **EXAMPLE 3.2**

Finding the probability of the intersection of independent events. The gender of the second child born to a couple is not influenced by the gender of their first child. The probability that a girl will be born is 0.5 for each birth. Hence the gender of the second child is independent of the gender of the first child. Using the simple multiplication rule, the probability that a girl will be born on the first birth *and* a girl will be born on the second birth is P[girl *and* girl] = P[girl] × P[girl] = 0.5 × 0.5 = 0.25.   ■

Nonindependent Events.

Let's say that 30% of all people who go to a hospital emergency room are later admitted to the hospital. In other words, P [Admitted] = 0.3. Let's also say that 20% of the people who go to the emergency room do not have health insurance, in other words, P [No health insurance] = 0.2. Applying the simple multiplication rule, the probability that a person who goes to the emergency room without health insurance *and* would be admitted to the hospital = 0.2 × 0.3 = 0.06. This probability is correct only if admission to the hospital is independent of whether or not the patient has health insurance, a premise that may or may not be true. (Example 3.3 shows an application of the independence assumption along with some illustrations of various probability rules.)

■ **EXAMPLE 3.3**

Applications of the rules of probability. In a standard deck of 52 cards, there are 2 colors (red and black, 26 cards each), 4 suits (heart, diamond, club, spade, 13 cards each), and 13 denominations in each suit (2 to 10, jack, queen, king, ace). The relative frequency of cards in a deck that meet criteria of color, suit, or denomination is the probability of drawing a card that meets those criteria. For example, the probability of drawing a red card = 26/52 = 0.5.

a. Find the probability that a single card drawn will be a red king (i.e., red *and* king). *Note:* Whenever you see a question about the occurrence of one event *and* another event, this indicates the intersection of events and the simple multiplication rule should be applied.

$$P \text{ [red and king]} = P \text{ [red]} \times P \text{ [king]} = 26/52 \times 4/52 = 2/52 = 0.03846$$

There are an equal number of kings for each color of card, so the probability of getting a king is independent of the color. Hence, the simple multiplication rule can be applied.

b. Find the probability that a single card drawn will be a king or an ace. *Note:* When the question involves whether one event *or* another event will occur, this indicates the union of events and the addition rule should be applied. Because a single card cannot simultaneously be a king and an ace, these two events are disjoint and the simple addition rule can be applied.

$$P \text{ [king } or \text{ ace]} = P \text{ [king]} + P \text{ [ace]} = 4/52 + 4/52 = 8/52 = 0.15385 = 0.57692$$

c. Find the probability that a single card drawn will be a king or will be red. "Getting a king or a red card" describes the intersection of two events. However, it is possible for a single card to meet both the criteria defining the two events in this example (i.e., be a king and be red). Therefore, these are *not* disjoint events. Let's see what happens when we apply the simple addition rule when the events violate the "disjoint events" assumption for the rule.

$$P \text{ [king } or \text{ red]} = P \text{ [king]} + P \text{ [red]} = 4/52 + 26/52 = 30/52$$

The true number of cards in a standard deck of 52 that meet this description include the 26 red cards plus the 2 black kings, for a probability of 28/52. The computed probability of 30/52 is incorrect because the two red kings were counted twice, once for being kings, and again for being red. This double-counting is why the events must be disjoint for the simple addition rule to provide correct probability statements.

d. What is the probability that two cards drawn from the same deck will both be face cards (jack, queen, king)? That is, what is the probability the first card drawn will be a face card *and* the second card will be a face card? Because this question involves the probability of one event *and* another event, the simple multiplication rule should be applied. However, this problem violates the "independence" assumption for the simple multiplication rule because removing cards from the deck changes probabilities for subsequent draws. Let's see what happens when the independence assumption is violated and we incorrectly use the simple multiplication rule of probabilities:

$$P \text{ [face } and \text{ face]} = P \text{ [face]} \times P \text{ [face]} = 12/52 \times 12/52 = 0.05325$$

This is not the correct probability for this experiment because if a face card is obtained on the first draw, the probability of getting a face card on the second card becomes 11/51. Hence, the correct probability of getting two face cards is computed as:

$$P[\text{face } and \text{ face}] = P[\text{face}] \times P[\text{face}] = 12/52 \times 11/51 = 0.04977 \quad ■$$

3.3

Probability Distributions for Random Variables

Random variables are quantitative representations of the outcomes of random phenomena. Because random sampling variation is always present in scientific studies, all sample statistics (means, medians, proportions, standard deviations) are random variables, and all statistical analyses are based on probability distributions for random variables.

One of the simplest kinds of random variables is the *count* of the number of times a specific outcome of a random phenomenon occurs out of some number of observations. For example, suppose a couple plans to have four children and they are interested in the gender mix of their children. One random variable could be X = number of girls out of four children. Another could be Y = number of boys out of four children. Each of these random variables has a sample space of {0, 1, 2, 3, 4}. Throughout this book, I will use uppercase letters from the end of the alphabet (X, Y, Z) to designate random variables. I use the notation $P[X]$ to indicate the probability of getting a specified value of X. In Example 3.4 I show how the probability distribution for a random variable can be used to determine probabilities for specific events defined by the value of that variable. The probability distribution in Example 3.4 can be derived based on logic and deduction, as will be shown in Chapter 4.

■ **EXAMPLE 3.4**

Probability distribution for the random variable X = number of girl children out of four births.

X	0	1	2	3	4	Sample space for X
$P[X]$	0.0625	0.25	0.375	0.25	0.0625	Probabilities

The probabilities associated with each value of the random variable X (Number of girls out of four births) allow us to determine which outcomes are more or less likely. The most likely outcome is that the couple will have two girls out of four births. It is also fairly unlikely that the couple would have all girls (X = 4) or all boys (X = 0).

What is the probability the couple will have at least one girl if they have a total of four children (i.e., what is $P[X \geq 1]$)? If they have 1 or 2 or 3 or 4 girls, any one of these outcomes would meet the criterion $X \geq 1$. Because these are disjoint events, the simple addition rule can be applied:

$$P[X \geq 1] = P[X = 1] + P[X = 2] + P[X = 3] + P[X = 4]$$

$$= 0.25 + 0.37 + 0.25 + 0.0625$$

$$= 0.9375$$

The complement rule provides an easier way to determine this probability:

$$P[X \geq 1] = 1 - P[X = 0] = 1 - 0.0625 = 0.9375$$

Once the probability distribution for a random variable has been determined, you can compute the probability associated with any one value or combination of values for that variable. ▪

3.4

Theoretical Probability vs. Empirical Probability

Theoretical probability values are based on assumptions about the nature of the random phenomenon and the application of the rules of probability. For example, when we flip a coin there are only two possible outcomes in the sample space {heads, tails}. If we can assume that the coin is fair, each of these two outcomes is equally likely to occur. Given the rules of probability, there is only one probability distribution for the random phenomenon of flipping a fair coin:

Outcomes:	Heads	Tails	
Probabilities:	0.5	0.5	← Two equal probabilities that sum to 1.0

The alternative way to determine probabilities for outcomes of a random phenomenon is to observe many repetitions of the random phenomenon and determine the long-term relative frequency for each distinct outcome in the sample space. A probability estimated from observed long-term relative frequencies is called an **empirical (data-based) probability**, and is computed as:

$$\text{Probability of event } A = \frac{\text{Number of occurences of event } A}{\text{Total number of observations}}$$

The empirical approach for determining probabilities is particularly appropriate when there is some question as to whether or not the assumptions about a random phenomenon are valid. Most of the probability problems and examples in this chapter deal with simple random phenomena like flipping a coin, rolling a die, drawing a card, or gender determination of humans. We were able to determine the theoretical probabilities of specific events by making assumptions like "the coin is fair," so all outcomes are equally likely. However, in many situations such assumptions are not correct. In these circumstances, it may not be possible to derive a theoretical probability distribution. We must observe many repetitions of the random phenomenon to learn the probabilities of its various possible outcomes.

A logical question to ask when you develop an empirical probability distribution is, "How many repeated observations of the random phenomenon do we need to obtain good empirical probabilities?" Let's take a simple example. The assumption that a coin is fair requires that the coin be exactly balanced so that each side is equally likely to land face-up. However, it is possible that by accident or design a coin is not balanced and that one side is heavier than the other. In this case, the heavier side will be more likely to land face-down and the lighter side face-up. Under such circumstances, the only way to determine the probability of the event ("Getting heads") is through empirical observation. How many times should the coin be flipped so that the relative frequency of heads represents the true probability of this outcome? Suppose you flip the coin 10 times and you get 6 heads. Is this evidence that the coin is not fair? You're not sure, so you continue to gather data by flipping the coin a total of 100 times, and count 45 heads. Is this evidence that the probability of a heads is not 0.5? You're still not sure, so you continue flipping the coin for a total of 1,000 times and you count 505 heads. Is this enough data, or do you need more? Figure 3.1 displays the outcome of a computer simulation of flipping a fair coin up to 128,000 times to determine the empirical value for P[Getting heads].

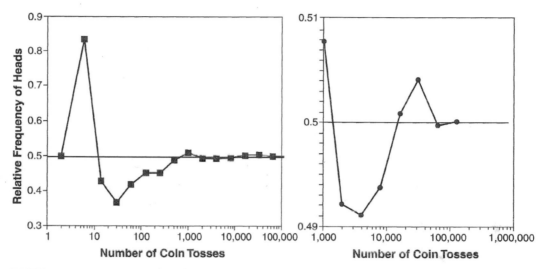

FIGURE 3.1   Computer simulation of flipping a fair coin, with the true probability P[Getting heads]= 0.5, to determine the relative frequency of heads as an empirical estimate of P[heads].   The left graph shows the results for 2 to 128,000 coin tosses. The right graph shows only the results for 1,000 to 128,000 coin tosses, and the Y-axis scale has been expanded to better see subtle variation in the relative frequency of heads as the number of observations increased. Note that as the number of observations increases, variation in the relative frequency of heads diminishes and the empirical probability approaches the true probability of 0.5 (horizontal line). However, a certain amount of deviation of the empirical from the true probability remains, even with more than 1,000 observations.

The question "How much data is enough to empirically estimate the probability of an event?" is answered by the **Law of Large Numbers:** *As the number of repeated, independent observations of a random phenomenon increases, the observed relative frequency of a specified event A will approach the true probability P[A].* Hence, the more data you have, the better will be your empirical estimate of the true probability of an event.

3.5
The General Addition Rule of Probability

The **general addition rule of probability** is used to determine the probability of the union of two events, *regardless of whether or not the events are disjoint.* Using this rule, the probability that event A or event B will occur, indicated as P[A or B], can be computed using the formula below:

General Addition Rule of Probability

$$P[A \text{ or } B] = P[A] + P[B] - P[A \text{ and } B]$$

The problem we saw with using the simple addition rule when the events were not disjoint was caused by double-counting the outcomes they have in common. If we subtract the probability associated with the union (overlap) of the nondisjoint events, P [A and B], the effect of this double-counting is eliminated. If events A and B are disjoint (have no outcomes in common), by definition the probability of A and B occurring will be zero. When P [A and B] = 0, this general formula reduces to the simpler formula given in Section 3.2 for disjoint events. Examples 3.5 and 3.6 allow you to compare the simple addition rule and the general addition rule in practice.

▪ **EXAMPLE 3.5**

Using the general addition rule. Determine the probability that a single card drawn from a standard deck will be a king *or* red. Note that a card can be both a king *and* red, so these are *not* disjoint events. If we use the simple addition rule we get the probability:

$$P \text{ [king } or \text{ red]} = P \text{ [king]} + P \text{ [red]} = 4/52 + 26/52 = 0.577 \text{ (30 out of 52 cards)}$$

However, there are only a total of 28 out of 52 (0.5385) cards that meet one or the other of these events (2 black kings + 26 red cards, including 2 kings). Applying the simple addition rule when events are not disjoint has produced an incorrect probability. However, if we use the general addition rule, the correct value for P[king *or* red] is computed as:

$$P \text{ [king]} + P \text{ [red]} - P \text{ [king and red]} = 4/52 + 26/52 - (4/52)(26/52)$$
$$= 0.5385 \text{ (28/52)} \qquad ▪$$

▪ **EXAMPLE 3.6**

Using the general addition rule. What is the probability that a single card drawn from a standard deck will be red or a heart? Let's try the general addition rule.

$$P \text{ [red]} + P \text{ [heart]} - P \text{ [red and heart]} = 26/52 + 13/52 - (26/52)(13/52)$$
$$= 0.625 \text{ (32.5 cards out of 52)}$$

This result is incorrect because the actual cards that would fulfill this description of an event are the 26 red cards, as all hearts are also red cards. The error in applying the general addition rule formula in this case is that whether or not a card is a heart is *not independent* of whether or not it is red. If a card is red, there is a 0.5 probability it is a heart. If a card is not red, there is a 0.0 probability that the card is a heart. When two events are not independent, it is not valid to use the simple multiplication rule to compute P [red *and* heart] = P [red] $\times$ P [heart]. When two events are not independent, it is necessary to use the general multiplication rule to determine the probability P[A and B]; this rule will be discussed next. ▪

3.6

The General Multiplication Rule of Probability

The General Multiplication Rule of Probability

The **general multiplication rule of probability** is used to determine the probability of the intersection of two events, *regardless of whether or not the events are independent*. The probability that both event A and event B will occur, indicated as P [A and B], can be computed using the formula below:

General Multiplication Rule of Probability
$$P \text{ [}A \text{ and } B\text{]} = P \text{ [}A\text{]} \times P \text{ [}B|A\text{]}$$

In this formula, the term $P[B|A]$, means "the probability of event B *given* that event A has happened," and is called the **conditional probability** of event B. If events A and B are independent, the probability of event B is not influenced by the occurrence of event A and $P[B|A] = P[B]$. However, if event B is somehow dependent on event A, then $P[B]$ will vary, depending on the outcome of event A. Examples 3.7 and 3.8 show the reasoning behind using the general multiplication rule.

■ **EXAMPLE 3.7**	**Using the general multiplication rule.** What is the probability that a card drawn from a standard deck is red *and* a heart? The two events "Getting a red" and "Getting a heart" are *not* independent. If a card is black, the probability it is a heart is zero. If the card is red, the probability it is also a heart is 13/26. The denominator term is the number of cards in the deck defined by the first event A "Getting a red." The numerator is the number of cards that meet criterion B, "Getting a heart." The value 13/26 is the conditional probability $P(\text{heart} \mid \text{red})$, that is, the probability that if a red card is drawn, it will be a heart.

$$P[\text{red } and \text{ heart}] = P[\text{red}] \times P[\text{heart} \mid \text{red}] = 26/52 \times 13/26$$
$$= 13/52 \ (0.25)$$

Going back to the question of Example 3.6, what is the probability that a single card drawn from a standard deck will be red *or* a heart? The correct computation is:

$$P[\text{red}] + P[\text{heart}] - P[\text{red and heart}] = 26/52 + 13/52 - 13/52$$
$$= 26/52 \text{ (or 0.5)} \quad ■$$

■ **EXAMPLE 3.8**	**Using the general multiplication rule.** Suppose that a university typically admits 50% of all applicants, and 10% of all applicants have combined SAT scores less than 600. Using the simple multiplication rule, the probability that a student with a combined SAT score less than 600 would be admitted by this university is computed as:

$$P[\text{SAT} < 600 \text{ and admitted}] = P[\text{SAT} < 600] \times P[\text{admitted}]$$
$$= 0.1 \times 0.5$$
$$= 0.05 \ (5\%)$$

However, any college student knows that admission to a university is *not* independent of their SAT scores, and only under very special circumstances would someone with combined SAT scores less than 600 be admitted.

How might we determine the **conditional probability** $P[\text{admitted} \mid \text{SAT} < 600]$? (I abbreviate "SAT scores" as SAT.) The *conditions* that determine whether or not a person with SAT less than 600 is admitted to a college are usually unknown, so the conditional probability can be determined only from an observed relative frequency of this combination of events. If a college receives 10,000 applications, of which 10% are from students with SAT less than 600, this would translate to 1,000 applications with SAT less than 600. Suppose that 10 of these 1,000 applicants with SAT less than 600 were admitted to the college; the relative frequency of students being admitted when they have SAT less than 600 is the estimate of the conditional probability: $P[\text{admitted} \mid \text{SAT} < 600] = 10/1000 = 0.01$ or 1%. The number of students who meet the condition SAT < 600 defines the denominator of the relative frequency calculation, and the number of students who are admitted "given" this condition is the numerator.

Given the conditional probability P [admitted | SAT < 600] = 0.01, we can now use the general multiplication rule to compute the probability that an applicant to this university would have a combined SAT score less than 600 and would be admitted:

$$P \text{ [SAT < 600 } and \text{ admitted]} = P \text{ [SAT < 600]} \times P \text{ [admitted | SAT < 600]}$$
$$= 0.1 \times 0.01$$
$$= 0.001$$

One in 1000 students who apply to this university would fulfill these two criteria. ▪

CHAPTER SUMMARY

1. **Probability** is a quantitative measure of how likely or unlikely are the possible outcomes of a random phenomenon Valid values for probabilities are within the range from 0 to 1 If probability is expressed as a percent, the valid range of values is 0 to 100%

2. **Random phenomena** are those for which specific outcomes cannot be predicted, but the likelihood of outcomes can be determined

3. **Random variables** are quantitative representations of random phenomena

4. **Probability distributions** for random variables are comprised of two parts (a) a list of all possible values of the variable, called the sample space, and (b) probabilities associated with all possible values The sum of the probabilities for all possible values in the sample space must equal 1 0 for the probability distribution to be valid

5. **Simple addition rule of probability:** If two events are disjoint (i e , events that have no outcomes in common), then the probability that one *or* the other event will occur can be computed as the sum of the probabilities of the two events

$$P \text{ [}A \text{ or } B\text{]} = P \text{ [}A\text{]} + P \text{ [}B\text{]}$$

6. **Simple multiplication rule of probability:** If two events are independent (i e , the probability that one event will occur is in no way influenced by whether or not the other event occurs), then the probability that both events will occur can be computed as the product of the probabilities of the two events

$$P \text{ [}A \text{ and } B\text{]} = P \text{ [}A\text{]} \times P \text{ [}B\text{]}$$

7. **Theoretical probabilities** for the values of a random variables are derived based on assumptions about the nature of the random phenomenon, the rules of probability, and logic If the assumptions and logic are valid, theoretical probabilities are exactly correct

8. **Empirical probabilities** for values of a random variable are derived from observations of many outcomes from the random phenomenon The *relative frequency* of a specific event out of the total number of observations is an estimate of the probability of that event The empirical estimate of a probability does not require any assumptions about the nature of the random phenomenon However, empirical estimates of probability are influenced by random variation and do not provide the exact value of the true probability

9. **The Law of Large Numbers** states that as the number of observations used to estimate an empirical probability increases, the value of the empirical probability estimate will approach the true probability of the event

10. **The general addition rule of probability:** When two events are *not* disjoint, the probability $P \text{ [}A \text{ or } B\text{]} = P \text{ [}A\text{]} + P \text{ [}B\text{]} - P \text{ [}A \text{ and } B\text{]}$

11. **The general multiplication rule of probability:** When two events are *not* independent, the probability $P \text{ [}A \text{ and } B\text{]} = P \text{ [}A\text{]} \times P \text{ [}B \text{ | } A\text{]}$ The term $P \text{ [}B \text{ | } A\text{]}$ is the **conditional probability** of the occurrence of event B, given that event A has occurred

GO!   ***Enhance your understanding of the material presented in this chapter by doing the practice problems and exercises found on pages 3–1—3–13 in the workbook companion to this textbook.***

4 Developing Probability Distributions for Binomial Random Variables

INTRODUCTION Even though it is not essential that a scientist be able to mathematically derive the probability distributions used in statistical analyses, I believe it is useful to understand the basic elements of this process. In this chapter I will present the logic and process used to derive the probability distribution for a simple sample statistic: the count of the number of individuals in a sample that meet some specified criterion. This probability distribution can be used in real-world statistical analyses, but its use is relatively uncommon. In later chapters of this book, you will learn about several other probability distributions that are the basis for widely used statistical analyses. Although I will not present the mathematical derivation for any of these other distributions, the basic logical process used to derive the simple probability distribution presented in this chapter is similar to that used to derive the more complex probability distributions introduced in later chapters. ▪

CHAPTER GOALS AND OBJECTIVES

In this chapter you will be able to derive, interpret, and use probability distributions for the count of X individuals in a sample of n that meet a specified criterion when the following conditions are fulfilled: (1) sample size (n) is fixed and (2) the true probability P that any randomly chosen individual meets the criterion is constant and independent of whether or not other individuals in the sample meet that criterion.

By the end of this chapter, you will be able to:

1. Identify when a count is a *Binomial random variable*.
2. Derive the probability distribution of a *Binomial count variable* using the following process:
   a. Use a systematic process for identifying all possible outcomes when a specified number of observations (n) are made on the outcomes of a random phenomenon.
   b. Determine the sample space for a Binomial count variable based on the list of possible outcomes of repeated observations of a random phenomenon.
   c. Use the rules of probability to determine probabilities for all values in the sample space of a Binomial count variable.
3. Describe the relationship between *theoretical* and *empirical probability distributions* for Binomial count variables.
4. Use the *Binomial formula* and the *Binomial table* as more efficient methods for obtaining the probability distribution for a Binomial count variable.
5. Use the *Binomial distribution* to determine the probability of specific events defined by a Binomial count variable.

4.1

Identifying When a Count Is a Binomial Random Variable

Binomial random variables are based on observations of random phenomena that are categorized based simply on whether or not a specified outcome occurs. For example, is the individual a male or not, does the individual have blood type AB or not, is the individual HIV positive or not. In the first chapter of this book I described several different types of variables (categorical, measured, discrete, continuous). By definition, a variable is a quantitative representation of some characteristic of an individual, a sample, or a population. Characteristics such as gender, blood type, or HIV status do not seem to meet this definition, as they are not represented by a number. However, scientists often analyze data for categorical variables based on the **count** of the number of individuals in a sample or population that fall into a specified category. These counts do meet the stipulation that variables are quantitative representations of a characteristic of a sample or population.

The count of the number of observations or individuals in a sample that belong to specified category is a **Binomial random variable** if the following four conditions are met:

1. The total number of observations or individuals in the sample taken from the population of interest is a fixed value, specified before the sample was taken. That is, the sampling design for a Binomial variable is *valid*. For example:

   - *A valid sampling design for a Binomial variable:* Randomly select 20 individuals from the population, count the number of individuals in the sample who have blue eyes.

   - An *invalid sampling design for a Binomial variable:* Randomly select individuals from the population until five blue-eyed individuals are obtained (indeterminate sample size). This design is invalid because the total number of individuals is not a fixed value, specified ahead of time.

2. The individuals or observations in the sample are randomly chosen from the population of interest such that all individuals in the population have an equal probability of being selected. Any deviation from random sampling violates this condition.

3. The true probability that any one randomly chosen individual or observation from the population will fall into the specified category (P) is constant. This condition is violated if the population size is small, such that the sampling of one individual changes the probability that the next individual will fall into the specified category.

4. The probability that a randomly chosen individual or observation will fall into the specified category (P) is independent of whether or not other individuals in the sample fall into that category. This condition can be violated two ways: (1) The population is small, so that whether or not one individual falls into the specified category influences the probability that the next individual sampled will fall into that category. (2) There are deviations from true random sampling.

For conditions 3 and 4 to be fulfilled, the size of the population must be much larger than the number of individuals in the sample. For example, the probability that the first card dealt from a standard deck of 52 cards will be a face card is 12/52. If the first card dealt is a face card, this changes the probability that the next card dealt from the same deck will also be a face card (11/51). Whenever there is a small number of individuals in a population, sampling one individual alters the probability that the next individual sampled will fall into a specified category, violating condition 3. However, if the population is very large, this change in probability is so small that it can be ignored. How much larger should the population be compared to the sample size for conditions 3 and 4 to be fulfilled?

Rule of thumb for population size: $N > 100n$ (where N is the population size and n is the sample size).

Example 4.1 illustrates the use of these four criteria to identify valid Binomial random variables.

Determining if counts are valid Binomial random variables.

a. X = the count of the number of times the one-dot side of a die comes up out of 100 tosses of a fair die. The sample size n is fixed at 100, $P[1 \text{ dot}] = 1/6$, and the outcome of one roll of the die is not influenced by the outcome of any other roll. **Valid.**

b. X = the count of individuals who are HIV positive in a random sample of $n = 100$ students taken from the student body of a large university with a population of 18,000 students. The sample size n is fixed at 100, $P[\text{HIV positive}]$ = true proportion of students who are HIV positive ($\boldsymbol{P}$), which is a constant at the point in time when the sample was obtained, and the population size is much larger than the sample size, so sampling has a negligible effect on $P[\text{HIV positive}]$ for each individual in the sample. **Valid.**

c. X = the count of individuals in a sample of $n = 45$ who have blood lead concentrations high enough to cause toxicity, with blood samples taken from all members of 10 randomly chosen families (parents and children living at home). **Invalid:** This design violates the independence condition. The two main causes of lead toxicity are lead leaching from water pipes where lead solder was used and lead in paint chips/dust. Parents and minor children share the same home, and are all exposed to whatever lead exists there. Hence, all individuals in families that live in older homes with lead in paint and water pipes are more likely to have high blood lead concentrations than individuals in families that live in newer homes that do not contain lead in paint and pipes.   ∎

4.2
Deriving the Probability Distribution for a Binomial Random Variable

This chapter demonstrates the logical deductive process for defining theoretical probability distributions. The derivation of probability distributions for Binomial random variables can be done through the following step-wise process: (1) Define all possible outcomes of the random phenomenon under study; (2) define the sample space of the Binomial random variable X, listing which random outcomes result in specific values for X; and (3) assign probabilities to each random outcome and each value of X.

Listing All Possible Outcomes of the Random Phenomenon

The first step in deriving the probability distribution for any random variable is to identify the sample space of possible values. This requires that you be able to describe all the possible outcomes of the random phenomenon. Although this could be done in a number of ways, the systematic "event-tree" approach described in Example 4.2 reduces the likelihood that you will overlook any possible outcomes.

Determining the Sample Space for the Binomial Count Variable (X)

Because a Binomial random variable is based on counts, this type of variable is a discrete quantitative variable. That is, it is possible to list all the possible values a Binomial random variable can take, and to assign probabilities to each value. For example, if you flip a fair coin three times and count X = the number of heads, this random variable has four possible values $\{0, 1, 2, 3\}$. If you take a random sample of 30 students at a large university and count X = the number who have consumed alcoholic beverages in the past week, this random variable can only have the values {0, 1, 2, 3, . . . , 29, 30}. Example 4.3 shows how the possible outcomes determined in Example 4.2 are translated into values of random variables.

■ **EXAMPLE 4.2**

An event-tree approach for listing all possible Binomial outcomes for a random phenomenon. The diagram below displays the application of the **event-tree approach** for determining the outcomes of flipping a fair coin three times On the first coin toss, the possible outcomes are H (heads) or T (tails). Given that the first toss comes up heads, the second toss could come up H or T. Given that the first toss was H and the second toss was H, the third toss could come up H or T. The lines connect the three independent events that produce each of the possible outcomes listed.

Coin Toss

1st	2nd	3rd	Outcome	P [Outcome]			
		H	HHH	0.5 × 0.5 × 0.5	=	0.125	= 1/8
	H	T	HHT	0.5 × 0.5 × 0.5	=	0.125	= 1/8
H		H	HTH	0.5 × 0.5 × 0.5	=	0.125	= 1/8
	T	T	HTT	0.5 × 0.5 × 0.5	=	0.125	= 1/8
		H	THH	0.5 × 0.5 × 0.5	=	0.125	= 1/8
	H	T	THT	0.5 × 0.5 × 0.5	=	0.125	= 1/8
T		H	TTH	0.5 × 0.5 × 0.5	=	0.125	= 1/8
	T	T	TTT	0.5 × 0.5 × 0.5	=	0.125	= 1/8

It is important to note that these eight outcomes are separate and distinct, even though some outcomes produce the same number of heads and tails. From the diagram, the outcomes THH, HTH, and HHT all result in two heads and one tail. However, the one tail happened on the first coin toss in one case, the second toss in another, and the third toss in the last. These are different **combinations** of outcomes that result in the same counts of heads and tails. The fact that there are three ways (combinations) to get two heads and one tail makes these count values more likely to occur than getting all heads or all tails, each of which can occur only in one way. ■

■ **EXAMPLE 4.3**

Translating multiple Binomial observations of a random phenomenon into values of a Binomial count variable. A Binomial count variable describes the number of times (X) that independent observations of a random phenomenon meet a specified criterion, out of a sample of n observations. For the example of flipping a coin three times, two different random count variables could be defined: (1) X = the number of heads and (2) Y = the number of tails. The table below lists the possible values for both of these random count variables, and the combinations of outcomes that produce these values.

Random Count Variables Combinations of Outcomes	X No. of heads	Y No. of tails
HHH	3	0
HHT HTH THH	2	1
HTT THT TTH	1	2
TTT	0	3

So from multiple Binomial observations of a random phenomenon, we now have values of a Binomial count variable. $X = 0, 1, 2, 3$, or $Y = 0, 1, 2, 3$. ▓

Determining Probabilities for All Values in the Sample Space of the Binomial Count Variable (X)

Step 1. *Determine probabilities for each of the possible outcomes of the Binomial random phenomenon.* If each of the observations of the outcome of a random phenomenon is independent of the others, then we can use the simple multiplication rule to determine the probability for the combination of observations in the outcome. For example, if we toss a fair coin $n = 3$ times, the probability that we will get the outcome HTH can be computed as:

$$P[H \text{ and } T \text{ and } H] = P[H] \times P[T] \times P[H] = 0.5 \times 0.5 \times 0.5 = 0.125$$

The probabilities for all possible outcomes of flipping a fair coin three times are listed in the last column of Example 4.2. Note that in cases where the probability for the two possible Binomial outcomes of a random phenomenon are equal (e.g., $P[\text{heads}] = 0.5$ and $P[\text{tails}] = 0.5$), all the possible outcomes of a sample of n independent observations of that Binomial phenomenon will be equally likely to occur. In Example 4.4, there were eight possible outcomes from flipping a fair coin $n = 3$ times. Because $P[\text{heads}] = P[\text{tails}] = 0.5$, all eight of these possible outcomes have the same probability of 0.125. This probability was computed using the simple multiplication rule. However, using the basic rules of probability (probability values must be between 0 and 1, sum of probabilities for all values in sample space $= 1.0$), we could also compute the probabilities for eight equally likely outcomes as $1/8 = 0.125$. (See Example 4.4.) If the probabilities of the two possible outcomes for the Binomial random phenomenon are not equal, the multiplication rule must be used to compute the probabilities for the outcomes of n observations of the random phenomenon. (See Example 4.5.)

Step 2. **Sum the probabilities for outcomes that produce the same value for the count variable X.** When there are multiple combinations of outcomes that result in the same value of the count variable X, we must use the simple addition rule to determine the overall probability of getting that particular value of X. For example, there were three different outcomes (combinations) of tossing a fair coin three times that produced the count $X = 2$ heads (HHT, HTH, THH). Each of these three outcomes has a probability of $1/8 = 0.125$. To determine the probability $P[X = 2]$, we must recognize that this value for X would occur if any of the outcomes HHT or HTH or THH occurs.

Therefore, the probability that $X = 2$ heads is computed using the simple addition rule this way:

$$P[X = 2] = P[\text{HHT or HTH or THH}]$$

$$= P[\text{HHT}] + P[\text{HTH}] + P[\text{THH}]$$

$$= 0.125 + 0.125 + 0.125$$

$$= 3 \times 0.125$$

$$= 0.375$$

The probability for any specific value of the count variable X can be computed by multiplying the probability associated with any one of the outcomes that produce the value of X by the number of combinations that result in that value of X. (See Example 4.4.)

■ **EXAMPLE 4.4**

Determining the probability distribution for the Binomial random count variable X. The probability for any specific value of X is equal to the probability of any one combination that produces the value X times the number of combinations that all produce the same value of X.

Combinations of Outcomes	X (= # heads)	Probability $P[X]$
TTT	0	$1 \times 0.125 = 0.125$
HTT THT TTH	1	$3 \times 0.125 = 0.375$
HHT HTH THH	2	$3 \times 0.125 = 0.375$
HHH	3	$1 \times 0.125 = 0.125$

■

■ **EXAMPLE 4.5**

Developing the probability distribution for a Binomial random variable when outcomes are *not* equally likely. A young couple are considering having children, but the woman's mother was a hemophiliac. This means that the woman carries the defective gene that causes hemophilia (a disease that results in uncontrolled bleeding, and often results in severe restrictions on activities and premature death). If this couple have children, there is a probability of 0.25 for each birth that the child will be a hemophiliac. The couple would like to have four children, but want to know the probability that at least one of these children would be a hemophiliac. In this example I develop the probability distribution for the Binomial random variable X = Number of hemophiliac children out of 4, with probability $P[\text{Hemophiliac}] = 0.25$. This is done in four steps:

1. List all possible outcomes of four children.
2. Determine the probability for each outcome.
3. Determine the number of combinations of outcomes that produce the same value for X (Number of hemophiliac children).
4. Multiply the probability of an outcome that produces a particular X value by the number of combinations that produce that value to obtain $P[X]$.

In the event-tree diagram that follows, H = a hemophiliac child is born, and N = a nonhemophiliac child is born.

Child's Birth Order

1st	2nd	3rd	4th	Outcome	X	P [Outcome]	
			H	HHHH	4	(.25)(.25)(.25)(.25)	= 0.00391
			N	HHHN	3	(.25)(.25)(.25)(.75)	= 0.01172
			H	HHNH	3	(.25)(.25)(.75)(.25)	= 0.01172
			N	HHNN	2	(.25)(.25)(.75)(.75)	= 0.03516
			H	HNHH	3	(.25)(.75)(.25)(.25)	= 0.01172
			N	HNHN	2	(.25)(.75)(.25)(.75)	= 0.03516
			H	HNNH	2	(.25)(.75)(.75)(.25)	= 0.03516
			N	HNNN	1	(.25)(.75)(.75)(.75)	= 0.10547
			H	NHHH	3	(.75)(.25)(.25)(.25)	= 0.01172
			N	NHHN	2	(.75)(.25)(.25)(.75)	= 0.03516
			H	NHNH	2	(.75)(.25)(.25)(.75)	= 0.03516
			N	NHNN	1	(.75)(.25)(.75)(.75)	= 0.10547
			H	NNHH	2	(.75)(.75)(.25)(.25)	= 0.03516
			N	NNHN	1	(.75)(.75)(.25)(.75)	= 0.10547
			H	NNNH	1	(.75)(.75)(.75)(.25)	= 0.10547
			N	NNNN	0	(.75)(.75)(.75)(.75)	= 0.31641

Probability distribution for X = Number of hemophiliac children out of four				
Sample Space for X	0	1	2	3	4
P[Outcome]	0.31641	0.10547	0.03516	0.01172	0.00391
# Combinations	1	4	6	4	1
P[X]	0.31641	0.42188	0.21096	0.04688	0.00391

Conclusion: Based on this probability distribution, if this couple have four children, the probability that none will be a hemophiliac, $P[X = 0]$, is approximately 0.32 or 32%. Using the complement rule, the probability that this couple will have at least one hemophiliac child can be computed as $1 - 0.32 = 0.68$ or 68%. The most likely outcome for this couple is that they would have one hemophiliac child out of four births.

Deriving Theoretical Probability Distributions

The **theoretical probability distributions** for Binomial random variables are derived based on assumptions, logic, and the rules of probability. The minimum necessary assumptions required are: (1) must assume some value for P (where P is the probability that any one observation of the random phenomenon will meet the specified criterion) and (2) must assume that the outcome for each observation is independent of the outcomes of the other observations.

The value of P is generally based on assumptions about the nature of the random phenomenon and the rules of probability. For example:

a. Assume a coin is fair (meaning each side is equally likely to land face-up). Since there are only two possible outcomes {head, tails}, each equally likely, $P = P[\text{heads}] = 1/2$.
b. Assume a die is fair (each side equally likely to land face-up). Since there are six possible outcomes {1, 2, 3, 4, 5, 6}, each equally likely, $P = P[\text{1 dot}] = 1/6$.
c. Assume that gender is determined by sex chromosomes (X and Y in the sperm that fertilize the ova). Assume that meiosis in the testes produces equal number of sperm that carry the X and Y chromosome. Hence, there are only two possible genders {male, female}, each equally likely to occur, and $P = P[\text{male}] = 1/2$.

The assumption that the outcome for each observation is independent of the outcomes of other observations is necessary if the simple multiplication rule is to be used to compute the probability associated with specific combinations of outcomes that define the value of a random count variable X. In Example 4.5 above, we were able to compute the probability $P[X = 0 \text{ hemophiliac children out of 4}] = 0.75 \times 0.75 \times 0.75 \times 0.75 = 0.3164$ only because the outcome of each birth was independent of the other births. If the outcomes were not independent, the value of P would vary from one observation to the next and we could not determine the probability of the combination of outcomes in this manner.

If the assumptions and the logic are sound, the probabilities for values of a Binomial random variable X derived by this method are exactly true. However, if the assumptions regarding the value of P and/or independent observations are invalid, probabilities derived by this method will be incorrect. If P is unknown and no specific value can be assumed, an empirical approach may be necessary to derive the probability distribution for values of the Binomial random variable X.

4.3

Theoretical vs. Empirical Probability Distributions for Binomial Random Variables

In Chapter 3 I introduced the concepts of theoretical and empirical probability. The derivation of theoretical probabilities for the values of a Binomial random variable was described in the previous section. **Empirical probabilities** for the outcomes of a random phenomenon are based on the relative frequencies of the outcomes in the sample space, determined from a large number of repeated observations of the phenomenon. Empirical probabilities are always approximate estimates of true probabilities. However, the Law of Large Numbers tells us that as the number of observations used to compute relative frequencies increases, the values of these empirical estimates of probability will approach the true probability values. If there is any question as to the value of P or the validity of the assumptions used to derive a theoretical probability distribution, the empirical approach is the only way to determine probabilities for outcomes of a random phenomenon.

In Figure 3.1, I presented a comparison of theoretical vs. empirical probability values for outcomes of the simple random phenomenon of flipping a fair coin. Even though the Binomial random variables described in Chapter 4 are more complex than a single coin toss, the concepts of theoretical and empirical probability distributions are equally applicable.

In Figure 4.1 I present a comparison of the theoretical vs. empirical probability distributions for the random variable X = the number of heads out of n = 4 tosses of a fair coin. These probability distributions are displayed using a **probability histogram**. This type of graphic displays the values in the sample space of the discrete random variable X on the X-axis, and the height of the histogram bars on the Y-axis displays the probability associated with each of the possible values of X. To demonstrate the Law of Large Numbers, three empirical probability distributions were derived using a computer simulation of tossing a fair coin four times and counting X = the number of heads. The first empirical probability distribution is based on 100 repetitions of tossing 4 coins, the second based on 1,000 repetitions, and the third based on 10,000 repetitions. The empirical probabilities are the relative frequencies for the outcomes X = 0, 1, 2, 3, or 4 heads out of 100, 1,000, or 10,000 repetitions of tossing four coins. For example, 5 out of 100 repetitions produced the outcome X = 0 heads out of n = four coin tosses. Hence, the empirical probability $P[X = 0]$ is 0.05. This deviates somewhat from the theoretical probability $P[X = 0] = 0.0625$ due to the random variation that is always associated with empirical studies. Nonetheless, with as few as 100 repetitions the general appearance of the theoretical and empirical probability distributions is similar; X = 2 was the most likely outcome and X = 0 and X = 4 the least likely. As the number of repetitions used to compute relative frequencies increases to 10,000, the empirical probability values get closer to the theoretical values, as described by the Law of Large Numbers.

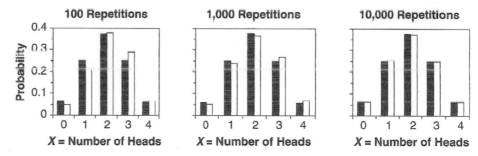

FIGURE 4.1   The Law of Large Numbers applied to probability distributions for Binomial random variables.   The histograms above display theoretical and empirical probability distributions for the Binomial random variable X = the number of heads out of four tosses of a fair coin. The black bars in each histogram display the theoretical probability distribution. The white bars display the empirical probabilities (relative frequencies) obtained from a computer simulation of tossing a fair coin four times and counting the number of heads. The number of repetitions above each probability histogram indicates the number of repeated trials (flipping four coins each) used to compute the empirical probabilities (relative frequencies).

4.4

Using the Binomial Formula and the Binomial Table

Either the Binomial formula or the Binomial table can be used to more efficiently determine probabilities for values of a Binomial random variable.

The Binomial Formula

The process described in Section 4.2 for determining probabilities associated with values of a Binomial random variable is quite laborious and is never used in actual practice. The Binomial formula provides a much more efficient method for computing these probabilities. To compute the probability that X out of n independent observations of a Binomial random phenomenon will meet some criterion, with a fixed probability P that each observation will meet the criterion, and $(1 - P)$ that the observation will not meet the criterion, insert the values for X, n, and P in the **Binomial formula**:

$$P[X] = \frac{n!}{X!(n - X)!} P^X (1 - P)^{n-X}$$

$$\underset{\text{Part 1}}{} \qquad \underset{\text{Part 2}}{}$$

The mathematical function indicated by the exclamation point "!" is called the **factorial** function. For any number n, the value of $n! = n\,(n - 1)(n - 2)\ldots((n - (n) + 1)$. For example: $4! = 4 \times 3 \times 2 \times 1$, and $6! = 6 \times 5 \times 4 \times 3 \times 2 \times 1$. Mathematicians define the special case of $0! = 1$. Many scientific calculators have an **x!** key to simplify this calculation.

Part 1 of the Binomial formula computes the number of combinations that result in X out of n observations that meet the specified criterion. This part performs the same function as the "event-tree" (left side) of Examples 4.2 and 4.5. For example, the number of combinations that produce two heads out of three tosses of a fair coin is $3!/2!(3 - 2)! = (3 \times 2 \times 1)/(2 \times 1)(1) = 3$ (HHT, HTH, THH).

Part 2 of the formula computes the probability associated with the combination of X out of n observations meeting the specified criterion. This part performs the same function as the P[Outcome] part of Examples 4.2 and 4.5. For example:

$$P[\text{2 heads and 1 tail}] = P[\text{heads}] \times P[\text{heads}] \times P[\text{tails}]$$

$$= 0.5 \times 0.5 \times (1 - 0.5)$$

$$= 0.5^2 \times (1 - 0.5)^1 \times P^X (1 - P)^{n-X}$$

$$= 0.125$$

The probability of getting two heads out of three tosses of a fair coin is computed as

$$\overset{\text{The number of combinations}}{\underset{}{\text{that produce X}=2\text{ heads}}}$$

$$3 \times 0.125 = 0.375$$

$$\underset{P[\text{2 heads and 1 tail}]}{}$$

See Example 4.6 for several illustrations of the use of the Binomial formula.

■ **EXAMPLE 4.6**

Using the Binomial formula to compute probabilities for Binomial variables.

a. Suppose X is the count of individuals in a sample of $n = 10$ college students who are infected with the herpes simplex II virus. The proportion of the U.S. population infected with this virus is $P = 0.2$ (U.S. Centers for Disease Control, 2001). What is the probability that $X = 4$ out of 10 students in the sample will test positive for this virus?

$$P[X = 4] = \frac{n!}{X!(n-X)!} P^X (1-P)^{n-X} = \frac{10!}{4!(10-4)!}(0.2)^4(1-0.2)^{10-4}$$

$$= \frac{10 \times 9 \times 8 \times 7 \times 6 \times 5 \times 4 \times 3 \times 2 \times 1}{(4 \times 3 \times 2 \times 1)(6 \times 5 \times 4 \times 3 \times 2 \times 1)}(0.0016)(0.2621)$$

$$= \frac{3628800}{17280}(0.0004194)$$

$$= 210 \times (0.0004194) = \mathbf{0.0881}$$

There are 210 combinations that produce four out of 10 infected individuals in a sample of 10. The probability associated with each one of these combinations is 0.0004194. To obtain the overall probability of getting $X = 4$ infected individuals out of $n = 10$, the number of combinations is multiplied by the probability associated with this outcome.

b. Suppose Y is the count of individuals in a sample of $n = 12$ college students who are African-American. The proportion of the U.S. population that self-identified as African-American is $P = 0.13$ (U.S. Census Bureau, 2001). What is the probability that $Y = 2$ out of 12 students in the sample will be African-American?

$$P[Y = 2] = \frac{n!}{Y!(n-Y)!} P^Y (1-P)^{n-Y} = \frac{12!}{2!(12-2)!}(0.13)^2(1-0.13)^{12-2}$$

$$= \frac{12 \times 11 \times 10 \times 9 \times 8 \times 7 \times 6 \times 5 \times 4 \times 3 \times 2 \times 1}{(2 \times 1)(10 \times 9 \times 8 \times 7 \times 6 \times 5 \times 4 \times 3 \times 2 \times 1)}(0.0169)(0.2484)$$

$$= \frac{479001600}{7257600}(0.00420)$$

$$= 66 \times (0.00420) = \mathbf{0.2772}$$

Some computer spreadsheet programs have a function that performs the Binomial formula calculation. In both Microsoft Excel and Corel QuattroPro, this function is:

BINOMDIST $(X, n, P, 0 \text{ or } 1)$.

Four values must be included in the parentheses after the BINOMDIST function: the value of X for which a probability is desired (or a spreadsheet cell address where the X value resides), the sample size n, the probability P an individual will fall in the specified category, and a 0 to indicate that you want the probability for a specific value of X. If you enter a 1 in the last position, the program will return the cumulative probability of getting an X value less than or equal to the specified X. Example 4.7 displays two implementations of this function in Microsoft Excel.

■ **EXAMPLE 4.7**

Using Microsoft Excel to Obtain Binomial Probabilities. The diagrams below show excerpts from a Microoft Excel spreadsheet in which the BINOMDIST function is used to compute the probabilities associated with the possible values of a Binomial count variable. The values in the sample space of X were manually entered into column A. The probability value $P[X]$ is listed in column B, and the spreadsheet function used to compute each probability value is listed as text in column C. In actual

Binomial Distribution for $n = 5$, $P = 0.5$. Probabilities listed are $P[X = \_\_]$.

Row# 1	A X	B P[X]	C Formula
2	0	0.0313	=BINOMDIST (A2, 5, 0.5, 0)
3	1	0.1563	=BINOMDIST (A3, 5, 0.5, 0)
4	2	0.3125	=BINOMDIST (A4, 5, 0.5, 0)
5	3	0.3125	=BINOMDIST (A5, 5, 0.5, 0)
6	4	0.1563	=BINOMDIST (A6, 5, 0.5, 0)
7	5	0.0313	=BINOMDIST (A7, 5, 0.5, 0)

Binomial Distribution for $n = 4$ and $P = 0.33$. Probabilities listed are $P[X \leq \_\_]$.

Row# 1	A X	B P[X]	C Formula
2	0	0.2015	=BINOMDIST(A2, 4, 0.33, 1)
3	1	0.5985	=BINOMDIST(A3, 4, 0.33, 1)
4	2	0.8918	=BINOMDIST(A4, 4, 0.33, 1)
5	3	0.9881	=BINOMDIST(A5, 4, 0.33, 1)
6	4	1.0000	=BINOMDIST(A6, 4, 0.33, 1)

practice, the BINOMDIST formula resides in the cells where the probabilities are listed (column B). I provide the text of these formulas in column C only to show you how to enter the formula. The first entry within the parentheses after BINOMDIST indicates the spreadsheet cell address where the X value resides (e.g., A2 indicates column A row 2).

The Binomial Table

Most statistics books provide tables of Binomial probability distributions for a range of sample sizes n and probabilities P. Look for the part of the table that has the appropriate sample size (n) listed in the left margin, then locate the column to the right of that location where the probability value (P) is listed at the top. The column of numbers lists the values $P[X]$ for each X value listed in the left margin. (See Example 4.8.)

▪ **EXAMPLE 4.8**

Reading a Binomial table. Listed below are probability distributions for a selection of Binomial random variables, obtained from Appendix Table 3. Given the specified values for n and P listed in bold type above each probability distribution, try to locate the appropriate distribution in the table. *Note:* The probabilities listed have only four significant digits. When the probability associated with a value of X is < 0.0001, no probability is listed. This does not mean that the probability for that

X value is zero. Rather, the probability is just so small that it is negligible and can be ignored.

n = 4, P = 0.01		n = 7, P = 0.09		n = 11, P = 0.35	
X	P[X]	X	P[X]	X	P[X]
0	0.9604	0	0.5168	0	0.0088
1	0.0388	1	0.3578	1	0.0518
2	0.0006	2	0.1061	2	0.1395
3		3	0.0175	3	0.2254
4		4	0.0017	4	0.2428
		5	0.0001	5	0.1830
		6		6	0.0985
		7		7	0.0379
				8	0.0102
				9	0.0018
				10	0.0002
				11	

4.5

Using the Binomial Distribution to Determine the Probability of Events

The Binomial count variable *X* has a discrete number of possible values, and each value in the sample space has a nonzero probability. The Binomial count variable *X* can only have integer values. When counting the number of times an event occurs, values such as 2.5 or 1.27 are not valid. If we have correctly specified the sample space of possible values for a Binomial count variable and if we have a valid probability distribution, the sum of the probabilities in the distribution should equal 1.0. Using the probability distribution for *X* and the rules for probability, we can determine the probability of any event defined by the value of *X*.

Determining the probabilities for events defined by Binomial random variables is a two-step process: (1) we list all values of *X* in the sample space that meet the specified description of an event and (2) we sum the probabilities of these values to obtain the overall probability for the event.

Suppose we flipped a fair coin three times and we wanted to know the probability *P*[*X* ≥ two heads]. First we would list the values of *X* that meet this criterion {*X* = 2 and *X* = 3}, and then we would sum the probabilities of these events (0.375 + 0.125) to obtain *P*[*X* ≥ 2] = 0.5.

We can use the Binomial probability distribution to determine the probability associated with any value or range of values for a Binomial variable. We illustrate with several different events, as shown in Example 4.9.

■ **EXAMPLE 4.9**

Using the two-step process to determine *overall* probabilities for events defined by a Binomial random variable. The Binomial probability distributions listed in Example 4.8 were used to determine probabilities for the following events.

a. Given $n = 7$ and $P = 0.09$, determine $P[X < 2]$.

Step 1: X 0   1
Step 2: $P[X]$ $0.5168 + 0.3578 = $ **0.8746**

b. Given $n = 7$ and $P = 0.09$, determine $P[X \geq 2]$.

Step 1: X 2   3   4   5   6   7
Step 2: $P[X]$ $0.1061 + 0.0175 + 0.0017 + 0.0001 + 0 + 0 = $ **0.1254**

Note: The event $X \geq 2$ is the complement of the event $X < 2$. That is, the union of these two events includes the entire sample space of X and there is no overlap between these two events. According to the Complement Rule, if events A and B are complements, then $P[B] = 1 - P[A]$. In this case, we could have computed the probability $P[X \geq 2] = 1 - P[X < 2] = 1 - 0.8746 = 0.1254$. Using the Complement Rule can sometimes reduce the amount of addition you must do to determine probabilities of events. This in turn reduces the opportunity for calculation errors.

c. Given $n = 11$ and $P = 0.35$, determine $P[3 < X < 6]$.

Step 1: X 4   5
Step 2: $P[X]$ $0.2428 + 0.1830 = $ **0.4258**

d. Given $n = 11$ and $P = 0.35$, determine $P[3 \leq X \leq 6]$.

Step 1: X 3   4   5   6
Step 2: $P[X]$ $0.2254 + 0.2428 + 0.1830 + 0.0985 = $ **0.7497**

Note: Whenever you use the Binomial table to determine probabilities you should be especially careful to distinguish $<$ from $\leq$. In parts (c) and (d) above, the first event did not include the values $X = 3$ and $X = 6$, but the second event did. This distinction in the descriptions of the events resulted in a substantial difference in the probability.

e. Given $n = 11$ and $P = 0.35$, determine $P[X \leq 1 \text{ or } X \geq 10] = P[X \leq 1] + P[X \geq 10]$

Step 1: X 0   1   10   11
Step 2: $P[X]$ $0.0088 + 0.0518 + 0.0002 + 0 = $ **0.0608**

If you scan the pages of Binomial probability distributions in Appendix Table 3, you will note that probability distributions are provided only for sample sizes $n \leq 20$ and for specific probability values P. If you must determine probabilities for a Binomial count variable for which the sample size or probability is not listed in Appendix Table 3, your only option is to use the Binomial formula. However, this can easily be implemented using most computer spreadsheet programs.

CHAPTER SUMMARY

1. **Binomial random variables** are counts of the number of observations of a random phenomenon that meet a specified criterion *if* the following conditions are true:
   a. The total number of observations (*n*) is fixed.
   b. The probability that any one observation will meet the specified criterion is constant.

   c. The outcomes of the multiple observations of the random phenomenon are independent.
   d. To fulfill conditions (b) and (c), the population size must be greater than 100× the sample size.
2. The probability associated with each value *X* of a **Binomial count variable** can be determined by the following method:

a. Determine all possible outcomes for a sample of n observations of the random phenomenon, where each of the n observations is simply the determination of whether or not a specified criterion was met.

b. Use the simple multiplication rule to determine the probability for each possible outcome of n observations of the random phenomenon.

c. For each possible outcome, determine X = count of the number of observations out of n that met the criterion.

d. Sum the probabilities for all outcomes (combinations) of events that produce the same value of X to determine the overall probability for each X value.

3. Approximate empirical probability distributions for Binomial count variables can be derived by repeatedly taking samples of n observations of the random phenomenon and counting X = number of observations out of n that meet the specified criterion. The relative frequency of the occurrence of each X value (= number of times that X value occurred divided by the total number of repetitions of n observations) is an empirical estimate of $P[X]$. The more repetitions used to compute such relative frequencies, the closer they will approximate the true probabilities for the X values in the sample space.

4. The **Binomial formula** can be used to directly compute the probability associated with any value of a Binomial count variable X, given the sample size n and the probability that an observation will meet the specified criterion **P**. Most computer spreadsheet programs have a function that simplifies this calculation.

5. The **Binomial table**, found in most statistics textbooks, provides the probability distributions for Binomial count variables for a selection of sample sizes (n) and probability (**P**) values.

6. The probability of any event defined by values of a Binomial count variable X can be determined by: (a) listing all values of X that meet the definition of the event and (b) summing their associated probabilities.

5 The Normal Distribution: A Probability Distribution for Continuous Random Variables

INTRODUCTION   The Binomial distribution described in Chapter 4 provides probabilities for values of random variables that are based on counting the number of observations in a sample that meet some criterion. Because they are based on counting, values for Binomial random variables can only be integers. However, many variables of interest to scientists are based on measurement (pH, temperature, mass, concentration). Such random variables can theoretically have an infinite number of possible values in their sample space, limited only by the number of significant decimal places provided by the measurement instrument. It is not possible to list all the possible values in the sample space, and it is impractical to base analyses of such variables on counting the number of times a specific value occurs in a sample. Hence, statistical analyses of data for measured variables must be based on a probability distribution other than the Binomial distribution.

For many different kinds of measured variables, values close to the middle of the range of the sample space occur most commonly, while values much smaller or larger than the average are relatively rare. For example, most people have heights close to the average, while very tall and very short people are much less common. Hence, there is a high probability associated with values close to the population mean, and probability associated with values away from the mean progressively decreases as the value deviates more from the mean. For measured variables that fit this description, a "bell-shaped" or Normal probability distribution, centered over the mean of the population, provides a useful description of probabilities associated with their values. Because such variables are common, Normal probability distributions are used for many of the most widely-used and familiar statistical analyses.

In this chapter I describe the characteristics of Normal probability distributions and the mechanics for assigning probabilities to values of Normal random variables. We will learn about applications of the Normal distribution in statistical analyses in later chapters. ■

CHAPTER GOALS AND OBJECTIVES

In this chapter you will be able to explain the differences between probability distributions for discrete and continuous random variables and to use the Normal probability distribution to determine probabilities associated with values of Normally distributed continuous random variables.

By the end of the chapter, you will be able to:

1. Determine when a random variable is continuous vs. discrete.
2. Describe how probability is represented as area under a probability density curve.
3. Describe a Normal probability distribution, given the mean and standard deviation of the variable of interest, sketch the distribution, and scale the X-axis.
4. Use the standard Normal distribution table to determine probabilities of events defined by values for any Normally distributed variable, given its mean and standard deviation.
5. Use Normal quantile plots of data values to determine if they are Normally distributed.

5.1

Characteristics of Continuous vs. Discrete Random Variables

The Binomial count variable X that was the subject of Chapter 4 is a discrete random variable. When we counted the number of times that a specified criterion was met out of n observations of a random phenomenon, the sample space included only integer values. In most real-world applications, the sample space for such a discrete random variable has a finite number of possible values, and each has its own associated probability. That is, it is possible to actually list all values in the sample space and to determine the probability of each. Scientists study Binomial count variables when the characteristic observed on each individual in their samples is a *categorical variable*, such as gender, blood type, or color.

Scientists often study characteristics that are based on measurements of *continuous quantitative variables* such as weight, height, concentration, and temperature. The number of decimal places reported for the values for such variables is determined by (1) the precision of the measuring instrument or protocol and (2) considerations of significant digits, that is, the number of digits that provide meaningful information. However, there is no theoretical limit on the number of decimal places for continuous measured variables. Therefore, such variables have an infinite number of possible values in their sample space. Given this distinction between discrete and continuous random variables, the probability distribution for continuous variables differs from those of discrete random variables in the following ways:

1. It is impossible to list all possible values in the sample space of a continuous variable.

2. Because the number of possible values is theoretically infinite, the probability of getting exactly any one specific value is approximately zero ($= 1/$ infinity).

Therefore, probabilities for continuous variables are expressed in terms of events defined by a range of values, e.g., $P[0 \leq X \leq 0.5]$, rather than any single value. Because the probability of getting any specific value of a continuous variable is approximately zero, the probability $P[X > y] = P[X \geq y]$ for continuous variables, unlike discrete variables.

5.2

Probability Distributions for Continuous Random Variables

In Chapter 4 I introduced the concept of a probability histogram for discrete random variables. This type of graphic lists all the possible values of X in the sample space on the x-axis, and above each value is a histogram bar that displays the probability associated with that value (e.g., Figure 4.1). Between each bar is a blank area, indicating that it is impossible (probability $= 0$) for the discrete random variable to have values that fall between the discrete values in the sample space. The height of each bar indicates the probability $P[X]$, but this probability is also indicated by the *area* of the bar. The base of each bar always has a width $= 1$ (one value of the discrete variable) and the height of the bar equals the probability $P[X]$. I make this connection between area and probability in probability histograms because we always represent probability with area when we work with the probability distributions for continuous random variables.

The graphic used to represent probability distributions for continuous random variable is called a **probability density curve**. The X-axis displays the *range* (minimum to maximum) of values for the continuous variable. Because there is an infinite number of possible decimal places, it is impossible to list all the possible values. The Y-axis scale does not represent probability. Rather, the probability of any event defined by a specific range of values within the sample space ($a < X < b$) is represented as *area under the curve* above the specified range of X-values (see Figure 5.1). Given the rules of probability, the total area under the curve is always 1.0. Probability density curves come in a variety of shapes, but all share the characteristics described here.

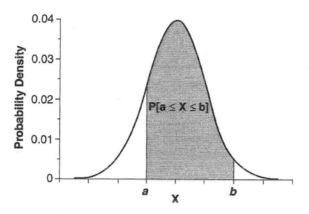

FIGURE 5.1   A probability density curve.   The probability associated with a range of values for the continuous random variable X ($a \leq X \leq b$) is represented by the area under the probability density curve above that range along the X-axis.

5.3
The Normal
Distribution

Normal distributions are a class of probability density curves that are: (1) symmetric, (2) single-peaked, and (3) bell-shaped (see Figure 5.2). The Normal distribution indicates there is a high probability associated with values of the random variable near the mean for the distribution, and there are progressively lower probabilities associated with values further from the mean. Because many measured variables studied by scientists have Normal distributions, the Normal distribution is one of the most widely used in statistical analyses. Normal distributions are defined based upon the **population mean (μ)** and the population **standard deviation (σ)** for the continuous random variable.

> **NOTE** In Chapter 2 I introduced the terms $\bar{x}$ and S for the *sample* mean and standard deviation. The terms μ and σ are the corresponding *population* parameters. The sample statistics $\bar{x}$ and S are used to estimate the population parameters μ and σ. ■

The mean μ determines the location of the center of the bell-shaped distribution along the X-axis. The standard deviation σ determines the spread of the bell-shaped distribution, e.g., short, squat bell-shaped vs. tall, skinny bell-shaped (Figure 5.2). The horizontal distance between the mean and the two points on the bell-shaped curve to either side of the mean where the curve changes from being convex-up to

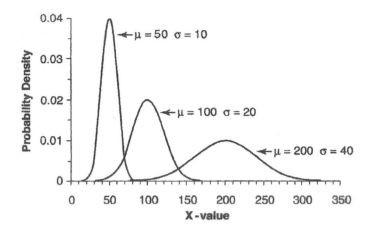

FIGURE 5.2   Examples of Normal probability distributions, as defined by different values for the population mean (μ) and the population standard deviation (σ).

concave-up (called the inflection points) is equal to the standard deviation σ (see Figure 5.3). Given these stipulations for the shape of all Normal distributions, the following Empirical Rule can be used to obtain approximate probabilities associated with ranges of values for any Normally distributed variable (also see Figure 5.3).

Empirical Rule for Normal Distributions

For any Normal distribution:

- Approximately 68% of the area under the curve falls within the range μ ± 1 σ. That is, $P\{\mu - 1\sigma \leq X \leq \mu + 1\sigma\} \approx 0.68$.
- Approximately 95% of the area under the curve falls within the range μ ± 2 σ. That is, $P\{\mu - 2\sigma \leq X \leq \mu + 2\sigma\} \approx 0.95$.
- Approximately 99% of the area under the curve falls within the range μ ± 3 σ. That is, $P\{\mu - 3\sigma \leq X \leq \mu + 3\sigma\} \approx 0.99$.

Given the difficulty associated with determining areas under curves (requires calculus), the empirical rule provides a useful approximation for determining probabilities associated with values for Normally distributed random variables (see the italics values in Figure 5.3).

Given the Rules of Probability, and that the Normal distribution is symmetric, the area under the curve for values greater than or equal to μ is equal to the area for values less than or equal to μ (= 0.5). The empirical rule states that 68% (0.68) of the area under a Normal curve is associated with values between (μ − 1σ) and (μ + 1σ). Because the Normal distribution is symmetric about the value of μ, half this area is associated with values between (μ − 1σ) and μ, and the other half is

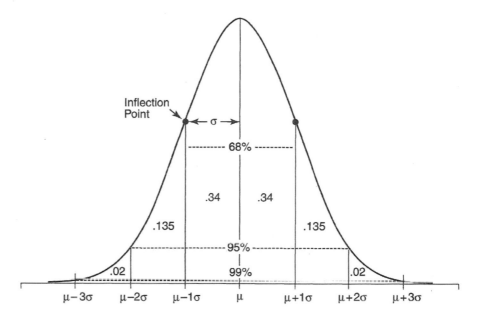

FIGURE 5.3   The empirical rule for Normal distributions.

associated with values between μ and (μ + 1σ). Hence the probabilities associated with values of μ in each of these two ranges is 0.34. (See Figure 5.3.)

If 95% (0.95) of the area under a Normal curve is associated with values within the range (μ − 2σ) to (μ + 2σ), and 68% of the area is associated with values in the range (μ − 1σ) and (μ + 1σ), then (95 − 68)/2 = 13.5% (0.135) of the area under the curve is associated with the range of values (μ − 2σ) to (μ − 1σ) and 13.5% is associated with the range of values (μ + 1σ) to (μ + 2σ). (See Figure 5.3.)

If 99% (0.99) of the area under a Normal curve is associated with values within the range (μ − 3σ) to (μ + 3σ), and 95% of the area is associated with values in the range (μ − 2σ) and (μ + 2σ), then (99 − 95)/2 − 2% (0.02) of the area under the curve is associated with the range of values (μ − 3σ) to (μ − 2σ) and 2% is associated with the range of values (μ + 2σ) to (μ + 3σ). (See Figure 5.3.)

Finally, if 99% (0.99) of the area under the curve is associated with the range of values (μ − 3σ) to (μ + 3σ), then 1% (0.01) of the area under the curve falls outside this range. Because the Normal distribution is symmetric, 0.5% (0.005) is associated with values less than or equal to (μ − 3σ) and 0.5% is associated with values greater than or equal to (μ + 3σ). Example 5.1 illustrates how the empirical rule can be applied.

■ **EXAMPLE 5.1**

Using the Empirical Rule to determine approximate probabilities for Normally distributed continuous random variables. In the figures that follow, the numbers positioned within the curve represent areas under the Normal curve (probabilities).

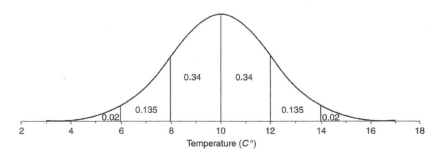

a. Suppose that average April temperature in Indianapolis, IN is $\mu = 10°C$, with standard deviation $\sigma = 2°C$. Using the Empirical Rule, you can determine approximate probabilities associated with ranges of $X =$ average April temperature values.

$P[8 \leq X \leq 12] = 0.68$	$P[X \geq 12] = 0.16$	$P[X \leq 8] = 0.16$
$P[6 \leq X \leq 14] = 0.95$	$P[X \geq 14] = 0.025$	$P[X \leq 6] = 0.025$
$P[4 \leq X \leq 16] = 0.99$	$P[X \geq 16] = 0.005$	$P[X \leq 4] = 0.005$
$P[X \geq 8] = 0.84$	$P[X \leq 14] = 0.975$	

b. Suppose that the weight of 10-yr-old children is Normally distributed with $\mu = 65$ lbs, with $\sigma = 10$ lbs. Use the Empirical Rule to obtain approximate probabilities associated with ranges of weight values.

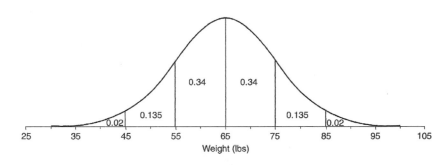

$P[55 \leq X \leq 75] = 0.68$	$P[X \geq 75] = 0.16$	$P[X \leq 55] = 0.16$
$P[45 \leq X \leq 85] = 0.95$	$P[X \geq 85] = 0.025$	$P[X \leq 45] = 0.025$
$P[35 \leq X \leq 95] = 0.99$	$P[X \geq 95] = 0.005$	$P[X \leq 35] = 0.005$
$P[X \geq 35] = 0.995$	$P(X \leq 75] = 0.84$	

5.4

The Standard Normal Distribution: Description and Applications

There are an infinite number of possible "Normal distributions," each defined by a unique combination of the mean μ and standard deviation σ values. The Empirical Rule provides approximate probabilities associated with values of Normally distributed random variables, but integral calculus is required to determine exact probabilities (areas under the Normal probability density curve). Because these calculations are tedious and difficult, mathematicians have performed the calculations and produced a table of probabilities for a single **Standard Normal Distribution**. This specific Normal distribution has a mean $\mu = 0$ and standard deviation $\sigma = 1$. Virtually all statistics books contain a table of areas under a standard Normal distribution curve, representing probabilities associated with specific ranges of this **standard Normal random variable** (given the symbol Z). This distribution can be found in Appendix Table 2 in this book.

How To Obtain Probabilities from the Standard Normal Distribution Table

1. The probabilities in the standard Normal table represent $P[Z \leq z]$. Because Z is a continuous random variable, $P[Z = z]$ is always 0. Hence, we can only determine probabilities for a range of Z-values. Although ranges can be stated in a number of ways, *the standard Normal table probabilities are given only for the range defined by $P[Z \leq z]$*. However, we can use the complement rule of probability to convert $P[Z \leq z_i]$ values into probabilities for events defined by $P[Z \geq z]$ and $P[z_a \leq Z \leq z_b]$.

2. The numbers on the left margin of the standard Normal table (Appendix Table 2), under the column titled z, are specific Z-values expressed with only one decimal place. To the right of the z-column title are a series of numeric column titles from .00 to .09. These values correspond to the second decimal place for the Z-value. To look up the probability associated with a Z-value specified with two decimal places, read down the column of z-values on the left margin of the table to the row that corresponds to the first two digits, then across on this row to the column with a header that corresponds to the number in the second decimal place. For example, to look up the probability $P[Z \leq -2.56]$, read down the leftmost column of z-values to the number -2.5, then read across on this row to the column with the header .06. The number at this position in the table, .0052, is the probability $P[Z \leq -2.56]$. The underlining has been added for emphasis.

z	.00	.01	.02	... ...	.05	.06	.07	... ...
−3.4	.0003	.0003	.0003	... ...	.0003	.0003	.0003	... . .
−3.3	.0005	.0005	.0005	... ...	.0004	.0004	.0004	... . .
−3.2	.0007	.0007	.0007	... ...	.0006	.0006	.0005	... . .
.	.	.	.		.	.	.	
.	.	.	.		.	.	.	
−2.6	.0047	.0045	.0044	... ...	.0040	.0039	.0038	... . .
<u>−2.5</u>	.0062	.0060	.0059	... ...	.0054	<u>.0052</u>	.0051	... . .
−2.4	.0082	.0080	.0078	... ...	.0071	.0069	.0068	... . .
.	.	.	.		.	.	.	
.	.	.	.		.	.	.	

3. The standard Normal table in this book spans two pages. The first pages give probabilities for negative Z-values, and the second page is for positive Z-values. *Note:* Because the Normal distribution is symmetric and centered on $Z = 0$, half the area under the curve is associated with Z-values ≤ 0, and half is associated with Z-values ≥ 0. The probability listed for $z \leq 0.0$ is .5000. Because the

probability values in the standard Normal table are for $P[Z \le z]$, all probability values for positive Z-values are greater than .5000, and all probability values for negative Z-values are less than .5000.

4. *To determine* $P[Z \ge z]$: Look up $P[Z \le z]$ in the standard Normal table. Given that the total area under the curve is 1.0, you can use the complement rule for probabilities:

$$P[Z \ge z] = 1 - P[Z \le z].$$ See Figure 5.4A.

5. *To determine* $P[z_a \le Z \le z_b]$: First, this only makes sense if z_b is larger than z_a. Look up $P[Z \le z_a]$ and $P[Z \le z_b]$ in the standard Normal table.

$$P[z_a \le Z \le z_b] = P[Z \le z_b] - P[Z \le z_a].$$ See Figure 5.4B.

Also see Example 5.2.

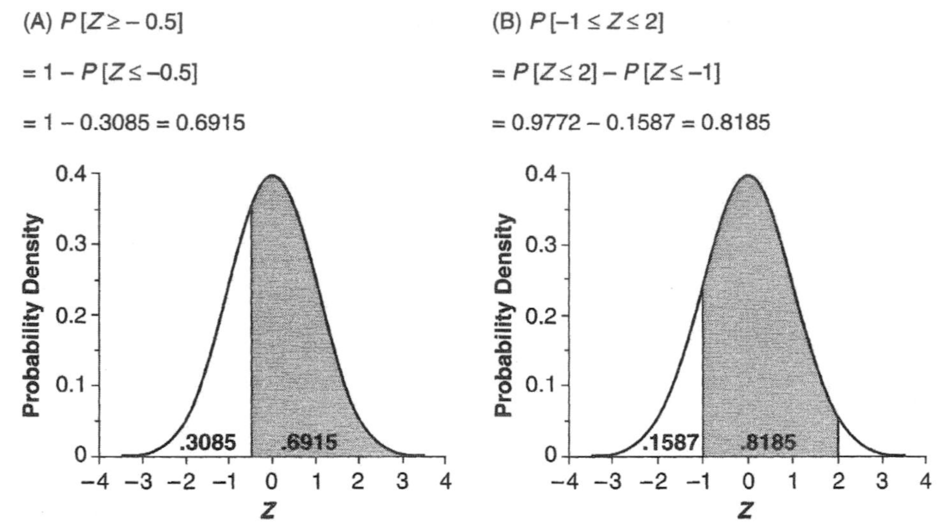
(A) P[Z≥−0.5]
= 1 − P[Z≤−0.5]
= 1 − 0.3085 = 0.6915
(B) P[−1 ≤ Z ≤ 2]
= P[Z≤2] − P[Z≤−1]
= 0.9772 − 0.1587 = 0.8185
Probability Density
0.4
0.3
0.2
0.1
0
.3085
.6915
.1587
.8185
−4 −3 −2 −1 0 1 2 3 4
Z

FIGURE 5.4 Examples of standard Normal curves with shaded areas corresponding to probabilities associated with specified ranges of Z-values.

■ EXAMPLE 5.2

Using the standard Normal table to assign probabilities to ranges of Z. What proportion of the area under a standard Normal curve is associated with the range of Z-score values $-1.00 \le Z \le +1.00$?

1. In Appendix Table 2, find -1.0 under the z-column and read across to the value under the .00 column (.1587 or 15.87% of the area under the curve is associated with Z-values ≤ -1.00).
2. Find $+1.0$ under the z-column and read across to the value under the .00 column (.8413 or 84.13% of the area under the curve is associated with Z-values $\le +1.00$).
3. If $P[Z \le -1.00] = 0.1587$ and $P[Z \le +1.00] = 0.8413$, then $P[-1.00 \le Z \le +1.00] = 0.8413 - 0.1587 = 0.6825$ (See accompanying figure.). Note that this value is very close to the approximate probability stated in the Empirical Rule that 68% of the area under a Normal curve falls within the range $\pm 1\ \sigma$ from the mean.

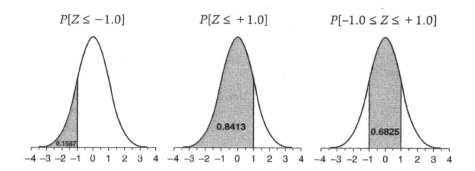

Using the Standard Normal Table to Determine Probabilities for Normally Distributed Random Variables

The standard Normal table can be used to determine probabilities associated with any normally distributed random variable.

The scale on the X-axis of the standard Normal curve is expressed in "standard deviation units." The standard Normal curve has a mean μ = 0 and a standard deviation σ = 1. Hence, a z-value of +1 corresponds to a value that is one "standard deviation unit" above the mean. Likewise, a z-value of −2.3 corresponds to a value that is 2.3 standard deviation units below the mean.

To determine probabilities associated with a range of values for any continuous, Normally distributed random variable X, transform the original x-value(s) to z-values on the standard Normal Z scale, using the formula:

Calculation of z-Values

$$z = \frac{(x - \mu)}{\sigma}$$

The values μ and σ are the population mean and standard deviation for the random variable X. This transformation re-expresses the value x of random variable X in terms of how many standard deviation units the value is away from μ, and whether the value x is below (−z) or above (+z) the population mean μ. The probability associated with the statement $P[X \leq x] = P[Z \leq z]$ can be found in the standard Normal table. Example 5.3 shows how the process is done.

■ **EXAMPLE 5.3**

Using the standard Normal distribution table to assign probabilities to Normal random variables. Suppose mean height for adult male Americans is 5.8 feet, with a standard deviation of 0.5 feet. If we can assume that height of adult male Americans is a Normally distributed random variable, we can determine the relative frequency of individuals in this population who fall within any specified range of heights.

a. A basketball coach wants to know the relative frequency of adult males with height ≥ 7 feet. This can be determined using the standard Normal distribution table if we re-express the question as, "How many standard deviation units above the mean of 5.8 is the value 7.0?"

Step 1. *Restate question in terms of standard Normal Z-values:*

$$P[X \geq 7] = P[Z \geq (7 - 5.8)/0.5] = P[Z \geq +2.40]$$

The height value 7 feet is 2.40 standard deviation units above the population mean $\mu = 5.8$ feet.

Step 2. *Look up the Z-value +2.40 in the standard Normal table:* Under the z-column, read down to the value 2.4 ($-Z$-values are on the first page of the table, + values on the second). Read across the +2.4 row to the .00 column. The value .9918 in the standard Normal table indicates the proportion of the total area under the curve that is to the left of (less than) the value +2.40 (i.e., 99.18% of the area under the curve). Hence, 99.18% of the American male population has height < 7 ft. However, we wanted to know the proportion of the population with height ≥ 7 feet ($Z \geq +2.40$). Using the complement rule of probability, this is computed as $1 - 0.9918 = 0.0082 = 0.82\%$ of American males are ≥ 7 ft tall.

b. What proportion of the population of adult male Americans have height ≤ 5 feet?

$$P[X \leq 5, \text{ given } \mu = 5.8 \text{ and } \sigma = 0.5] = P[Z \leq (5 - 5.8)/0.5]$$

$= P[Z \leq -1.60]$ The height value 5 feet is 1.60 standard deviation units below the population mean $\mu = 5.8$.

$= 0.0548$ 5.48% of the male population has height ≤ 5 feet.

Assumption for Using the Standard Normal Table of Probabilities

To determine the probability associated with values for a random variable X by using the standard Normal table, we must be able to assume that the true distribution of X is Normal. If the true population distribution for a random variable X is *not* Normal, but we use the procedures described in this chapter to assign probabilities to values of X, these probabilities will be inaccurate to the point of being meaningless. See Examples 5.4 and 5.5.

■ **EXAMPLE 5.4**

Using the standard Normal distribution to assign probabilities to values of a Normally-distributed variable. The histogram that follows is the population distribution for Math SAT scores of the student body at Watsamata University ($N = 3,791$ students). This population distribution is Normal, with a mean SAT score of $\mu = 516$ and standard deviation $\sigma = 107$. Of the 3,791 students, 190 (5% or 0.05) have Math SAT scores ≤ 340. Because the population distribution is Normal, we can use the standard Normal distribution to estimate this true relative frequency (probability).

$$P[\text{Math SAT} \le 340 \text{ if } \mu = 516, \sigma = 107] = P[Z \le (340 - 516)/107]$$

$$= P[Z \le -1.64]$$

$$= 0.0505$$

Note how this probability (or relative frequency) is very close to the true relative frequency of students in this population who have a Math SAT score ≤ 340.

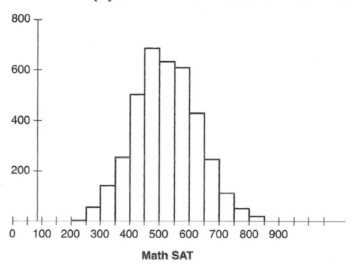

■ **EXAMPLE 5.5**

Using the standard Normal distribution to assign probabilities to values of a random variable that is *not* Normally-distributed. The histogram that follows is the population distribution for Cumulative Hours Earned for students at Watsamata University. This population distribution is multimodal (for freshmen, sophomores, juniors, and seniors), with $\mu = 70$ and $\sigma = 46$. Of the 3,791 students, 948 (25% or 0.25) have earned ≥ 117 credit hours. Because the population distribution is not Normal, we should *not* use the standard Normal distribution to estimate this true relative frequency (probability). The calculations below demonstrate the consequence of misapplying the standard Normal distribution calculations.

$$P[\text{CumHrE} \ge 117 \text{ if } \mu = 70, \sigma = 46] = P[Z \ge (117 - 70)/46]$$

$$= P[Z \ge 1.02]$$

$$= 1 - P[Z \le 1.02] = 1 - 0.8461$$

$$= 0.1539$$

NOTE This estimated probability (relative frequency) is quite different from the actual relative frequency of 0.25, because the distribution of this variable is not Normal.

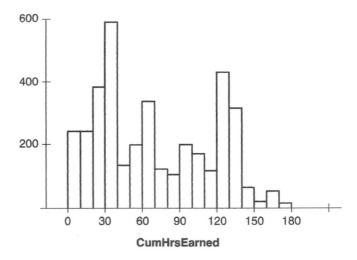

Normal quantile plots.   Although it is important that the population distribution be Normal for us to use the standard Normal table to assign probabilities to ranges of values of a random variable, we very rarely know the shape of the population distribution. To know with certainty that the population distribution is Normal, we would have to measure all the individuals in the population and produce a histogram of the values. This histogram would have to be symmetric, unimodal, and bell-shaped. Because we are rarely able to obtain data for an entire population, we must evaluate whether or not this assumption of Normality is valid by considering the data available. If a histogram or stem-leaf plot of the individual data values in a sample is approximately symmetric and bell-shaped, this would provide a basis for assuming the population distribution from which the data were obtained is also Normal. However, when sample size is small, the distribution of data values obtained from a truly Normal population may not appear bell-shaped in a histogram or boxplot.

A **Normal quantile plot** (also called a **Normal probability plot**) is a specialized type of graphic that is used to determine whether or not data for a variable are Normally distributed. Since these plots will usually be produced by a computer program, I will not go into how these graphics are produced. It is sufficient to know that when the sample data value points in the Normal quantile plot lie close to a straight line with a slope of 1, this indicates that the data are Normally distributed. Systematic deviations of the cloud of data points from a straight line indicate the data have a non-Normal distribution. Individual points far away from the other points indicate outliers. See Figure 5.5 for examples of Normal quantile plots and their interpretations regarding Normality of the distribution of data values. The variable names Normal, −Skew, +Skew, Flat, and Peaky describe how the distribution deviates from Normal. The *X*-axis labeled "nscores" in the Normal quantile plot refers to computed values of these variables if they were Normal.

FIGURE 5.5   Examples of distributions of various shapes and their associated Normal quantile plots (called Normal probability plots in some statistics programs).

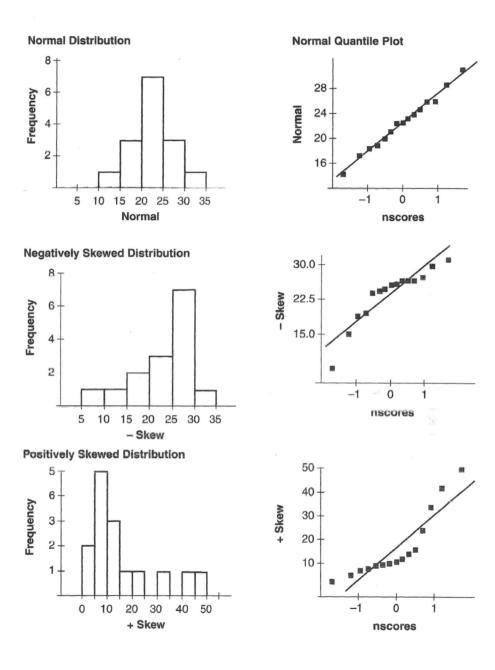

"Flat" (Platykurtic) Distribution

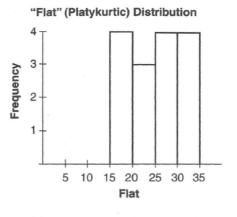

Normal Quantile Plots

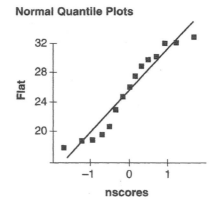

"Peaky" (Leptokurtic) Distribution

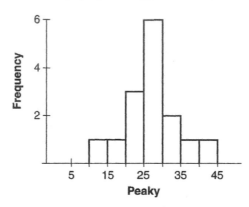

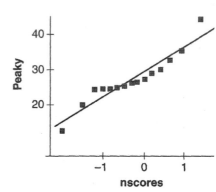

CHAPTER SUMMARY

1. Measured quantitative variables such as length, area, mass, concentration, and temperature are *continuous random variables* with a sample space that includes an infinite number of possible values.

2. The probabilities for continuous random variables are determined only for ranges of values ($a \le X \le b$). Because there are an infinite number of values in the sample space, the probability associated with any one exact value is approximately zero.

3. Probability distributions for continuous random variables are represented by **probability density curves**. The probability associated with any range of values ($a \le X \le b$) is represented by the *area under the curve* above the length of the X-axis that spans this range of values.

4. The **Normal distribution** is a family of probability density curves that are symmetric, unimodal, and bell-shaped, defined by the mean μ and standard deviation σ of a continuous random variable. The **mean** defines the location of the center of the distribution along the X-axis. The **standard deviation** defines the spread of the bell-shaped distribution.

5. The **empirical rule** provides approximate areas under Normal probability density curves:
   a. 68% of the area under the curve is above X-values in the range ($\mu - 1\sigma$) to ($\mu + 1\sigma$).
   b. 95% of the area under the curve is above X-values in the range ($\mu - 2\sigma$) to ($\mu + 2\sigma$).
   c. 99% of the area under the curve is above X-values in the range ($\mu - 3\sigma$) to ($\mu + 3\sigma$).

6. The **standard Normal distribution** is a special Normal probability density curve defined by $\mu = 0$ and $\sigma = 1$. Most statistics textbooks include tables

of probabilities for values of the standard Normal distribution variable Z.

7. The standard Normal distribution table can be used to determine probabilities associated with values for any Normally distributed variable:

    a. Convert the values of the original X variable to standard Normal Z-values using the formula:

    $$z = \frac{(x - \mu)}{\sigma}$$

    Look up the Z-value in the standard Normal table. The probability of the event defined in terms of the Z-value will equal the probability of the event defined by the value(s) of the original Normally distributed variable X.

8. The standard Normal table can be used to assign probabilities to values of a random variable only if the values of that variable are Normally distributed. To evaluate whether or not the available data values for a quantitative random variable X support this assumption of a Normal distribution, the data should be plotted in a **Normal quantile (probability) plot**. If the array of points in the plot form a straight line, the data values are Normally distributed. Deviations from a straight line indicate a non-Normal distribution.

6 Sampling Distributions: Probability Distributions for Sample Statistics

INTRODUCTION A fundamental concept that you must understand before you can correctly use sample data to draw conclusions about the nature of the real world is that *sample statistics are random variables*. In Chapter 1 I defined a **statistic** as a number computed from data obtained by a sampling or experimental study that is used to estimate the value of a population parameter. Even if a truly random, unbiased study design is used to obtain multiple independent, large, representative samples *from the same population*, the values for any sample statistic will differ among these samples. Each of these repeated samples would include different individuals from the population, and so the values of sample statistics would differ. In Chapter 1, I referred to this as **random sampling variation**. If we assume that the study design is valid, all of these different values for the sample statistic are equally valid estimates of the one true value of the corresponding population parameter. Unless we obtain data for every individual in the population (something that virtually never happens), we will never know the true value of the parameter. This concept can be discomforting for students who believe that there is always a "right answer," and that "facts" are certain. In truth, most progress in science has required that we draw conclusions and make decisions based on the very *un*-certain evidence provided by sample statistics. The topics presented in this chapter are fundamental concepts that will help you understand how to correctly draw conclusions from sample data, rather than being paralyzed into inaction by doubts about the uncertainties of your estimates.

In Chapter 3, I described how probability and probability distributions can be used to understand the behavior of random phenomena. Even though the individual outcomes of these phenomena are not predictable, the likelihood of possible outcomes can be determined.

For example, we cannot predict whether a toss of a fair coin will come up heads or tails, but we know these are the only two possible outcomes and assume each outcome is equally likely to occur. To understand a random phenomenon, we could observe it in operation many times, listing the outcomes and their associated relative frequencies. For example, we could flip a coin thousands of times to determine the relative frequency (empirical probability) of the outcome "Getting heads". To apply this empirical approach to understanding the random behavior of sample statistics, you would need to: (1) repeatedly take independent samples of size n from exactly the same population, (2) compute the sample statistic for each, and (3) determine the relative frequency for each unique value of the sample statistic. However, very few people have the time or interest to flip a coin ten thousand times to determine if P [heads] = 0.5. Likewise, scientists never repeatedly sample the same real population thousands of times to determine the probability distribution of a sample statistic.

How can we know what to expect regarding the value of a sample statistic? How can we use sample statistics that exhibit random variation to test hypotheses about the true value of an unknown population parameter? In fact, we can use the same approaches to assign probabilities to specified values of a sample statistic that we used in Chapter 4 to assign probabilities to simple random variables like X = number of heads out of four tosses of a fair coin. Theoretical probability distributions for the values of many different sample statistics can be derived by mathematical proofs, based on probability theory and logical deduction (theoretical approach). With the development of computers, these theory-based probability distributions for statistics can be demonstrated using simulations of repeated sampling from well-defined "virtual populations" (simulated empirical approach). Although any one value of a sample statistic never provides certain knowledge about the value of a population parameter, we can describe probability distributions that give us a basis for evaluating the uncertain evidence provided by statistics. ■

CHAPTER GOALS AND OBJECTIVES

In this chapter you will be able to describe probability distributions for sample proportions and sample means, and to use these distributions to determine probabilities associated with ranges of values for these sample statistics.

By the end of this chapter, you will be able to:

1. Describe the concept of a *sampling distribution*.
2. Describe the center, spread, and shape of the *sampling distribution of the sample proportion* $\hat{p}$, which is the proportion of individuals in a sample of n individuals that meet some criterion.
   a. Explain the link between randomized unbiased sampling and the assumption that the center of the sampling distribution of $\hat{p}$ is the value of the population proportion P.
   b. Compute the standard deviation of $\hat{p}$ to quantify the spread of the sampling distribution of $\hat{p}$, and explain how sample size and the value of the population proportion P influences the random sampling variation of $\hat{p}$.
   c. Describe the combinations of sample size n and values for P or $\hat{p}$ required for the sampling distribution of $\hat{p}$ to be approximately Normal.

d. Use the sampling distribution of $\hat{p}$ to determine the probability associated with a specified range of values for $\hat{p}$.

3. Describe the center, spread, and shape of the *sampling distribution of the sample mean* $\bar{x}$.
   a. Explain the link between randomized unbiased sampling and the assumption that the center of the sampling distribution of $\bar{x}$ is the value of the population mean μ.
   b. Compute the standard deviation of $\bar{x}$ to quantify the spread of the sampling distribution of $\bar{x}$, and explain how sample size n and the amount of variation among individuals in the population influences the random sampling variation of $\bar{x}$.
   c. Describe the circumstances under which it is appropriate to assume that the sampling distribution of $\bar{x}$ is Normal, as defined by the Central Limit Theorem.
   d. Use the sampling distribution of $\bar{x}$ to determine the probability associated with a specified range of values for $\bar{x}$.
4. Describe how the *Law of Large Numbers* applies to sample statistics as estimates of population parameters.

6.1

What Is a Sampling Distribution?

The **sampling distribution of a statistic** is a probability distribution for the values of a sample statistic. Sampling distributions describe the range of possible values for a sample statistic and display the probabilities associated with those values. The sampling distribution of a sample statistic might be visualized as a probability histogram or probability density curve. The list or range of possible values for the statistic is on the X-axis, and the height of the bars on the Y-axis, or area under a probability density curve, represents the relative frequency (probability) of obtaining specific values of the statistic from random samples.

The theoretical derivations of probability distributions for sample statistics are based on a fundamental assumption that sample units are selected by a random, unbiased procedure. If this assumption is fulfilled, simple logic dictates that the value of the sample statistic is equally likely to fall above or below the true parameter value. If we compute many values of a sample statistic, derived from repeated, independent samples from the same population, and we plot a frequency histogram of these values, the true value of the parameter should be at the center of this histogram. This center of the sampling distribution is the **expected value** of the sample statistic. If the study design is random and unbiased, *the expected value of the statistic is the true value of the population parameter*.

A second characteristic of sampling distributions is that the variation among values of a sample statistic computed from repeated samples from the same population should decrease if the sample size n of each sample is increased. The values of a statistic computed from large samples are less variable because they are less affected by unusual observations (outliers) than are values of a statistic computed from small samples.

If we combine random, unbiased sampling with large sample sizes, the values of a sample statistic obtained by repeated independent samples from the same population will tend to be closely bunched around the true value of the population parameter. That is, we can expect that the relative frequency (probability) of sample statistic values close to the true parameter value will be high, and the probability of getting a sample statistic value far from the parameter value will be low. Hence, by randomly sampling a large number of sample units, we can maximize the probability that we will get a **representative sample**, which provides a precise and accurate estimate of the true population parameter value. Sample statistics computed from representative samples provide a sound basis for describing the real world, drawing conclusions, and making informed decisions.

6.2

Sampling Distribution for the Sample Proportion $\hat{p}$

When the characteristic of interest recorded for each study subject is **categorical** (e.g., belonging to a category such as gender, blood type, or color), statistical analyses of these data are based on the count (frequency) or the proportion (relative frequency) of individuals that fall into particular categories. The count of individuals in the sample of n units that fall in a specified category (denoted by the symbol X) is the Binomial random variable I described in Chapter 4. However, the statistic that is more commonly analyzed in studies of categorical variables is the *proportion* of individuals in the sample of n units that fall into a specified category. The **sample proportion** is denoted by the symbol $\hat{p}$ (pronounced "p-hat") and is computed as X/n (*Note:* The sample proportion is a relative frequency.) This sample proportion is an empirical estimate of the proportion of individuals in the entire population that fall into the specified category. This population proportion parameter is denoted by the symbol P.

Determining the Center of the Sampling Distribution of $\hat{p}$

If individuals are sampled from the population of interest by a random, unbiased procedure, the value of the sample proportion $\hat{p}$ is equally likely to fall above or below the value of the true population proportion P. Therefore, the center of the sampling distribution of $\hat{p}$ is the true value of P, assuming a representative sample is obtained. Hence:

Expected Value of $\hat{p}$

$$E(\hat{p}) = P$$

Determining the Spread of the Sampling Distribution of $\hat{p}$

The spread of the sampling distribution of $\hat{p}$ reflects the amount of random sampling variation that would be exhibited by this statistic if independent samples of n observations were repeatedly taken from the same population. The sample space of possible values for $\hat{p}$ can be determined from knowledge of the center (expected value of $\hat{p}$) and spread of the sampling distribution. In Chapter 1 I described in a conceptual manner that random sampling variation of a sample statistic decreases with increasing sample size n. This general rule applies to sample proportions as well as sample means. However, sample size is not the only factor that controls random sampling variation of $\hat{p}$. The value of the population proportion P also influences how much random variation is observed in the value of $\hat{p}$. For any given sample size, random sampling variation of $\hat{p}$ will be greatest if the population proportion $P = 0.5$, and will decrease as the value of P becomes either smaller or larger than 0.5. This is because the value of a proportion must be within the range of 0 to 1. There is more room for variation when the value of P is in the middle of this range. If P is close to 0 or 1, there is a fixed "floor" ($\hat{p} \geq 0$) or a "ceiling" ($\hat{p} \leq 1$) that constrains random sampling variation. Table 6.1 displays some examples of how this equation reflects the influence of the value of the population proportion P on random sampling variation of $\hat{p}$. Based on a theoretical proof, the amount of random sampling variation that is expected in the value of $\hat{p}$ can be computed as the standard deviation of the sample proportion ($\sigma_{\hat{p}}$, pronounced "sigma sub p-hat"), computed as follows:

Standard Deviation of the Sample Proportion $\hat{p}$

$$\sigma_{\hat{p}} = \sqrt{\frac{P(1-P)}{n}}$$

Note that as sample size n increases, the value of $\sigma_{\hat{p}}$ will decrease.

TABLE 6.1   The standard deviation of the sample proportion ($\sigma_{\hat{p}}$) for a fixed sample size of $n = 100$ and a range of values for the population proportion P.

P	$\sigma_{\hat{p}}$	$=$	$\sqrt{P(1-P)/n}$
0.01	0.00995	$=$	$\sqrt{0.01(0.99/100}$
0.10	0.0300	$=$	$\sqrt{0.1(0.9)/100}$
0.25	0.0433	$=$	$\sqrt{0.25(.75)/100}$
0.50	0.05	$=$	$\sqrt{0.5(0.5)/100}$
0.75	0.0433	$=$	$\sqrt{0.75(0.25)/100}$
0.90	0.0300	$=$	$\sqrt{0.9(0.1)/100}$
0.99	0.00995	$=$	$\sqrt{0.99(0.01)/100}$

The Shape of the Sampling Distribution of $\hat{p}$

The sample proportion $\hat{p}$ is computed from the count X of individuals in a sample of size n who meet some specified criterion ($\hat{p} = X/n$). The count variable X is a discrete random variable that can have only integer values. Since the sample proportion is computed from the count X, the sample proportion $\hat{p}$ is also a discrete random variable that can take only particular values. For example, if the sample size is

$n = 4$, the sample space of the count variable X is $\{0, 1, 2, 3, 4\}$ and the sample space for $\hat{p}$ is $\{0.0, 0.25, 0.5, 0.75, 1.0\}$. The sample proportion $\hat{p}$ cannot take values between those listed. However, as the sample size increases, the values in the sample space of $\hat{p}$ look more like a continuous variable, with a range of 0 to 1.0. For example, with a sample size of 100, the sample space for $\hat{p}$ would be $\{0, 0.01, 0.02, 0.03, \ldots, 0.97, 0.98, 0.99, 1.0\}$. The consequence of this change with regard to the shape of the sampling distribution is that the probability histogram for $\hat{p}$ looks less like a "staircase" and more like a smooth curve (see Figure 6.1). As the sample size increases, the sample space of values for $\hat{p}$ becomes approximately continuous.

If the population proportion deviates from **P** = 0.5, the sampling distribution for the sample proportion $\hat{p}$ will be skewed when the sample size is small. Figure 6.2 displays sampling distributions for a selection of **P** values with the small sample size of $n = 5$. Note that the more the value of **P** deviates from 0.5, the more skewed is the sampling distribution of $\hat{p}$. However, as the sample size used to compute $\hat{p}$ increases, the sampling distribution of $\hat{p}$ becomes more symmetric. *If the sample size is sufficiently large, the sampling distribution of $\hat{p}$ becomes approximately Normal.*

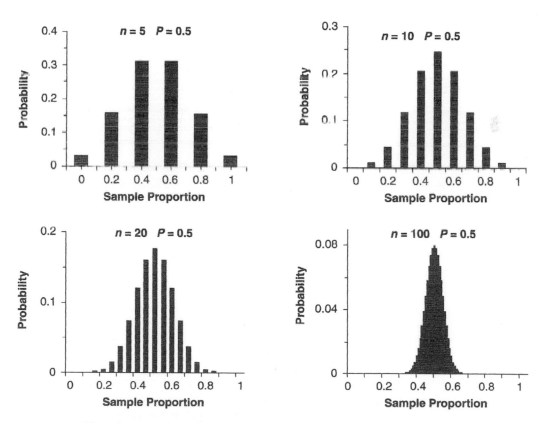

FIGURE 6.1   Effect of sample size on the sampling distribution of $\hat{p}$.

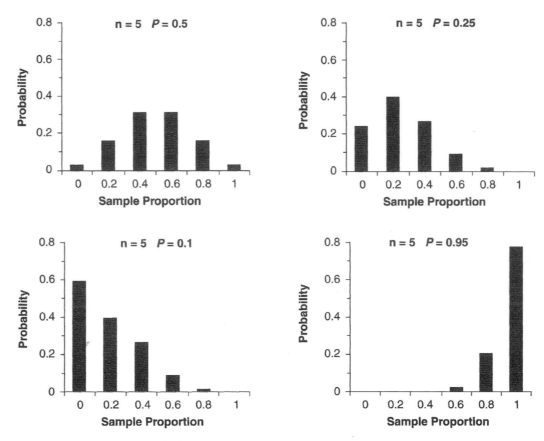

FIGURE 6.2 **Effect of the population proportion _P_ on the shape of the sampling distribution for _p̂_.**

To demonstrate this phenomenon, Figure 6.3 displays sampling distributions for $\hat{p}$ for selected combinations of population proportion **_P_** and sample size _n_. This suggests the question, "How large a sample size is required for the sampling distribution for $\hat{p}$ to become approximately Normal?" The necessary sample size depends on how far the value of **_P_** deviates from 0.5, and so, how skewed is the sampling distribution at small sample sizes.

Rule of Thumb: If $n\textbf{P} \geq 10$ and $n(1 - \textbf{P}) \geq 10$ the sampling distribution of $\hat{p}$ will be approximately Normal.

To summarize: The sampling distribution of the sample proportion can be characterized by its center, shape, and spread.

Description of the Sampling Distribution for $\hat{p}$

Center: $E(\hat{p}) = \textbf{\textit{P}}$

Spread: $\sigma_{\hat{p}} = \sqrt{\dfrac{\textbf{\textit{P}}(1 - \textbf{\textit{P}})}{n}}$

Shape: Normal *If $n\textbf{P} \geq 10$ and $n(1 - \textbf{P}) \geq 10$*

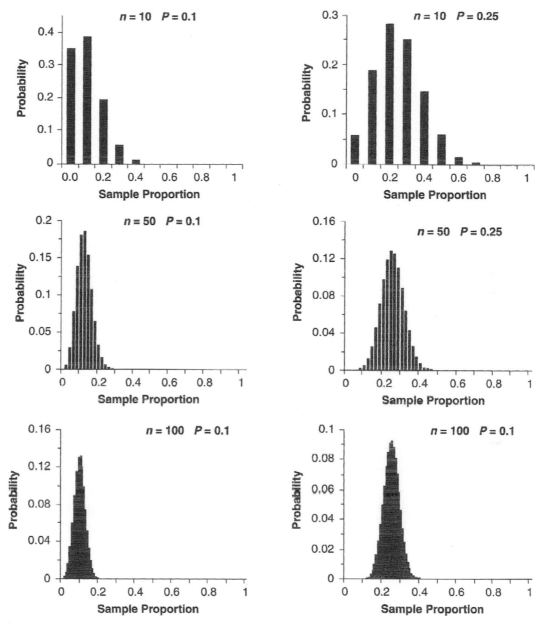

FIGURE 6.3   Interactive effects of population proportion *P* and sample size *n* on the shape of the sampling distribution of $\hat{p}$. When *P* is further from 0.5, larger sample sizes are required for the sampling distribution to become approximately Normal.

Simulating the Sampling Distribution of $\hat{p}$

In Section 3.4 I first presented the concept of *empirical probability* and the use of simulations to determine the probability distribution for simple random phenomena (e.g., flipping a fair coin). The basic premise was that it is possible to approximate the probabilities associated with the outcomes of a random phenomena by observing thousands of outcomes and computing the relative frequency of each specific type of outcome. In Figure 3.1 I presented the results from a simulation of flipping a fair coin and showed that an empirical probability (relative frequency) for P[heads] approximates the theoretical probability $P = 0.5$ when the relative frequency is based on thousands of observations. This simulation approach for describing probability distributions can be extended to more complex random phenomena, including the random variables we call sample statistics.

The description of the sampling distribution of $\hat{p}$ derived from theoretical proofs can be demonstrated empirically using computer simulations. In these simulations, a "virtual population" of known characteristics is specified; for simulations of the sampling distribution of $\hat{p}$ we need only specify the population proportion P. The computer simulates taking thousands of repeated, independent samples from this virtual population. The sample proportion $\hat{p}$ is computed for each of these thousands of independent samples. Relative frequencies are determined for each specific value of $\hat{p}$, based on the thousands of values from the simulation. The relative frequency of each $\hat{p}$ value is an empirical estimate of the probability that a single sample from this population would result in that specific value of the sample proportion, given P and sample size n. A histogram of the relative frequencies for the various observed $\hat{p}$ values is an empirical sampling distribution for $\hat{p}$. The mean of the $\hat{p}$ values from thousands of simulated samples is an empirical estimate of $E(\hat{p})$. The standard deviation of the simulated $\hat{p}$ values is an empirical estimate of $\sigma_{\hat{p}}$. The shape of the relative frequency histogram for $\hat{p}$ values from thousands of simulated samples is an empirical picture of the shape of the sampling distribution of $\hat{p}$.

Computer simulations are used in this chapter to allow you to visually assess the influence of population characteristics and sample size on the center, spread, and shape of sampling distributions. Because the computer randomly samples from the virtual population, we expect the simulated sample proportions to be unbiased estimates of P, and the center of the sampling distribution to be located at the true value of this population parameter. This true value is stipulated as part of the simulation, so we can actually determine if this expectation is fulfilled (something we rarely are able to do in real-world studies). Because the precision of sample statistics increases with larger sample sizes (Chapter 1), we expect the spread of sampling distributions for these statistics to decrease when larger sample sizes are specified for the simulation. These simulated sampling distributions also allow us to see for ourselves that the shape of the sampling distribution becomes Normal if the sample size is sufficiently large.

Results of a computer simulation.

Figure 6.4 and Table 6.2 display the results of a simulation to assess the effect of sample size on the center, spread, and shape of the sampling distribution for $\hat{p}$. The true value of the population parameter P was set at 0.10. The computer simulated taking 5,000 repeated samples from this population, the value of $\hat{p}$ was computed for each sample, relative frequencies were determined for each unique value of $\hat{p}$, and a histogram of these relative frequencies was produced to display the empirical sampling distribution for $\hat{p}$. This simulation procedure was performed for each of a

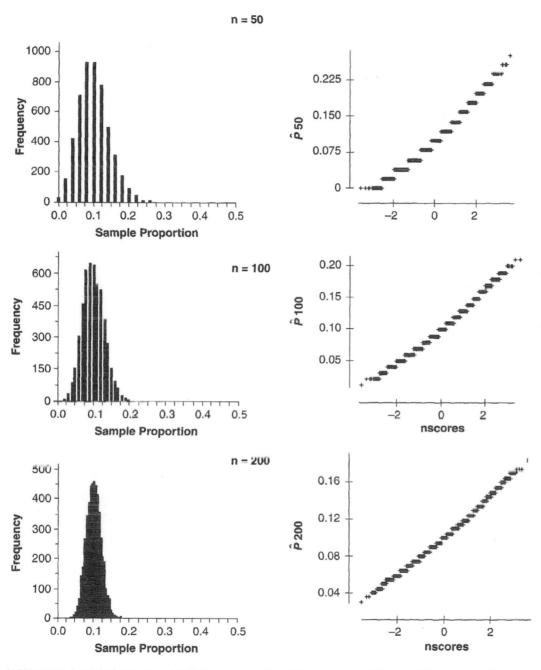

FIGURE 6.4 **Empirical sampling distributions for $\hat{p}$.**   The histograms (left side) and Normal quantile plots (right side) display the distribution of values for $\hat{p}$ obtained from a simulation of taking 5,000 repeated samples of the specified sample size n from the same population with **P** = 0.1.

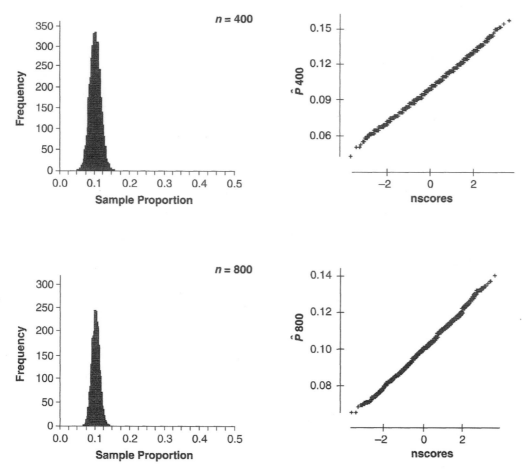

FIGURE 6.4   (continued)

range of sample sizes from $n = 50$ to $n = 800$. The histograms on the left side of Figure 6.4 display the empirical sampling distributions for $\hat{p}$.

The Normal quantile plots to the right of each histogram provide a clear indication of how closely each empirical sampling distribution approximates a Normal distribution. When the sample size is $n = 50$, the empirical sampling distribution histogram for $\hat{p}$ is positively skewed. The points in the Normal probability plot occur in separate clusters, indicating that the sample space of $\hat{p}$ includes only specific discrete values, and so, differs from a continuous Normal distribution. As the sample size is increased from 50 to 800, the shape of the sampling distribution for $\hat{p}$ becomes progressively more symmetric and bell-shaped. Also, as sample size increases, the number of possible values for $\hat{p}$ increases, and the points in the Normal probability plot begin to approximate a continuous straight line. Hence, as sample size increases, the sampling distribution of $\hat{p}$ more closely approximates a Normal distribution.

According to the "rule of thumb" for the Normal approximation, if $P = 0.1$, then $n = 100$ is the smallest sample size that allows us to consider the sampling distribution for $\hat{p}$ to be "approximately Normal." Note that the sampling distribution for $n = 100$ is still slightly skewed and slightly "discrete" looking. However, the distribution is sufficiently close to being Normal that we can use the Normal distribution (Appendix Table 2) to determine reasonably accurate probabilities of obtaining specified values of $\hat{p}$.

TABLE 6.2 The mean and standard deviation for 5,000 $\hat{p}$ values obtained from simulations of repeated sampling from a population with $P = 0.1$. Separate simulations were done for each of a range of sample sizes $n = 10$ to $n = 800$. These summary statistics are empirical estimates for $E(\hat{p})$ and $\sigma_{\hat{p}}$.

	Empirical Estimates from Simulation		Theoretical Values	
Sample Size	Mean of 5,000 $\hat{p}$ Values	Standard Deviation of 5,000 $\hat{p}$ Values	$E(\hat{p}) = P$	$\sigma_{\hat{p}} = \sqrt{\dfrac{P(1-P)}{n}}$
10	0.10026	0.09507	0.1	0.09487
20	0.10058	0.06738	0.1	0.06708
50	0.09974	0.04274	0.1	0.04243
100	0.10013	0.02950	0.1	0.03000
200	0.09967	0.02111	0.1	0.02121
400	0.09979	0.01516	0.1	0.01500
800	0.10021	0.01078	0.1	0.01061

We can also use the results from this computer simulation to demonstrate the theoretical expected value $E(\hat{p})$ and spread $\sigma_{\hat{p}}$ of the sampling distribution of $\hat{p}$. Because the 5,000 repeated samples of size n were obtained by random sampling from the virtual population, the values of the sample proportion $\hat{p}$ should be unbiased estimates of **P**. This expectation is fulfilled if equal numbers of the $\hat{p}$ values fall above and below the true value **P** $= 0.1$, such that the mean of the 5,000 $\hat{p}$ values is 0.1. Likewise, if we apply the formula for computing a sample standard deviation (Chapter 2) to the 5,000 $\hat{p}$ values, this "standard deviation of $\hat{p}$ values" is an empirical estimate of $\sigma_{\hat{p}}$.

Table 6.2 presents a comparison of theoretical versus empirical values for the center and spread of the sampling distribution of $\hat{p}$. Note that the theoretical and empirical values are very similar; they differ only because of the random variation inherent in any empirical study. As sample size increases, both the spread of the empirical sampling distributions and $\sigma_{\hat{p}}$ decrease, indicating greater precision with larger sample sizes.

In real-world science, we never take repeated samples from our population of interest to document the sampling distribution of our statistics. Rather, we depend on the theoretical descriptions of these sampling distributions to perform statistical analyses. Simulations such as this one allow us to reassure ourselves that the theoretically derived probability distributions actually describe the real-world behavior of sample statistics. These theoretical sampling distributions are powerful tools that allow us to describe a large population based on data from a single sample.

Using the Sampling Distribution of $\hat{p}$ to Determine Probabilities of Events

In Chapter 5 you learned how to use the table of probabilities for the standard Normal distribution (Appendix Table 2) to determine the probability associated with any range of values for any Normally distributed variable. As described above, the sample proportion $\hat{p}$ is a Normally distributed random variable, with a mean (or expected value) equal to the population proportion **P**, and a standard deviation $\sigma_{\hat{p}}$. Therefore, we can determine the probability associated with any range of values for $\hat{p}$ by converting the proportion value to a standard Normal Z-value and using the standard Normal distribution to obtain the probability.

Using the formula below and the standard Normal table, you can easily determine the probability of obtaining $\hat{p}$ values less than or equal to (or greater than or equal to) any observed value $\hat{p}_{obs}$ if the sample size is sufficiently large for the Normal approximation to apply.

Determining Probabilities for Values of $\hat{p}$

$$P[\hat{p} \leq \text{ or } \geq \ \hat{p}_{obs}] = P[Z \leq \text{ or } \geq \frac{(\hat{p}_{obs} - E[\hat{p}])}{\sigma_{\hat{p}}}]$$

Examples 6.1 through 6.3 illustrate the use of this formula.

■ **EXAMPLE 6.1**

Using the sampling distribution of $\hat{p}$.   Suppose a coin is flipped 40 times and the number of heads is $X = 22$; the observed value of the sample proportion is $\hat{p}_{obs} = 0.55$. What is the probability that a fair coin ($P = P[\text{heads}] = 0.5$) would produce $\hat{p}_{obs} \geq 0.55$ heads out of $n = 40$ tosses?

Description of the Sampling Distribution for $\hat{p}$

Center: $E(\hat{p}) = P = 0.5$ If the coin is fair

Spread: $\sigma_{\hat{p}} = \sqrt{\dfrac{P(1 - P)}{n}} = \sqrt{\dfrac{0.5(1 - 0.5)}{40}} = 0.079.$

Shape: Normal, because $(0.5)(40) \geq 10$ *and* $(1 - 0.5)(40) \geq 10$

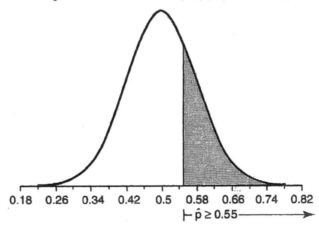

Therefore:

$$P[\hat{p} \geq \hat{p}_{obs}] = P\left[Z \geq \frac{(\hat{p}_{obs} - E[\hat{p}])}{\sigma_{\hat{p}}} \right]$$

$$P[\hat{p} \geq 0.55] = P\left[Z \geq \frac{(0.55 - 0.5)}{0.079} \right]$$

$$= P[Z \geq 0.63]$$

$$= 1 - 0.7357$$

$$= 0.2643$$

■

EXAMPLE 6.2

Using the sampling distribution of $\hat{p}$. What is the probability of obtaining 28 heads or more out of 40 flips of a fair coin ($\hat{p}_{obs} \geq 0.7$ if $P = 0.5$ and $n = 40$)?

$$P[\hat{p} \geq 0.70] = P\left[Z \geq \frac{(0.7 - 0.5)}{0.079}\right]$$

$$= P[Z \geq 2.53]$$

$$= 1 - 0.9943$$

$$= 0.0057$$

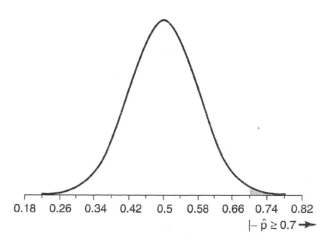

0.18   0.26   0.34   0.42   0.5   0.58   0.66   0.74   0.82

$\vdash \hat{p} \geq 0.7 \rightarrow$

EXAMPLE 6.3

Using the sampling distribution of $\hat{p}$—Healthcare application. Suppose an auditor is charged with determining whether or not a hospital discriminates against patients who do not have health insurance in decisions regarding whether or not to admit patients who come to its emergency room. Complaints about bias have been raised since the hospital came under new management. Based on a survey of hospital records for the past 10 years, 28% of patients who came to the emergency room did not have medical insurance. If the hospital did not discriminate, the proportion of emergency room patients admitted to the hospital who did not have medical insurance should also have been 28% (or $P = 0.28$). The auditor reviews records for $n = 500$ people who went to the hospital emergency room under the new management, and determines that 24% ($\hat{p} = 0.24$) of patients admitted did not have health insurance. What is the probability that the auditor would get a value $\hat{p} \leq 0.24$ if the admission policy of the new management did not discriminate against people without health insurance?

Description of the Sampling Distribution of $\hat{p}$

Center: $E(\hat{p}) = P = 0.28$ If there is no discrimination

Spread: $\sigma_{\hat{p}} = \sqrt{\frac{P(1 - P)}{n}} = \sqrt{\frac{0.28(1 - 0.28)}{500}} = 0.02.$

Shape: Normal, because $(0.28)(500) \geq 10$ *and* $(1 - 0.28)(500) \geq 10$

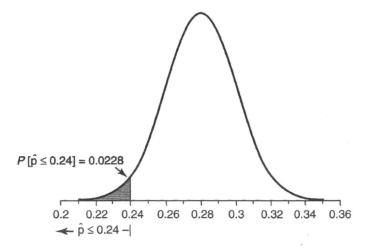

$P[\hat{p} \leq 0.24] = 0.0228$

$\leftarrow \hat{p} \leq 0.24 \dashv$

Therefore:

$$P[\hat{p} \leq \hat{p}_{obs}] = P\left[Z \leq \frac{(\hat{p}_{obs} - E[\hat{p}])}{\sigma_{\hat{p}}}\right]$$

$$P[\hat{p} \leq 0.24] = P\left[Z \leq \frac{(0.24 - 0.28)}{0.02}\right]$$

$$= P[Z \leq -2.00]$$

$$= 0.0228$$

Interpretation: If the new management at this hospital did not have a discriminatory admissions policy against individuals without health insurances, there would be a 0.0228 probability of obtaining a sample proportion less than or equal to 0.24. ∎

6.3
Sampling Distribution for the Sample Mean $\overline{x}$

When the characteristic of interest recorded for each study subject is *measured* or *quantitative* (e.g., mass, size, concentration, temperature, energy), statistical analyses often focus on the **average (mean) of the data values**, denoted by the symbol $\overline{x}$ (pronounced "x-bar"). The sample mean is computed by dividing the sum of the data values for all individuals in the sample by the sample size, as described in Chapter 2. The **sample mean** is used to estimate the value of the true mean for the entire population of interest, denoted by the symbol μ (pronounced "mew"). However, the value of the sample mean is influenced by random sampling variation; the sample mean rarely is exactly equal to the population mean. Again, we use probability distributions for the value of the sample mean to arrive at appropriate conclusions based on the uncertain evidence provided by sample data. The sample mean is a continuous random variable, and theoretically the mean has an infinite number of possible values. Therefore, the sampling distribution for sample means is a probability density curve.

Determining the Center of the Sampling Distribution of $\overline{x}$

If individuals are sampled from the population of interest by a random, unbiased procedure, the value of the sample mean $\overline{x}$ is equally likely to fall above or below the true value of the population mean μ. Therefore, the center of the sampling distribution of $\bar{x}$ is the true value of the population mean μ, assuming a representative sample is obtained. Hence:

Expected Value of $\bar{x}$

$$E(\bar{x}) = \mu$$

Determining the Spread of the Sampling Distribution of $\bar{x}$

The spread of the sampling distribution of $\bar{x}$ reflects the amount of random sampling variation that would be exhibited by this statistic if independent samples of n observations were repeatedly taken from the same population. The amount of random variation exhibited by $\bar{x}$ is influenced by: (1) the amount of variation among individuals in the population and (2) the sample size n.

The amount of variation among individuals in the population is quantified by the **population standard deviation** σ (pronounced "sigma"). This is a characteristic that differs both between variables and between different populations. For example, there is more individual variation in body weight of humans (from 2 to more than 400 pounds) than in number of teeth (0 to 32). There is also more variation in body weight among adults (60–600 lbs) than among infants (2–12 lbs). The greater the variation among individuals in the population (σ), the greater the amount of random sampling variation we can expect to see in values of $\bar{x}$ computed from independent samples.

In Chapter 1, I described at a conceptual level how random sampling variation in the value of sample statistics decreases when sample size (n) is increased. The larger the sample size, the less influence a small number of unusual individuals in the sample can have on the value of the mean. This statement is based on a mathematical proof and can be demonstrated by empirical simulations (e.g., see Figures 6.5 and 6.6). The amount of random sampling variation in the value of $\bar{x}$ is quantified by the **standard deviation of the sample mean** ($\sigma_{\bar{x}}$, pronounced "sigma sub x-bar"), computed as the population standard deviation divided by the square root of the sample size:

Standard Deviation of the Sample Mean $\bar{x}$

$$\sigma_{\bar{x}} = \sigma/\sqrt{n}$$

We rarely know the true population standard deviation σ, and will usually quantify the spread of the sampling distribution of $\bar{x}$ by replacing σ in this equation with an estimated standard deviation (S), computed from data values in a sample. The resulting measure of spread for the sampling distribution of the mean is given the symbol $S_{\bar{x}}$, and is called the **standard error of the mean**. I describe this statistic in more detail in Chapter 9.

The Shape of the Sampling Distribution of $\bar{x}$

The shape of the sampling distribution of $\bar{x}$ is determined by: (1) the shape of the population distribution for variable X and (2) sample size n. The **population distribution** is the probability distribution for individual values of variable X that would be obtained if the entire population were measured. This distribution displays the range of possible values of X for individuals and the relative frequency (probability) associated with these values within the population. *If the shape of the population distribution is Normal, the sampling distribution of $\bar{x}$ will always be Normal.* To assess if the population distribution is Normal, investigators generally look at histograms, boxplots, or Normal quantile plots of the individual data values in a sample (called the **data distribution**). *If the data distribution is approximately Normal, then we infer the population distribution is Normal, as is the sampling distribution of $\bar{x}$.* If the data distribution (and by inference the population distribution) is *not* Normal, (e.g., skewed or multimodal), the shape of the sampling distribution of $\bar{x}$ depends on the sample size n. When the sample size is small, the shape of the sampling distribution of $\bar{x}$ will be similar to the shape of the population distribution. However, with increasing sample size, the sampling distribution of $\bar{x}$ will become approximately Normal, as described by the *Central Limit Theorem:*

The Central Limit Theorem

The **Central Limit Theorem** says: *When sample size n is sufficiently large, the shape of the sampling distribution of the mean $\bar{x}$ will be Normal, no matter what the shape of the population distribution.* The question is, "How large a sample is sufficiently large?" The answer will depend on how far the data distribution is from Normal (and by inference how far the population distribution is from Normal). The more skewed and multimodal the population distribution, the larger the sample size that is required before the sampling distribution of $\bar{x}$ will be Normal.

Rules of Thumb for Applying the Central Limit Theorem

1. If $n < 15$, the sampling distribution of $\bar{x}$ can be assumed Normal only if the data distribution (and by inference the population distribution) is Normal.
2. If $15 < n \leq 40$, the sampling distribution of $\bar{x}$ can be assumed approximately Normal unless the data distribution has outliers or extreme skewness. (Look at a boxplot or Normal quantile plot for the sample data values to identify outliers and skew.)
3. If $n \geq 40$, the sampling distribution of $\bar{x}$ can be assumed approximately Normal unless there are extreme outliers or multimodal population distributions.
4. If $n \geq 100$, the sampling distribution of $\bar{x}$ will always be approximately Normal.

The sampling distribution for the sample mean $\bar{x}$ can be characterized by its center, spread, and shape.

To summarize:

Description of the Sampling Distribution for $\bar{x}$

Center: $E(\bar{x}) = \mu$

Spread: $\sigma_{\bar{x}} = \sigma/\sqrt{n}$

Shape: Normal if either one of the following conditions is met:

1. The population distribution is Normal (evaluated based on the distribution of individual sample data values, as displayed by a boxplot or Normal quantile plot).
2. The sample size is sufficiently large that the Central Limit Theorem applies.

Simulating the Sampling Distribution of $\bar{x}$

The Central Limit Theorem is based on a theoretical proof, but can also be demonstrated by computer simulations. Figures 6.5 and 6.6 show the results of two such simulations. Part (A) in each figure displays a simulated population distribution that is *not* Normal; the population distribution is extremely positively skewed in Figure 6.5, and the population is multimodal in Figure 6.6. The graphs in part (B), below

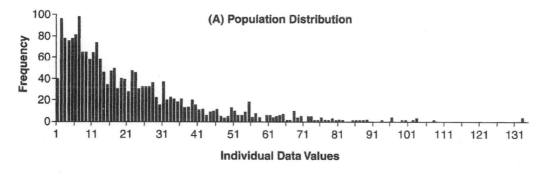

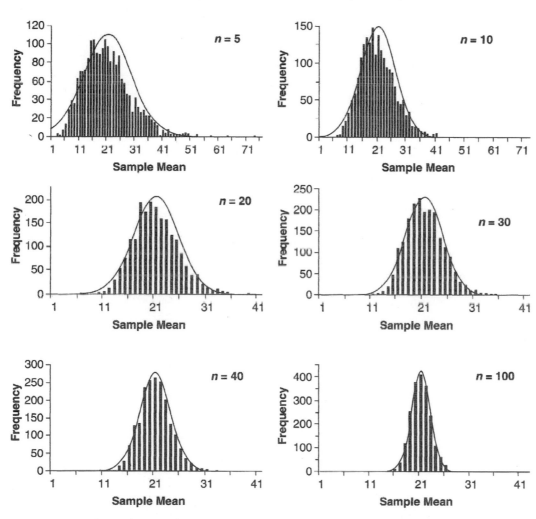

FIGURE 6.5   Simulating the Central Limit Theorem.   This figure displays a simulated positively skewed population distribution (part A) and simulated empirical sampling distributions of $\bar{x}$ for sample sizes $n = 5$ to 100 taken from this population. The effect of sample size on the spread and shape of the sampling distribution of $\bar{x}$ can be seen by comparing the various empirical sampling distributions.

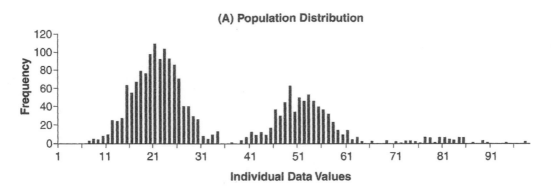

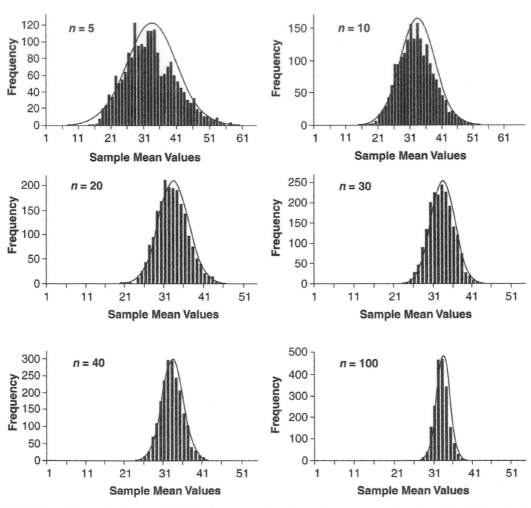

FIGURE 6.6   Simulating the Central Limit Theorem.   This figure displays a simulated multimodal population distribution (part A) and the simulated empirical sampling distributions of $\bar{x}$ for a range of sample sizes (part B). The interpretation of this figure is the same as described for Figure 6.5.

each population distribution, display empirical sampling distributions for $\bar{x}$. They are derived by: (1) taking 2,000 repeated samples of the indicated size n, (2) computing the mean for each sample, and (3) plotting the relative frequencies for different values of $\bar{x}$ (histogram bars). To help you assess how "Normal" these empirical sampling distributions are, I have also plotted a smooth Normal curve over each, based on the known μ and σ for the simulated population distributions, and the sample size n. Note that when sample size is small, the shape of the sampling distribution is distinctly not Normal. However, when sample size is at least 40, the sampling distribution of $\bar{x}$ is approximately Normal, even for these very non-Normal population distributions.

The simulated empirical sampling distribution, also demonstrate the center and spread of the sampling distribution of $\bar{x}$ ($E(\bar{x})$ and $\sigma_{\bar{x}}$). Tables 6.3 and 6.4 present a comparison of empirical estimates and theoretical values, computed using the formulae presented in this chapter, for a range of sample sizes.

TABLE 6.3. Theoretical vs. Empirical Sampling Distributions for $\bar{x}$.   The mean and standard deviation for the 2,000 sample means computed from repeated, independent samples from the population distribution in Figure 6.5 are empirical estimates for $E(\bar{x})$ and $\sigma_{\bar{x}}$. Even for relatively small sample sizes, these empirical values approximate the theoretically derived values computed using the known values for μ and σ and the formulae in this chapter.

| Sample Size | Empirical Estimates from Simulation | | Theoretical Values | |
	Mean of 2,000 $\bar{x}$ Values	Standard Deviation of 2,000 $\bar{x}$ Values	$E(\bar{x}) = \mu$	$\sigma_{\bar{x}} = \sigma/\sqrt{n}$
5	19.7	8.64	20	8.47
10	19.9	6.13	20	5.99
20	20.0	4.29	20	4.24
30	20.0	3.85	20	3.46
40	20.0	3.06	20	3.00
100	20.0	1.90	20	1.89

TABLE 6.4. Theoretical vs. Empirical Sampling Distributions for $\bar{x}$.   The mean and standard deviation for the 2,000 sample means computed from repeated, independent samples from the multimodal population in Figure 6.6 are empirical estimates for $E(\bar{x})$ and $\sigma_{\bar{x}}$.

| Sample Size | Empirical Estimates from Simulation | | Theoretical Values | |
	Mean of 2,000 $\bar{x}$ Values	Standard Deviation of 2,000 $\bar{x}$ Values	$E(\bar{x}) = \mu$	$\sigma_{\bar{x}} = \sigma/\sqrt{n}$
5	31.6	7.69	32.4	7.81
10	31.6	5.49	32.4	5.52
20	31.8	3.86	32.4	3.91
30	31.7	3.14	32.4	3.19
40	31.9	2.75	32.4	2.76
100	31.7	1.72	32.4	1.75

Applying the Central Limit Theorem

Applying the Central Limit Theorem in real-world situations requires a certain amount of judgment that comes only with experience. Students often wonder what constitutes "extremely skewed" versus "moderately skewed," and how they can distinguish a regular outlier from an "extreme outlier." I offer the following to guide beginners as they make these distinctions: Boxplots of *skewed* distributions will have one whisker longer than the other, and the median line will usually not be located in the middle of the box. A *moderately skewed* distribution might have one whisker 2 to 5× longer than the other, but with no outliers. *Extremely skewed* distributions might have one whisker more than 10× longer than the other, and usually include outliers off the end of the longer whisker. If there is a twofold difference in the length of the boxplot whiskers, a sample size of $n = 15$ is likely be sufficient to ensure the sampling distribution of $\bar{x}$ is Normal. If there is a fivefold difference, a sample size of $n = 20$ might be required, and a sample size of $n = 30$ might be needed for an eightfold difference in whisker length.

The population distribution in Figure 6.5 is extremely positively skewed by anyone's definition, as indicated in the boxplot and Normal quantile plot for this population distribution in Figure 6.7. The software used to produce the boxplot in Figure 6.7 uses an * plot symbol to indicate "extreme" outliers (defined as values more than 3× the IQR above the 75th percentile or below the 25th percentile). The population distribution in Figure 6.5 also included extreme outliers; the largest data value was almost 10× larger than the median. The simulation results in Figure 6.5 demonstrate that even for this extremely skewed population distribution, a sample size of $n = 40$ is sufficient to produce a empirical sampling distributions for $\bar{x}$ that is approximately Normal. So how much more skewed or "extreme" an outlier would provide cause for concern that a sample size of $n = 40$ would not be sufficient to ensure a Normal sampling distribution? There is no hard and fast rule. However, if the largest data value is more than 10×× the median, I would be sufficiently concerned that I would either obtain a larger sample ($n \geq 100$) or I would not assume the sampling distribution of $\bar{x}$ is Normal.

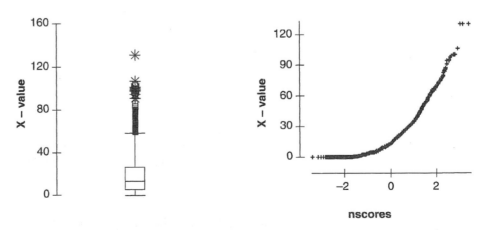

FIGURE 6.7   **Boxplot and Normal quantile plot for the population distribution in Figure 6.5.**

Why Is It So Important That the Sampling Distribution of $\bar{x}$ Be Normal?
Many statistical analyses are based on assumptions about the characteristics of the sampling distributions of statistics. The most powerful statistical analyses for sample means are based on a Normal sampling distribution for $\bar{x}$. If these procedures are used under circumstances in which it is likely the sampling distribution is *not* Normal (skewed data distribution and small sample size), the results will be inaccurate, and conclusions based on those results may be inappropriate. There are alternative statistical analyses that are not based on a Normal sampling distribution, but these procedures are often less powerful and less familiar to many scientists. Hence, it is important to recognize when we *can* assume the sampling distribution is Normal, even if the data distribution is not. For these reasons, it is very important that you be able to accurately assess whether or not the sampling distribution is Normal. In real-world situations, you will not be able to make this assessment by direct observation of the sampling distribution. You will typically have only a single sample from which you compute a single value for the sample mean. However, the simulations presented here, and the theoretical proofs used to describe the sampling distribution of $\bar{x}$, provide a basis for you to make this assessment based on data from a single sample.

Using the Sampling Distribution of $\bar{x}$ to Determine Probabilities of Events

We can use the description of the sampling distribution for $\bar{x}$ and the standard Normal distribution (Appendix Table 2) to determine probabilities associated with events defined by the value of a sample mean. The center (expected value) of the sampling distribution is equal to the population mean μ, and the standard deviation of the sampling distribution of $\bar{x}$ can be computed from the population standard deviation σ and the sample size (n). Therefore, we can determine the probability associated with any range of values for $\bar{x}$ by converting the observed value of the sample mean $\bar{x}_{obs}$ to a standard Normal Z-value and then using the standard Normal distribution to obtain the probability. Using the formulae below and the standard Normal table (Appendix Table 2) you can easily determine the probability of obtaining any specified range of values for the sample mean, if the conditions of the Central Limit Theorem are met.

Probability of a Range of Values for the Sample Mean

$$P[\bar{x} \leq \text{ or } \geq \bar{x}_{obs}] = P[Z \leq \text{ or } \geq \frac{\bar{x}_{obs} - E[\bar{x}]}{\sigma_{\bar{x}}} = \frac{\bar{x}_{obs} - E[\bar{x}]}{\sigma/\sqrt{n}}]$$

Examples 6.4 and 6.5 apply this formula.

■ **EXAMPLE 6.4**

Using the sampling distribution of $\bar{x}$. A taxonomic description for an Iris species states that mean petal length is $\mu = 10$ cm, with a standard deviation $\sigma = 2$ cm. A new population of Iris is found and a random sample of $n = 16$ plants is collected. While the Iris plants appear to be very similar to other populations, they have a smaller flower, with mean petal length $\bar{x} = 9.2$ cm. A Normal quantile plot of the 16 petal length data values indicates the data distribution is Normal. If this new population is the same species of Iris, the expected value for the sample mean would be $E(\bar{x}) = 10$ cm, the value in the taxonomic description, and the spread of the sampling distribution $\sigma_{\bar{x}} = 2/\sqrt{16} = 0.5$. If this new population is the same species, what is the probability of getting $\bar{x}_{obs} \leq 9.2$ cm?

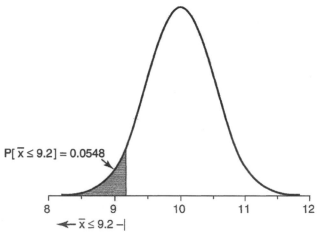

Sampling Distribution of $\bar{x}$

$P[\bar{x} \leq 9.2] = 0.0548$

$\leftarrow \bar{x} \leq 9.2 -|$

$$
\begin{aligned}
P[\bar{x} \leq 9.2] &= P[Z \leq (9.2 - 10)/0.5] \\
&= P[Z \leq -1.60] \\
&= 0.0548
\end{aligned}
$$

■ **EXAMPLE 6.5**

Effect of sample size. If the sample had included only $n = 9$ Iris plants, how would the probability $P[\bar{x} \leq 9.2]$ be altered? The sampling distribution would have a greater spread ($\sigma_{\bar{x}} = 2/\sqrt{9} = 0.67$). If the new population was the same species, the expected value of the sample mean would still be $E(\bar{x}) = 10$ cm. Given that the data values are Normally distributed, we can assume the sampling distribution of the mean is Normal regardless of sample size. Therefore, the probability of getting $\bar{x}_{obs} \leq 9$ cm is:

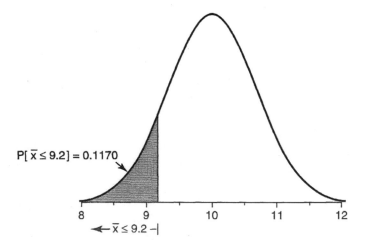

Sampling Distribution of $\bar{x}$

$P[\bar{x} \leq 9.2] = 0.1170$

$\leftarrow \bar{x} \leq 9.2 -|$

$$P[\bar{x} \le 9.2] = P[Z \le (9.2 - 10)/0.67]$$
$$= P[Z \le -1.19]$$
$$= .1170$$

Conclusion: The probability of getting $\bar{x} \le 9.2$ cm is greater when the sample size is reduced from 16 to 9 (due to increased random sampling variation with the smaller sample size).

6.4
Extension of the Law of Large Numbers to Sample Statistics

In Chapter 3, I presented the Law of Large Numbers with regard to the relationship between empirical vs. theoretical probabilities: "As the number of observations of a random phenomenon becomes very large, the observed relative frequency of an event will approach the theoretical or 'true' probability of the event." In this chapter I have described sample statistics as random variables that estimate the theoretical or "true" values of population parameters.

The Law of Large Numbers can be expanded to describe the relationship between sample size, sample statistics, and population parameters: *If study subjects are selected from the population of interest by a random, unbiased method, as the size of the sample used to compute the sample statistic increases, the value of the sample statistic will tend to approach the true population parameter value.*

This application of the Law of Large Numbers to sample statistics is based on two major features of good study design: randomization to eliminate bias and replication to increase the precision of estimates. The results of the simulations displayed in Figures 6.5 and 6.6 demonstrate that as sample size increases, the values of sample means from repeated, independent samples tend to fall closer to the true value of the population mean. The results from the simulation displayed in Figure 6.4 demonstrate that the same holds true for the value of the sample proportion.

CHAPTER SUMMARY

1. The **sampling distribution** (of a **statistic**) is a probability distribution that displays the range possible values for the statistic and provides probabilities associated with particular values or ranges of values of the statistic.
2. The characteristics of a sampling distribution (center, spread, and shape) depend on:
   a. The nature of the population distribution:
      - The **center** of the population distribution is the true parameter value, and this determines the center of the sampling distribution (i.e., the expected value of the statistic, but also see item (b) below).
      - The **spread** and **shape** of the population distribution influence, but do not entirely deter-

   mine, the spread and shape of the sampling distribution.
   b. The characteristics of the study design:
      - *Randomization.* Used to eliminate bias, randomization ensures that the value of a sample statistic is equally likely to fall above or below the true parameter value. Randomization is necessary if we are to assume that the center of the sampling distribution is equal to the true value of the population parameter.
      - *Replication.* Increasing the sample size decreases the spread of the sampling distribution (decreases variability of the statistic in repeated samples). Also, if the sample size is sufficiently large, the shape of sampling

distributions for means and proportions will be approximately Normal.

3. The **sampling distribution for the sample proportion** $\hat{p}$ can be described in terms of:

   a. Center: $E(\hat{p}) = \boldsymbol{P}$ (the true population proportion)

   b. Spread: $\sigma_{\hat{p}} = \sqrt{\boldsymbol{P}(1 - \boldsymbol{P})/n}$

   c. Shape: If the sample size is sufficiently large, such that $n\boldsymbol{P} \geq 10$ and $n(1 - \boldsymbol{P}) \geq 10$, the shape of the sampling distribution of $\hat{p}$ will be approximately Normal

4. The **sampling distribution for the sample mean** $\bar{x}$ can be described in terms of:

   a. Center: $E(\bar{x}) = \mu$ (the true population mean)

   b. Spread: $\sigma_{\bar{x}} = \sigma/\sqrt{n}$ (where σ is the population standard deviation)

   c. Shape: If the sample size n is sufficiently large, the shape of the sampling distribution of $\bar{x}$ will

be Normal, regardless of the shape of the population distribution. However, the more the population distribution deviates from Normal, the greater the sample size must be to ensure that the sampling distribution of $\bar{x}$ is Normal (**Central Limit Theorem**).

5. When the sampling distributions of $\hat{p}$ and $\bar{x}$ are approximately Normal, the standard Normal distribution can be used to assign probabilities to values of these statistics.

6. The **Law of Large Numbers** applies to the relationship between sample statistics and the true value of the population parameter. So long as data are obtained by a randomized, unbiased method, the values of sample statistics computed from progressively larger samples will tend to become less variable and to approach the true value of the population parameter.

7 Hypothesis Testing, Statistical Inference, and Estimation

INTRODUCTION In the preceding chapters, you have acquired the tools necessary to apply statistical analyses to address scientific questions. You have learned the importance of randomization and replication for obtaining samples that are representative of a larger population of interest. You have learned that the sample statistics you use to estimate characteristics of populations are random variables; different samples from the same population will give different estimates of the same population parameter. This **random sampling variation** can be described by **sampling distributions**, which provide probabilities associated with specified values of the sample statistic. You have learned how to determine the center, spread, and shape of these sampling distributions, given information regarding study design and a specific hypothesis about the true nature of the population. Using these tools, you can determine the probability associated with obtaining the observed value of a sample statistic, if some hypothetical value for the population parameter is true. In this chapter, you will learn how these concepts are applied in the process of scientific inquiry.

Important activities in the process we call science include collecting, analyzing, and evaluating data for the purposes of (1) testing competing hypotheses regarding the nature of a population of interest, and (2) estimating the value of a population parameter. For example, biologists long believed that gender of animals was determined by sex chromosomes, such that the proportion of male or female offspring was $P = 0.5$. This constituted a hypothesis about the nature of gender determination. Substantial deviation of observed proportions of male or female offspring from the expected value 0.5 would constitute evidence that some other mechanism might be involved in gender determination. If there was evidence that the true proportion of males or females deviated from the hypothesized value of 0.5, scientists might use a sample to estimate the true proportion of males or females in a population.

Inference can be defined as drawing a conclusion about the nature of a population based on empirical observations (data) from a sample. That is, we observe or measure individuals in a representative sample taken from the population, and based on this information we infer something about the nature of the entire population. However, we know that the data we obtain from samples or experiments exhibit random sampling variation. Although sampling variability can be minimized by making consistent measurements on a large sample, it can never be completely eliminated. Therefore, scientists must recognize that *sample statistics are always "uncertain evidence" regarding the true nature of the universe*. Hence, all inferences have some probability of error. There is always a possibility that a sample statistic value may be quite different from the true population parameter value, due only to random sampling variation. The probability of this unfortunate outcome is directly related to the precision and accuracy of sample estimates (controlled by study design and implementation). Different studies may use different samples sizes, have different study designs, or differ in the training and skill of their participants. All of these factors can affect the accuracy of statistics, and influence the confidence that we can place in the estimates and inferences that result from a scientific study.

A central role of statistics in the practice of science is to quantify the probability of drawing an erroneous conclusion based on empirical data that are affected by random variation. This determination is based on probability distributions of the sample statistics (i.e., sampling distributions). If the probability of an error is very small, we place a high degree of confidence in the estimates and inferences produced by a study, and these results may be widely published and provide useful insights. But if there is a relatively large probability that the estimates or inferences from a study may be incorrect, they will likely never be reported and the effort spent on the study will have been largely wasted. Careful implementation of a valid study design is the best way to avoid this unfortunate outcome.

This chapter will explain two modes of using sampling distributions to evaluate sample statistics: (1) performing *statistical tests of significance* to obtain probabilities that are used to assess whether or not sample data support a hypothesis and (2) computing *confidence intervals* that describe the precision of statistical estimates of population parameters. In this chapter I introduce these concepts only in the context of the sample proportion $\hat{p}$. However, these concepts can be applied to all types of sample statistics, as described in Section 2 of this book. ▪

GOALS AND OBJECTIVES

In this chapter you will gain an understanding of the basic concepts and logic used to draw inferences about the nature of the universe based on sample statistics that are subject to random sampling variation. You will also realize that all inferences based on sample data are uncertain, and you will be able to express the level of uncertainty as probabilities of making specific types of error. Based on these probabilities, you will be able to draw appropriate inferences from sample data and represent the precision of sample estimates of population parameters.

In particular, by the end of this chapter you will be able to:

1. Explain the reasoning for *statistical tests of significance:*
   a. Explain the concepts of Null hypothesis and Alternative hypothesis.
   b. Explain the interpretation of a test statistic and its associated *p*-value.
   c. Explain the meaning of statistical "significance," how this differs from "practical importance," and how "effect size" is used to assess importance.
   d. Explain the circumstances under which one-tailed and two-tailed tests are appropriate.
   e. Explain the concepts of Type I error, Type II error, and power, and be able to describe the trade-offs inherent in efforts to minimize Type I and Type II errors.
   f. Compute the power of one-sample *Z*-tests for hypotheses regarding proportions.
2. Explain the concept of a *confidence interval*, including the meaning of the following terms: *confidence level*, *standard error of the statistic*, and *margin of error*.
   a. Compute confidence intervals for the population proportion **P**.
   b. Compute the sample size necessary to obtain a confidence interval for the population proportion **P** with a specified width and confidence level.

7.1

Tests of Significance

Introduction to Basic Concepts

A circus psychic claims she has ESP (extrasensory perception) that allows her to "read your mind." Of course, you are skeptical about such a claim and propose a simple test. You will look at a card from a standard deck of playing cards, and the psychic will try to read your mind to "see" what is the suit of the card (heart, diamond, club, spade). You will do this 100 times, replacing the card back in the deck each time, and count the number of times the psychic correctly identifies the suit of the card. If the psychic is just guessing, she has a 1 in 4 chance of guessing correctly (only four possible suits, each guess is equally likely to be correct since each suit accounts for 25% of the cards). Hence, even if the psychic doesn't have ESP, you expect her to correctly guess the suit of the playing card 25% of the time (**P** = 0.25), just by random chance.

Suppose the psychic correctly identifies the suit of the card 28 times out of 100 ($\hat{p}$ = 0.28). Based on this result, she claims she must have some ESP because she did better than expected if she were just guessing. You respond, "Not so fast!" You have studied statistics and know how to determine the probability that the psychic just guessed correctly 28 times out of 100, without any help from ESP. If the psychic is just guessing, the true probability of making a correct guess is **P** = 0.25 for each. The psychic made n = 100 attempts to identify the suit of the card. Using the sampling distribution for $\hat{p}$, with $P = 0.25$ and $n = 100$, you determine the probability that a person could correctly guess the suit of the card 28% of the time, without any ESP, is 0.2451. That is, if you repeated this critical test many times with a person who did not have ESP, in approximately 25% of those tests the person would get 28 or more correct guesses out of 100 just by random chance. What do you conclude about the psychic's claim of having ESP?

Tests of significance.

Statistical "tests of significance" are used to evaluate sample data as evidence that either supports or refutes a hypothesis or claim about the real world. Using *The Scientific Method*, scientists must apply a critical test to any hypothesis they offer to explain some observed phenomenon. They make a prediction based on the proposed hypothesis, and then collect data from the real world to determine if the prediction comes true. If the prediction comes true, the test supports the proposed hypothesis, *but does not prove the hypothesis is true*. If the prediction does not come true, the hypothesis or explanation is modified or rejected entirely. For example, a person claiming to have ESP is making a statement about the existence of a phenomenon that most scientists do not believe exists. To support this claim, the person must submit to a critical test similar to that described above. If the person's claim is true, we would expect (predict) that individual would be able to correctly identify the suit of a card more often than would be expected for a normal person who was just guessing. In the critical test described above, the person must attain greater than 25% correct guesses. Does this mean that the psychic can claim to have some level of ESP if they correctly identify the suit on 28 out of 100 cards? Would a normal person, who does *not* have ESP and just guessed, always get exactly 25% correct guesses? Random variation would cause a normal person to sometimes correctly guess more than 25 out of 100, and sometimes less than 25. So even if the expected proportion of correct guesses for a nonpsychic is $P = 0.25$, we don't expect to get *exactly* this proportion of correct guesses in every test. Hence, the appropriate predictions for our critical test are: (1) If a person does not have ESP, the proportion of correct guesses of the suit of the playing card should be close to 0.25 and (2) if the person does have ESP, the proportion of correct guesses should be substantially higher than 0.25. This suggests the question, "What is *substantially higher*?"

The first step in any test of significance is to state a prediction in terms of the expected value of the sample statistic *and* the expected amount of random sampling variation. If the study design provides for a sample that is randomized and unbiased, we can expect that the value of our sample statistic is equally likely to fall above or below the true population parameter value. Hence, the expected value for the sample proportion $E(\hat{p})$ is the true value of the population proportion P. The expected amount of random sampling variation is quantified by the standard deviation of the sample proportion $(\sigma_{\hat{p}})$, and depends on the sample size and the value of population proportion $(\sigma_{\hat{p}} = \sqrt{P(1 - P)/n})$. If the sample size is sufficiently large, the sampling distribution for $\hat{p}$ will be approximately Normal. This description of the sampling distribution of $\hat{p}$, given a predicted value of P, is the best prediction of the outcome from a critical test involving counting the occurrences of an event (e.g., number of correct guesses out of 100 tries by the circus psychic).

If a study is designed and implemented correctly, there should be only two possible explanations for why the value of a sample statistic differs from the hypothesized parameter value:

1. *Random variation only:* The difference is due only to random sampling variation (i.e., the hypothesized parameter value actually does equal the true parameter value).

2. *Real difference:* The difference between the statistic and the hypothesized parameter value occurs because the hypothesized parameter value is really not equal to the true population parameter value.

The study must be correctly designed and implemented to eliminate any other possible explanations for the observed difference between the sample statistic and the hypothesized parameter value. For example, one way that a circus psychic might be able to correctly guess the card you are looking at is by cheating. There might be a mirror, closed-circuit TV, or some other device that allows her to actually see the cards. This would be a form of bias in the study to test her claim of ESP. The phrase "correctly designed and implemented study" means that the critical test is explicitly designed so that at the end there will be only two possible explanations for any observed difference, as described above. Hence, if you can eliminate one of these two possible explanations, the other is the only remaining explanation.

Assuming a randomized, unbiased study design, the values of the sample statistic that are most likely to occur are close to the true population parameter value. If the hypothesized parameter value is correct, the sampling distribution is centered over this value and there is a high probability that sample statistic values will be close to the hypothesized value. Alternatively, if the hypothesized population parameter value is not correct, the sampling distribution will not be centered on this hypothesized value. Rather, it will be centered on whatever is the true parameter value. In this case, there is a high probability that the sample statistic will have a value different from the hypothesized parameter value.

Based on our knowledge of the sampling distribution of the sample statistic, **a test of significance** *determines the probability that the observed difference between the value of the sample statistic and the hypothesized value of the population parameter is due only to random sampling variability*. If the difference between the sample statistic and the hypothesized population parameter value is large relative to the expected amount of random sampling variation, the probability that this difference is simply due to random variation is very small. If the *random-variation* explanation is very unlikely to be true, then the *real-difference* explanation is the only likely explanation remaining. This outcome would provide empirical evidence that whatever we thought might cause a difference in some variable actually does cause the difference (e.g., the psychic actually does have ESP).

Procedure for a statistical test of significance.

I present the procedure for a statistical test of significance by further exploring the gender determination hypothesis mentioned earlier.

One explanation (hypothesis) for the phenomenon of gender determination in many animal species is that the gender of an organism is determined by a sex chromosome in the sperm that fertilizes the egg. If this explanation for gender determination is true, the expected proportions of male and female offspring are both $P = 0.5$.

A scientist working on conservation efforts to save the endangered loggerhead turtle has organized a program to dig up turtle nests and bring the eggs into an incubator to protect them from destruction by predators and human disturbance. Based on observations of hatchlings from the first group of eggs, the proportion of female hatchlings coming out of the incubator is 0.36, less than the expected 0.5. This observation concerns the scientist because the number of reproductive females in the population is critical to its recovery. He wonders if something other than sex chromosomes influences gender determination of loggerhead turtles. The scientist collects data for a number of environmental factors (temperature, relative humidity, etc.) at natural nest sites and in his incubators. He determines that the average temperature in the incubators was 3°C less than the average temperature in natural nests. Based on this observation, the scientist can ask a more specific scientific question, "Does lowering incubation temperature of loggerhead turtle eggs reduce the proportion of females hatched below that expected if gender is determined by sex chromosomes?"

Notice that this more specific question is more amenable to study and experimentation than the scientist's earlier, more general question. Application of the formal scientific method to the question of gender determination in loggerhead turtles might include:

Scientific Hypothesis: Lowering incubation temperature reduces the proportion of females hatched from loggerhead turtle eggs below that expected if gender is determined entirely by the sex chromosome in the sperm that fertilizes the egg.

Prediction: When incubation temperature of loggerhead turtle eggs is lower than in natural nests, the proportions of females hatched will be less than 0.5 (the expected proportion if gender is determined by sex chromosomes and not affected by temperature).

Critical Test: Incubate a clutch of eggs at a temperature below the average for natural nests and determine the proportions of female hatchlings. *Note:* A more complex study design that incubated eggs at multiple temperatures would be more realistic, but let's keep the analysis simple for now. Suppose the scientist incubated a sample of $n = 50$ loggerhead turtle eggs at a temperature 3°C cooler than natural nests, and 17 female turtles were hatched ($\hat{p} = 0.34$).

Test of Significance: The decision the scientist must make is whether or not this sample proportion $\hat{p} = 0.34$ is sufficiently different from 0.5 to conclude that lowering incubation temperature reduced the proportion of female hatchlings. We will assume that the scientist correctly implemented an appropriate study design. Hence, there are only two possible explanations for the difference between the observed proportion of females and the expected proportion of 0.5: (1) The difference between $\hat{p} = 0.34$ and $P = 0.5$ is due only to random sampling variation, *or* 2) the difference occurred because lower incubation temperature affects gender determination of the loggerhead turtles.

A statistical test of significance is used to determine the probability that the observed difference between $\hat{p} = 0.34$ and $P = 0.5$ is due only to random sampling variation. If this probability is sufficiently small, the scientist will have enough evidence to reject the random-variation-only explanation. The remaining real-difference explanation is that the lower incubation temperature affected gender determination in a manner that reduced the proportion of females hatched.

The procedure for a test of significance is highly formalized, starting with the specification of two competing *statistical* hypotheses that describe values for the population parameter that would be consistent with the two possible explanations for outcomes of the critical test.

- **Null Hypothesis [H₀]:** This is the hypothesis of "no difference"; it states the expected value of the population parameter if groups are *not* different or if a treatment does *not* cause a change.
- **Example:** If lowering incubation temperature does *not* influence gender determination, then the Null hypothesis states H_0: P_{female} = 0.5.
- **Alternative Hypothesis [Hₐ]:** This hypothesis states the expected manner in which the parameter value differs from the value stated in the Null hypothesis if the groups under study actually differ, or if the treatment does cause a change.
- **Example:** If lower incubation temperature actually does reduce the probability that the eggs will develop into females, then the Alternative hypothesis states H_a: P_{female} < 0.5.

Important: These hypotheses are *always* stated in terms of *population parameter values, not* values for the sample statistics that estimate the parameters. We can actually *know* the value of statistics, and so, do not have to hypothesize about these values. It is the populations parameters that are unknown.

The next step in a test of significance (as shown in Example 7.1) is to describe the sampling distribution of the sample statistic *if the Null hypothesis is true.*

▪ **EXAMPLE 7.1**

Describing the sampling distribution for $\hat{p}$. If gender of loggerhead turtles is entirely determined by sex chromosomes and unaffected by incubation temperature, then $P_{females}$ = 0.5 and the sampling distribution for $\hat{p}$ with n = 50 has the following characteristics:

Description of the Sampling Distribution

Center: $E(\hat{p})$ = P = 0.5
Spread: $\sigma_{\hat{p}}$ = $\sqrt{P(1 - P)/n}$
  = $\sqrt{0.5(1 - 0.5)/50}$
  = 0.07
Shape: Normal
  nP = 50 (0.5) = 25 > 10
  $n(1 - P)$ = 50 (0.5) = 25 > 10

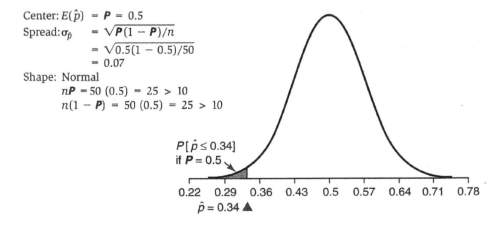

$P[\hat{p} \leq 0.34]$
if P = 0.5

0.22 0.29 0.36 0.43 0.5 0.57 0.64 0.71 0.78
$\hat{p}$ = 0.34 ▲

The next step in a test of significance is to compute an appropriate **test statistic** that represents the size of the difference between the observed sample statistic value and the population parameter value proposed by the Null hypothesis. This difference is always expressed relative to the expected amount of random sampling variation for the sample statistic. The general formula for a test statistic is:

$$\text{Test Statistic} = \frac{\text{Sample Statistic} - \text{Expected Value of Parameter under H}_0}{\text{Standard Deviation of the Sample Statistic}}$$

Test statistics always have a known probability distribution that allows us to assign a probability to the observed value of the sample statistic if the value of the population parameter stated in the Null hypothesis is true. When the sample statistic is a proportion, and the sample size is sufficiently large that the Normal approximation is valid, the test statistic is a standard Normal Z-value, as in Example 7.2.

■ **EXAMPLE 7.2**

Computing the test statistic. The scientist incubated 50 loggerhead turtle eggs at a temperature 3°C below the average temperature for natural nests, and 17 eggs produced female hatchlings ($\hat{p}$ = 0.34). Under the Null hypothesis (H$_0$), incubation temperature would have no effect on gender, so the true proportion of females is P = 0.5 and $E(\hat{p})$ = 0.5. Given this expected value, the expected amount of random sampling variation for the value of $\hat{p}$ is $\sigma_{\hat{p}}$ = 0.07. The test statistic is computed as:

$$Z_{\text{test}} = \frac{(\hat{p} - P_0)}{\sigma_{\hat{p}}} = \frac{(0.34 - 0.5)}{0.07} = -2.29$$

The difference between the observed sample proportion of female hatchlings ($\hat{p}$ = 0.34) and the predicted proportion if incubation temperature has no effect (P_0 = 0.5) is 2.29 times larger than the expected amount of random sampling variation ($\sigma_{\hat{p}}$). ▨

The final step in a test of significance (shown in Example 7.3) is to determine the probability of obtaining the observed sample statistic value *if the true population parameter is equal to the value specified by the Null hypothesis.*

■ **EXAMPLE 7.3**

Using the test statistic to determine the probability.

$$P[\hat{p} \leq 0.34, \text{ if } P = 0.5] = P[Z \leq -2.29]$$

$$= 0.0110$$

The probability of getting a proportion of females ≤ 0.34 in a sample of 50 hatchlings due only to random sampling variation (if gender is determined by sex chromosomes and P = 0.5) is 0.0110, as determined using the Standard Normal table.

▨

Drawing a conclusion from the critical test and test of significance.

Using the results of the test of significance, the scientist can make an informed inference about the nature of the real world based on the uncertain evidence of data derived from his experiment. If gender is determined entirely by sex chromosomes, the probability of getting 17 or fewer female hatchlings out of 50 eggs is 0.0110, or about 1 time out of 100 repeated experiments. Most people would consider this an unlikely outcome. With this information, the scientist must now choose between the two possible conclusions: (1) an unlikely random event occurred during his experiment, *or* (2) incubating loggerhead turtles at temperatures below the average for natural nests reduces the proportion of female hatchlings. Faced with this evidence, most scientists would conclude that there is sufficient evidence to conclude that the observed difference was *not* due only to random variation. The only remaining likely explanation is that incubation temperature influences gender determination of loggerhead turtles.

Tests of significance are always critical tests of the Null hypothesis (i.e., that there is *no difference*). To many people, this seems odd. If the scientist in Examples 7.1–7.3 did not expect incubation temperature to affect gender determination in loggerhead turtle hatchlings, why bother doing the experiment? The reason tests of significance are always critical tests of the Null hypothesis is that we can predict an exact expected value for the sample statistic *only for the Null hypothesis*. In Examples 7.1–7.3, if the Null hypothesis is true (gender determination is due entirely to sex chromosomes), the expected proportion of female hatchlings, regardless of incubation temperature, is $E(\hat{p}) = 0.5$. On the other hand, if temperature affects gender determination, we have no idea what the expected proportion of female hatchlings should be, and so, cannot propose an exact expected value for the sampling distribution. Since H_0 and H_a are complementary, tests that fail to support the Null hypothesis can be interpreted as providing evidence that supports the Alternative hypothesis.

Interpreting the Results from a Test of Significance

The probability value obtained from a test of significance is called a "*p*-value." The **p-value** is the probability that the observed difference between the value of the sample statistic and the Null hypothesis parameter value is *due only to random sampling variation*, if the Null hypothesis is really true. This probability is a measure of the strength of the evidence against the Null hypothesis. That is, if the observed value of the sample statistic is very unlikely to occur if the Null hypothesis was true, perhaps the Null hypothesis is *not* true.

If we use the *p*-value as evidence to reject the Null hypothesis (i.e., draw a conclusion that the true population parameter value differs from the value stated in the Null hypothesis), the *p*-value has another important interpretation. *The p-value is also the probability that a conclusion to reject the Null hypothesis value is incorrect.*

Remember, the *p*-value is the probability of getting the observed statistic value due only to random sampling variation *if the Null hypothesis parameter value is true*. If the Null hypothesis is true, concluding it is false would be an error, and the *p*-value is the probability of this error. In Example 7.3, the *p*-value 0.0110 is the probability of getting $\hat{p} \leq 0.34$ female hatchlings due only to random sampling variation, if gender is determined by sex chromosomes and $P = 0.5$. If the scientist used this evidence to conclude that lowering incubation temperature reduced the proportion of females hatched from loggerhead turtle eggs, there is a probability 0.0110 that this conclusion is wrong. If the scientist published a paper that claimed to have found a new mechanism for gender determination, it would likely result in substantial controversy. His credibility among his peers could be questioned. Hence, he would want to minimize the probability of making an incorrect conclusion.

The *p*-value from a test of significance is also called the **significance level**. In this context, the *p*-value is a quantitative representation of the strength of evidence against the Null hypothesis. The difference between a sample statistic and the Null hypothesis parameter value is called **statistically significant** if the value of the sample statistic represents strong evidence that the true population parameter value differs from the Null hypothesis parameter value. The phrase "strong evidence" means that the observed difference between the sample statistic and the Null hypothesis value is unlikely to be due only to random variation. This outcome is associated with conclusions like, "The treatment had an effect."

How small must a p-value be to infer a sample statistic is significant?

1. *Using the decision approach* (Figure 7.1): By tradition, most scientists use the standard $p \leq 0.05$ as the critical threshold for inferring that the evidence from a test of significance is sufficiently strong to reject the Null hypothesis and claim that there is a real difference. Based on this criterion, we will incorrectly reject the Null hypothesis value when it is true in 5 out of 100 tests of significance. That is, 5 times out of 100 we would obtain an apparently "significant" difference between the sample statistic and Null hypothesis value due only to random variation, when there really is no difference. Using the traditional approach, the exact *p*-value from a test of significance is not reported, only the decision as to whether or not the statistic was significant, with $p \leq 0.05$. For example, "the proportion of female turtles hatched in nests with lower incubation temperature was significantly less than 0.5."

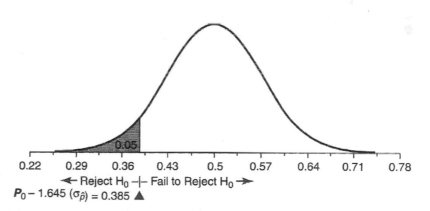

$P_0 - 1.645\,(\sigma_{\hat{p}}) = 0.385$ ▲

FIGURE 7.1   **Sampling distribution of $\hat{p}$, with *rejection region* of $\hat{p}$ values defined by all values that are sufficiently less than $P_0 = 0.5$ to reject the Null hypothesis and conclude that the true population proportion P is < 0.5.**   The sampling distribution corresponds to the study described in Examples 7.1–7.3. The boundary of the rejection region is defined by the critical value of $\hat{p}$ that has a probability $P[\hat{p} \leq \hat{p}_{\text{critical}}] = 0.05$. This region corresponds to all $\hat{p}$ values that are less than $P_0 - 1.645\,(\sigma_{\hat{p}})$ or $\hat{p} \leq 0.385$. The value -1.645 is the critical Z-value corresponding to a probability of 0.05 in the lower (left) tail of the Z-distribution (see Appendix Table 2). Using the traditional inference approach, any $\hat{p}$ value that falls within the rejection region is evidence of a "significant difference" from $P_0 = 0.5$.

2. *Using the inference approach* (Figure 7.2): The exact *p*-value from the test of significance is reported. The smaller the *p*-value, the stronger the evidence that the unknown true value of the population parameter differs from the value proposed in the Null hypothesis. The investigator states the conclusion without using the word "significant," but instead states the exact p-value as a measure of the strength of evidence against the Null hypothesis. For example, "the proportion of female turtles hatched in nests with lower incubation temperature was less than 0.5 ($p = 0.011$)."

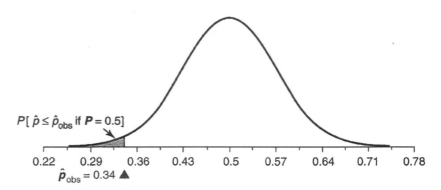

FIGURE 7.2   Sampling distribution of $\hat{p}$, with shaded area under the curve representing the exact probability $P[\hat{p} \leq \hat{p}_{obs}, \text{ if } P = 0.5]$.   This sampling distribution corresponds to the study described in Examples 7.1–7.3. The boundary of the shaded area is defined by the observed sample statistic value $\hat{p} = 0.34$. The shaded area represents the exact probability of getting a sample proportion $\hat{p} \leq 0.34$ if the Null hypothesis value for the population proportion $P_0 = 0.5$ is true (the *p*-value $= 0.011$).

Rules of Thumb for Interpreting *p*-Values Using the Inference Approach

$p \leq 0.05$	In the community of scientists, *p*-values in this range are widely considered to be "strong evidence" that the unknown true population parameter value differs from the Null hypothesis value.
$0.05 < p \leq 0.10$	There is not widespread consensus in the scientific community regarding how to interpret *p*-values in this "gray zone." A middle-of-the-road approach is to conclude, "The data suggest there is a difference, but more data are needed to verify this inference."
$p > 0.10$	Given *p*-values in this range, almost all scientists would conclude there is "insufficient evidence" to claim that a difference exists (i.e., not enough evidence to reject the Null hypothesis value for the population parameter).

Important: Even if the *p*-value is large, it is *not* appropriate to conclude that the parameter value proposed by the Null hypothesis is "true." As we will see later, there is more than one reason why a *p*-value might be large. Only one of these reasons is that the parameter value proposed by the Null hypothesis is correct. When the *p*-value is large, we can conclude only that there is not enough evidence to reject the Null hypothesis parameter value.

The Effect of Sample Size on Tests of Significance

Suppose the scientist working with loggerhead turtle conservation in Examples 7.1–7.3 was the first to "discover" that gender was not determined entirely by sex chromosomes in a vertebrate animal species. Based on the experiment and statistical test of significance described in this example, the scientist has "strong evidence" that the proportion of female hatchlings differs from the proportion expected if gender is determined entirely by sex chromosomes ($P = 0.5$). However, he knows that this evidence is contrary to the prevailing understanding about gender determination, and it will likely generate controversy when it is reported in the scientific literature. His research will be intensely scrutinized, and his credibility may be questioned. He wants to minimize criticism of his analysis and conclusions.

The scientist could strengthen the evidence in support of his conclusion if he increases his sample size (number of eggs incubated at a temperature 3°C below natural nest temperature). As a general rule, the more data used to arrive at a conclusion, the more confident we are in that conclusion (again, assuming an appropriate, unbiased study design). Common sense tells us that an unexpected outcome based on a small sample of observations could be just a "fluke" (i.e., due to random variation). The more observations used to document the unexpected outcome, the less likely it is due merely to random variation. If you flip a coin 2 or 3 times and get all heads, you wouldn't be surprised. But if you flip a coin 20 times and get 20 heads, you would have reason to doubt the fairness of the coin. How does this commonsense faith in larger sample sizes enter into statistical tests of significance? Example 7.4 explores this question.

■ EXAMPLE 7.4

Effect of increasing the sample size. To strengthen the evidence in support of his observation of a temperature effect on gender determination, the scientist decides to increase his sample size to $n = 100$ eggs incubated at the lower temperature. Let's suppose that he gets the same sample proportion of $\hat{p} = 0.34$ female hatchlings from these 100 eggs. How has increasing the sample size strengthened the evidence in support of a temperature effect on gender determination?

To perform a test of significance with the larger data set, we must first describe the sampling distribution of $\hat{p}$ based on the larger sample size. Remember, the sampling distribution is specified under the assumption that the Null hypothesis is true.

If you compare this sampling distribution with the one specified in Example 7.1, the only difference is that with a sample size of 100, $\sigma_{\hat{p}} = 0.05$ is smaller than it was with a sample size of 50 ($\sigma_{\hat{p}} = 0.07$). We expect sample statistics computed from larger samples to exhibit less random sampling variation than statistics computed from smaller samples.

Center: $E(\hat{p}) = P = 0.5$
Spread: $\sigma_{\hat{p}} = \sqrt{P(1-P)/n}$
$= \sqrt{0.5(1-0.5)/100}$
$= 0.05$

Shape: Normal
$nP = 100\,(0.5) = 50 > 10$
$n(1-P) = 100\,(0.5) = 50 > 10$

The next step in the test of significance is to compute the Z_{test} statistic and determine the p-value, as shown below:

$$P[\hat{p} \le 0.34, \text{ if } P = 0.5] = P[Z \le (0.34 - 0.5)/0.05]$$

$$= P[Z \le -3.20]$$

$$= 0.0007$$

The probability of getting a proportion of females ≤ 0.34 in the sample of 100 hatchlings if gender is determined entirely by sex chromosomes is about 7 times out of 10,000 repeated experiments.

Because the observed difference $(\hat{p} - P_0) = 0.34 - 0.5 = -0.16$ is very unlikely to occur due to random variation alone, we can conclude there is strong evidence that gender determination is influenced by incubation temperature. If we made such an inference based on our sample data, there is only a 0.0007 probability that we would be wrong.

> **NOTE** When the sample size was increased from $n = 50$ to 100 eggs, the probability of getting the same difference between $\hat{p}$ and P_0 due to random sampling variation alone decreased from 0.011 to 0.0007. As sample size n increases, random sampling variation (i.e., $\sigma_{\hat{p}} = \sqrt{P(1-P)/n}$) decreases, and the probability of getting any specific difference between $\hat{p}$ and P_0 due only to random variation becomes smaller. With larger sample sizes, a test of significance is more likely to reject the Null hypothesis parameter value, if the true population parameter value really is different. ■

The difference between "statistically significant" and "practically important."

When a statistical test of significance is based on a large sample size, even small differences between the sample statistic and the Null hypothesis parameter value can have small p-values and be deemed statistically significant. For example, suppose the scientist studying sex determination in loggerhead turtles collected data for 10,000 eggs incubated at 3°C lower than in natural nests, and the proportion of females was $\hat{p} = 0.49$. A test of significance would determine that the probability of getting $\hat{p} \leq 0.49$ if $P = 0.5$ is $p = 0.0228$. (Do the calculations for yourself and see if you can get the same answer.) This result provides strong evidence that the proportion of females hatched is less than the expected 0.5 if gender was determined by sex chromosomes alone. That is, this difference between 0.49 and 0.5 is statistically significant. However, this simply means that with a large sample size we are confident this difference reflects a real, albeit small, difference between the true P and the Null hypothesis value of 0.5.

The example above demonstrates that in some situations there may be a distinction between **statistically significant** (which means "not due to random sampling variability alone") and "**practically important**." Even though the difference between $\hat{p} = 0.49$ and $P = 0.5$ was unlikely to be due only to random variation when $n = 10,000$, the practical question that must now be addressed is whether or not this difference is sufficiently large to justify doing something about it. A statistically significant result is "important" only if it provides a reason for us to change the way we think about something, or do something, that has wide implications for a human enterprise (business, scientific discipline, or society).[1] The magnitude of the difference between the observed statistic and the Null hypothesis value must be big enough to have real-world consequences. The difference must also be unexpected (contrary to prevailing understanding), what Abelson (1995) calls "interesting." Finally, the more connections there are between the "interesting" result and major theories or societal concerns, the more "important" the result. In the example above, a finding that something other than sex chromosomes influences gender determination in a vertebrate species would be considered "interesting" to biologists. However, the magnitude of the difference between $\hat{p} = 0.49$ and $P = 0.5$ is small, and this result would probably not be considered important.

Effect size is the difference between an observed sample statistic and the parameter value proposed by the Null hypothesis. This is a measure of the "magnitude" of the effect (difference) documented by a study. Effect size can be expressed in absolute terms or in relative terms, computed as shown below:

Absolute Effect Size = Statistic Value − Null Hypothesis Parameter Value

$$\text{Relative Effect Size (\%)} = 100 \times \frac{\text{Statistic Value} - \text{Null Hypothesis Parameter Value}}{\text{Null Hypothesis Parameter Value}}$$

For example, the absolute effect size documented in the experiment described above with $n = 10,000$ loggerhead turtle eggs was $(\hat{p} - P_0) = 0.49 - 0.5 = -0.01$. The relative effect size would be the difference $(\hat{p} - P_0)$, expressed as a percent of the Null hypothesis value. For the preceding example, $100 (-0.01/0.5) = -2\%$. This relative effect size means that there were 2% fewer females than expected if gender was determined solely by sex chromosomes.

Relative effect size provides an easily understood measure of the effect, but this means of expressing effect size can be misleading if the Null hypothesis parameter value is a very small number. Expressing the effect size as a percent of a very small number tends to overstate the true effect. In one real-world example, a study widely reported in the news media of Britain estimated that a new type of birth control

---

[1]Abelson, R .P. 1995. *Statistics as Principled Argument.* Lawrence Erlbaum Associates, Publishers, Hillsdale, NJ.

pill was twice as likely to cause potentially fatal blood clots (relative effect size = $+100\%$). The result of this news report was that large numbers of women stopped taking the new pill, resulting in an estimated 8,000 extra abortions plus an unknown number of unplanned pregnancies. The absolute effect size for the increased number of fatalities due to blood clots associated with this new birth control pill was an increase from 2 to 4 fatalities per year for all of Britain.

Bottom Line: Whenever you report the results of a statistical test of significance, you should *always* report either the sample statistic value(s) or the effect size along with the *p*-value. Readers can then make their own judgment as to whether or not a statistically significant difference is of practical importance.

"One-Tailed" vs. "Two-Tailed" Tests of Significance

The way that a scientific question is stated influences how the test of significance is done. In some situations, the researcher has reason to believe that the true value of the population parameter might differ from the Null hypothesis value in a specific direction ($<$ or $>$). For example, the scientist studying loggerhead turtle hatchlings had reason to believe lower incubation temperature reduced the proportion of female hatchlings, based on prior observations. However, it is often the case that researchers do not have sufficient knowledge to specify the expected type of difference. Rather, the question is simply whether or not the population parameter is *different* from some Null hypothesis value. For example: "Does the proportion of females produced from eggs incubated at lower temperatures differ from the value expected if gender is determined by sex chromosomes ($P = 0.5$)?" This difference in the manner that the scientific question is posed is reflected in the statistical Alternative hypothesis (H_a). If the question indicates a specific kind of expected difference ("less than" or "greater than"), the Alternative hypothesis should indicate this as $P < P_0$ or $P > P_0$, respectively. If the question simply asks whether or not a difference exists, the Alternative hypothesis is stated as $P \neq P_0$.

1. *One-tailed test:* If H_a is $P < P_0$, then only values of $\hat{p}$ much less than P_0 will cause us to reject H_0 and conclude there is a real difference between P and P_0. That is, all possible values of $\hat{p}$ that would cause us to reject H_0 are in the left tail of the sampling distribution. (See Figure 7.3.)

2. *Two-tailed test:* If H_a is $P \neq P_0$, then values of $\hat{p}$ much greater than P_0 *or* $\hat{p}$ much less than P_0 will cause us to conclude there is a real difference between the true population P and the Null hypothesis value P_0. There are two ranges of $\hat{p}$ values, located in the extreme left and right tails of the sampling distribution, that would cause us to reject the Null hypothesis value. When we state H_a as $P \neq P_0$, we must determine the overall probability of getting an absolute difference $|(\hat{p} - P_0)|$, either positive or negative, due only to random sampling variation. To do this, we simply multiply the *p*-value obtained in the test of significance by 2 to obtain the two-tailed probability of getting the observed absolute difference $|(\hat{p} - P_0)|$ due only to random variation. (See Figure 7.4.)

Which test to choose?

Because one-tailed tests of significance always have smaller *p*-values than two-tailed tests applied to the same data, one-tailed tests are always more likely to support a

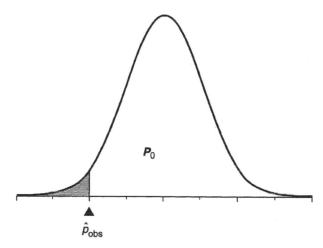

FIGURE 7.3   Sampling distribution of $\hat{p}$.   The shaded area represents the p-value for a one-tailed test of significance corresponding to the probability $P[\hat{p} \leq \hat{p}_{obs}$, if $P = P_0]$.

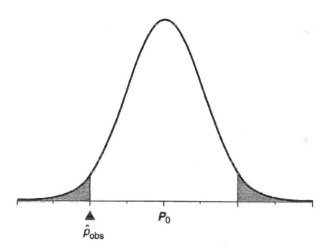

FIGURE 7.4   Sampling distribution of $\hat{p}$.   The shaded area represents the p-value of a two-tailed test of significance. The area in the left tail below $\hat{p}_{obs}$ was shaded first, then the "mirror image" of that area was shaded in the right tail.

conclusion that a significant difference exists. In most situations, it is in the scientist's personal interest to reject the Null hypothesis, as this conclusion leads to "Interesting" claims that are more likely to be published. A scientist's reputation, job opportunities, and salary are often based on publishing the results of their research. To avoid even the appearance that personal interests influence conclusions drawn from scientific research, most editors of scientific publications require rigorous documentation of the basis for using one-tailed tests of significance. The most common standard that must be met for a one-tailed test to be considered valid is that there is a strong, theory-based reason, well documented in the published scientific literature, for predicting a specific type of difference.

A second, though less widely accepted reason for performing a one-tailed test is in situations where the investigator is only interested in a difference in a particular direction; a significant difference in the other direction would be ignored. For example, if a new drug is supposed to reduce blood pressure, the investigator performing the drug trial would only recommend production of the drug if the treated subjects had *lower* blood pressure than untreated subjects. Even in this situation, some scientists might wonder if the investigator would really ignore the fact that the drug significantly increased blood pressure.

Bottom Line: The use of one-tailed tests is rare in real-world scientific research. Unless you have an iron-clad justification, supported by a large body of published research, using a one-tailed test is almost certain to generate criticism of your analysis. The distinction between using a one-tailed versus two-tailed test often does not affect the nature of your final conclusion as to whether or not there is a "significant difference." Only in cases where the evidence against the Null hypothesis is borderline (*p*-value close to 0.05) will the choice of one-tailed versus two-tailed test make a difference. Example 7.5 gives a comparison of the use of one-tailed and two-tailed tests of significance.

▪ **EXAMPLE 7.5**

One-tailed vs. two-tailed tests of significance. In Example 7.1, the scientist studying gender determination of loggerhead turtles asked the specific question, "Does lower incubation temperature reduce the proportion of female hatchlings?" Given this scientific question, a one-tailed test of significance was used to evaluate the evidence that $P < P_0 = 0.5$. Figure 7.5 below is a diagrammatic representation of that test of significance. The *p*-value computed in Examples 7.1–7.3 represents the probability $P[\hat{p} \leq 0.34$, if $P = 0.5]$. However, if this scientist had been studying the effect of temperature on gender determination for the first time, and there was no existing literature on this phenomenon, the more appropriate scientific question would have been, "Does incubation temperature influence gender determination in loggerhead turtles?" In this case, the scientist does not have a sufficiently strong basis to predict a specific temperature effect, so the test of significance must be two-tailed. Using the results described in Examples 7.1–7.3, now the test of significance must determine the probability of getting a difference of $(\hat{p} - P_0)$ of $+0.16$ *or* -0.16 due to random variation. Figure 7.6 is a diagrammatic representation of this test of sig-

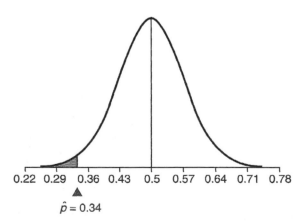

FIGURE 7.5   Sampling distribution with shaded area = *p*-value for one-tailed test of significance = $P[(\hat{p} - P_0) \leq -0.16]$.

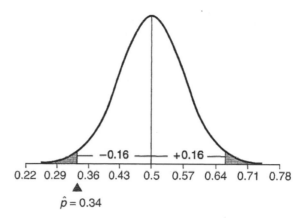

FIGURE 7.6 Sampling distribution with shaded area = *p*-value for two-tailed test of significance = $P\left[\left|(\hat{p} - P_0)\right|\right] \geq 0.16$.

nificance, and the total shaded area represents the two-tailed *p*-value. The calculations for the Z_{test} statistic would be exactly the same as in Example 7.2, and you would look up this value in the standard Normal table in the same manner. However, this probability would only account for the area in the left tail. Because the Normal distribution is symmetric, $P[Z_{test} \leq -2.29] = P[Z_{test} \geq +2.29]$, and the *p*-value is the total area in both tails = $0.011 \times 2 = 0.022$.

Probability of Erroneous Conclusions from Tests of Significance

Using a statistical test of significance to evaluate competing hypotheses about the nature of the universe is similar to serving on a jury in a criminal proceeding that is based entirely on circumstantial evidence. In such cases, no one actually saw who committed the crime, so the "truth" cannot be known with certainty. Rather, the evidence consists of such things as whether or not the defendant had a motive to commit the crime, the likelihood that the defendant could have been on the scene at the time of the crime, and the plausibility of the prosecution's case versus the defendant's alibi. In such cases, it is impossible to determine without the possibility of error whether or not the defendant is guilty. Nonetheless, the jury must come to a conclusion/judgment. If they bring a guilty verdict, there is a possibility they will send an innocent person to prison. If they bring a not guilty verdict, they may release a dangerous person back into society. If their deliberations result in a hung jury, then the entire court proceedings must be repeated with a new jury.

In scientific investigations, we cannot know the true population parameter value because sampling variability causes the "evidence" (data) to be uncertain. Nonetheless, the scientist must draw a conclusion regarding the scientific question of interest. For example, "Does the new protocol for extracting DNA produce more DNA than the one currently in use?" or "Is there evidence that a new chemical is toxic or will cause cancer or birth defects?" If the scientist concludes the new process increases DNA extraction, when it really does not, then an entire lab may be renovated and money spent on new equipment, all to no benefit. If the scientist concludes that the new chemical does not cause cancer or birth defects, when the chemical actually *does* cause such effects, then human health is put at risk. If the results of the scientific study are inconclusive, so that the scientist cannot draw any conclusion, then the entire study must be redone. In all of these situations, the jury and the scientist must come to a conclusion with full awareness of the probabilities and relative costs of making different types of error.

In scientific studies, statistical analyses provide the basis for determining the probabilities of various errors associated with each of the possible conclusions. In both jury and science situations, personal and societal values provide the context for weighing the relative costs of different kinds of error. Because we would rather release a criminal than jail an innocent person, juries convict only when the evidence meets the "beyond a reasonable doubt" criterion. In scientific research, we usually do not reject the Null hypothesis of "no difference" unless $p \leq 0.05$. Most scientists want to minimize the risk that they will erroneously conclude a new process/theory is better than the existing process/theory and spend resources making a change that will not produce the anticipated benefits.

There are four possible outcomes from a test of significance. The outcome that occurs depends on two conditions:

1. *The actual nature of the population:* For example, $P = P_0$ or P is less than (or greater than) P_0.

2. *The result of the test of significance:* Based on the data, the investigator concludes there is a difference *or* concludes there is insufficient evidence to claim that a difference exists.

These four outcomes are displayed in Table 7.1.

Type I error rate (α) is the probability that a statistical test will provide evidence for a conclusion that "there is a difference" (reject H_0) when there actually is none. This probability is estimated by the p-value (significance level) from a test of significance. This type of error occurs simply as a result of random sampling variability. For example, occasionally a large difference between $\hat{p}$ and P_0 will occur due only to random variation, even when the true population proportion is $P = P_0$. Scientists control the probability of making a Type I error simply by choosing the critical probability (α) required to reject the Null hypothesis (usually $p \leq 0.05$) and conclude that a dif-

TABLE 7.1 Possible outcomes from a test of significance and associated terminology.   The probabilities associated with outcomes are indicated in the table below by α and β. *Note: The probabilities within each column sum to 1.0. In fact, a difference either does or does not exist. There is only one real-world fact. The probabilities are associated with the investigator's inferences.*

Inference from Test	Real-World Fact (Not Known)	
	Difference Really Exists	There Is No Difference
Reject H_0 (Conclude there is a difference)	Correct conclusion $(1 - \beta)$ = **Power**	**Type I Error** (Error rate = α or the *p*-value)
Fail to Reject H_0 (Insufficient evidence that a difference exists)	**Type II Error** (Error rate = β)	Correct conclusion $(1 - \alpha)$

ference exists. In Example 7.3, the p-value from the one-tailed test was $p = 0.011$. Most scientists would feel that this probability of a Type I error was sufficiently small to conclude that lower incubation temperature reduced the proportion of female hatchlings. If the results from this study had produced a p-value $= 0.11$, most scientists would feel this probability of making a Type I error was too high to warrant a claim that incubation temperature affects gender determination of turtles.

Type II error rate (β) is the probability that a statistical test will result in a conclusion that there is no difference (fail to reject H_0) when the true population parameter value actually is different from the value stated by the Null hypothesis. This type of error occurs when the spread of the sampling distribution is large relative to the size of the difference between the true value of P and the Null hypothesis value P_0. This means that even a fairly large difference has a high probability of occurring due to random sampling variability (has a high p-value). Because the Type II error rate is determined in large part by the spread of the sampling distribution (e.g., $\sigma_{\hat{p}}$), the probability of this kind of error can be reduced by increasing sample size (remember $\sigma_{\hat{p}} = \sqrt{P(1 - P)/n}$). As sample size increases, smaller and smaller differences between $\hat{p}$ and P_0 will be found to be significant (have p-values ≤ 0.05).

Statistical power $(1 - \beta)$ is the probability that a statistical test of significance will result in a conclusion that a difference exists ($P \neq P_0$) when the true population parameter value really is different from the value stated in the Null hypothesis. In other words, the probability the test will "detect" a difference that really does exist. Power is determined by : (1) the Type I error rate (α) the investigator is willing to accept, (2) the size of the "important" difference that the investigator wants to detect if it actually exists, and (3) the spread of the sampling distribution, as influenced by the amount of variability in the population and the sample size. These issues are covered in more detail in the next section that describes how to compute power.

Why should you be concerned about power?

There are two possible reasons why a test of significance fails to "detect" a difference between P and P_0, that is, fails to reject the Null hypothesis parameter value (Table 7.1):

1. There really is no difference. (The true population parameter value P equals the value stipulated by the Null hypothesis P_0.)
2. There really is a difference between the true population parameter value and the value given in the Null hypothesis, *but* the power of the test was insufficient to detect the difference (usually due to inadequate sample size and low precision).

When a study fails to find a difference, it is often difficult to publish the results because of this ambiguity as to *why* the study failed to find a difference. By computing the power of your test of significance, you can document whether or not your study had sufficient power to detect some "minimum important difference," if one actually existed. If the study had sufficient power to detect an important difference, the failure to reject the Null hypothesis might be interpreted as strong evidence that there really isn't an important difference. Alternatively, estimating the power of your study design *before* implementation can save you time and money. If you know a study design has low power and little chance of success beforehand, you can either modify the design (increase sample size) or simply decide not to do a study with little chance for success.

Balancing the relative costs of Type I and Type II errors.

A nasty fact of life is that for any given study, trying to decrease the Type I error rate by requiring a smaller p-value to reject the Null hypothesis necessarily increases the probability of a Type II error. Anything that reduces the probability of rejecting the Null hypothesis (like requiring a smaller p-value) necessarily decreases power and increases Type II error rate (β). Scientists must always balance the relative costs of the two types of error when they make judgments about how small the p-value must be for them to conclude that a difference exists between the population parameter and the Null hypothesis value.

If we conclude there is a difference when there really is not (Type I error), the consequence is that we take an action that is unnecessary. For example, suppose the scientist in Example 7.1 was the first to claim that incubation temperature influenced gender determination in turtles. This claim would have challenged a widely accepted understanding of gender determination. The likely consequence is that many scientists would perform research to confirm the validity of the new claim, at a cost of time, effort, and money. If the scientist had made a Type I error, his colleagues would be wasting their time to verify a temperature effect that did not exist. Likewise, if an experiment on the efficacy of a new drug incorrectly concluded that people who took the drug did better than the control group, much money would be wasted producing a drug that did not relieve disease symptoms. Because the costs of replacing old understanding with new information or of producing new drugs are often high, most studies try to minimize the chance of a Type I error by not rejecting the Null hypothesis unless the p-value is very small (≤ 0.05).

If we fail to reject the Null hypothesis when in fact there is a difference (Type II error), we may fail to take necessary actions. For example, suppose a study of a potentially carcinogenic chemical failed to detect an increased rate of cancer in a lab experiment. The investigator would conclude that there is no cancer risk associated with the chemical, and government regulators might allow the chemical to be released into the environment. The cost of failing to detect that the chemical actually is carcinogenic might be increased cancer rates among people exposed to the chemical. Environmental damage or human deaths would result from this inaction. Environmental research is one area in which balancing Type I and Type II errors is critically important and controversial. If a study concludes a chemical is hazardous to people or natural ecosystems, then millions of dollars may be spent to control this chemical. If the chemical is actually *not* harmful, this money is wasted. If the chemical *is* harmful, but a study concludes it does not cause harmful effects (Type II error), then the costs to affected persons and natural resources could be devastating. This is but one of many aspects of scientific inquiry where considerations of societal values and ethics must enter into what is widely perceived as objective science.

Computing Power

The definition of power can be broken into two parts, defined by the row and column descriptions in Table 7.1. **Power** is the probability that a test of significance will (1) reject the Null hypothesis (claim there is a difference) when (2) a difference really exists. The computation of power is a two-step process based on these two parts of the definition.

Computing Power

1. Determine the **minimum significant difference (MSD)**, which corresponds to the **smallest effect size** ($\hat{p}^* - P_0$) that would provide sufficient evidence to re-

ject the Null hypothesis with a specified Type I error rate α. $\hat{p}^*$ is the critical value of the sample proportion that defines the range of values for $\hat{p}$ that would provide sufficient evidence to reject the Null hypothesis (rejection region). The value for $\hat{p}^*$ that corresponds to the MSD is computed as:

Computing the critical value of $\hat{p}$ to reject the Null hypothesis.

$$\hat{p}^* = Z_{\text{critical}} \, (\sigma_{\hat{p}}) + \boldsymbol{P}_0$$

Z_{critical} is determined by the following: (1) the acceptable probability of a Type I error (usually set at 0.05), (2) whether the test of significance is one-tailed (Z-value corresponding to $0.05 = 1.645$) or two tailed (Z-value corresponding to $0.05/2 = 0.025 = 1.96$), and (3) for one-tailed tests, whether H_a stipulates $\boldsymbol{P} < \boldsymbol{P}_0$ (Z_{critical} is negative) or $\boldsymbol{P} > \boldsymbol{P}_0$ (Z_{critical} is positive). The value of $\sigma_{\hat{p}}$ is computed using the Null hypothesis value $\boldsymbol{P}_0$.

2. Power is the probability of obtaining an effect size greater than or equal to the minimum significant difference ($\hat{p}^* - \boldsymbol{P}_0$) if the true population parameter has a value that is some **minimum important difference** (**MID**) from the Null hypothesis value. This Alternative hypothesis value is given the symbol $\boldsymbol{P}_a$. The value for $\boldsymbol{P}_a$ associated with the MID is defined by the judgment of the investigator about what true effect size ($\boldsymbol{P}_a - \boldsymbol{P}_0$) would be "important," as defined earlier. This probability is determined as follows:

Determining the probability of rejecting H_0 if a specified minimum important difference exists.

$$P[\hat{p} \leq \text{or} \geq \hat{p}^* \text{ if } \boldsymbol{P}_a = \boldsymbol{P}] = P[Z \leq \text{or} \geq (\hat{p}^* - \boldsymbol{P}_a)/\sigma_{\hat{p}}]$$

The value of $\sigma_{\hat{p}}$ is recomputed using the Alternative hypothesis value $\boldsymbol{P}_a$. The direction of the inequality symbol is the same as stated in the Alternative hypothesis.

The use of these two steps in computing power is illustrated in Example 7.6.

EXAMPLE 7.6

Computing power—The gender determination experiment.   Suppose the scientist conducting the study of gender determination in loggerhead turtles decided that if the proportion of females was ≤ 0.4, this would be an important enough difference from the expected $\boldsymbol{P} = 0.5$ to warrant making a claim that a mechanism other than sex chromosomes was at work. Let us also suppose that the scientist wants to conform to the most widely accepted standards of evidence, so he will only conclude that $\boldsymbol{P} < 0.5$ if the p-value is ≤ 0.05 (i.e., the maximum acceptable probability of a Type I error is $\alpha = 0.05$). What is the power of the scientist's original one-tailed test (Examples 7.1 to 7.3), with $n = 50$ eggs, to detect that the true proportion of female turtles hatched from eggs incubated at temperatures 3°C below natural nests is $\boldsymbol{P} < 0.5$ *if* the true value for $\boldsymbol{P}$ was 0.4?

1. Compute the value of $\hat{p}^*$ that corresponds to the minimum significant difference (the smallest *negative* difference $\hat{p}^* - \boldsymbol{P}_0$ that would be statistically significant at the $p \leq 0.05$ level) (see Figure 7.7A, Step 1).

   Note 1: For a one-tailed test of significance with maximum acceptable Type I error rate $\alpha \leq 0.05$, the Z-value that corresponds to an area of 0.05 in the left tail is -1.645.

Note 2: Because Alternative hypothesis indicates the proportion of female hatchlings is predicted to be less than the value expected if gender is determined by sex chromosomes, the corresponding Z-value should be negative (in the left side of the Normal distribution).

$$\hat{p}^* = -Z_{0.05}\, \sigma_{\hat{p}} + P_0 = -1.645\,(0.07) + 0.5 = 0.385$$

Interpretation: The critical value of the sample proportion needed to conclude that eggs incubated at the lower temperature produce $P < 0.5$ female hatchlings is $\hat{p}^* = 0.385$. That is, any value for $\hat{p}$ that was ≤ 0.385 would cause the scientist to reject the Null hypothesis and claim that lower incubation temperature reduces the proportion of female hatchlings below 0.5.

2. Determine the probability of obtaining a sample proportion $\hat{p} \leq 0.385$ *if* the true proportion of female hatchlings produced by the lower incubation temperature is $P_a = 0.4$ (defined by the investigator's minimum important difference) (see Figure 7.7A, Step 2).

$$\begin{aligned} P[\hat{p} \leq 0.385 \text{ if } P_a = 0.4] &= P[Z \leq (0.385 - 0.4)/0.0693] \\ &= P[Z \leq -0.22] \\ &= 0.4129 \end{aligned}$$

$$\begin{aligned} \text{where } \sigma_{\hat{p}} &= \sqrt{P_a\,(1 - P_a)/n} \\ &= \sqrt{0.4\,(1 - 0.4)/50} \\ &= 0.0693 \end{aligned}$$

Conclusion: There is only a 41.29% probability that the study based on a sample size of $n = 50$ eggs would detect that lowered incubation temperature reduced the true proportion of female hatchlings below 0.5 if the true proportion of female hatchlings at this lower temperature was $P = 0.4$. This would be considered unacceptably low power.   ■

How does increasing sample size affect power? Example 7.7 explores this question.

■ **EXAMPLE 7.7**

Changing the study design to increase the power of a test of significance.   In Example 7.4, the scientist used a larger sample size of $n = 100$ to obtain more evidence to support his claim that incubation temperature influences gender determination in loggerhead turtles. How does this change in the study design affect the power of the test of significance?

1. Compute the value of $\hat{p}^*$ that corresponds to the minimum significant difference, with the standard deviation of the sampling distribution $\sigma_{\hat{p}} = \sqrt{0.5\,(1 - 0.5)/100} = 0.05$.

$$\hat{p}^* = -Z_{0.05}\, \sigma_{\hat{p}} + P_0 = -1.645\,(0.05) + 0.5 = 0.418$$

Given the "one-tailed" research question, sample proportion values less than $\hat{p}^* = 0.418$ are required to conclude that eggs incubated at the lower temperature reduce the true proportion of female hatchlings to $P < 0.5$. (See Figure 7.7B, Step 1.)

Step 1

Compute $\hat{p}^*$ that corresponds to a minimum significant difference with $\alpha = 0.05$.

Step 2

Determine $P[\hat{p} \le \hat{p}^*]$ if $P = 0.4$.

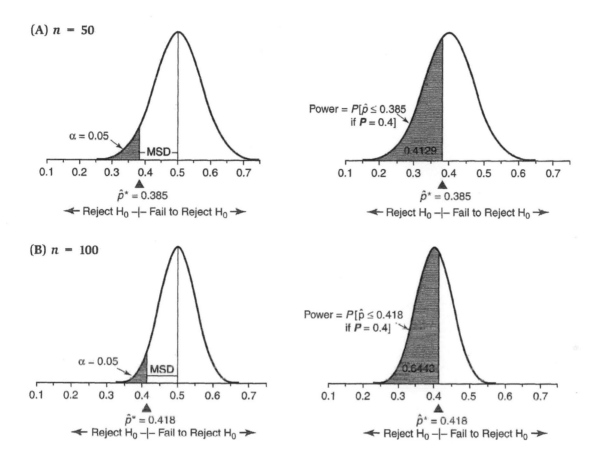

FIGURE 7.7   Effect of sample size on sampling distributions and the power of statistical tests.   Sampling distributions in part A correspond to the power computations in Example 7.6 ($n = 50$) and those in part B correspond to the power computations in Example 7.7 ($n = 100$). The shaded area under the curve in Step 1 (left side) is the acceptable Type I error rate ($\alpha = 0.05$). The shaded area under the curve in Step 2 (right side) represents power.

2. Determine the probability of obtaining a sample proportion $\hat{p} \le 0.418$ *if* the true proportion of female hatchlings produced by the lower incubation temperature is $P = 0.4$, as defined by the investigator's judgment regarding what constitutes a minimum important difference. (See Figure 7.7B, Step 2.)

$$P[\hat{p} \le 0.418 \text{ if } P_a = 0.4] = P[Z \le (0.418 - 0.4)/0.049]$$

$$= P[Z \le +0.37]$$

$$= 0.6443$$

where: $\sigma_{\hat{p}} = \sqrt{P_a(1 - P_a)/n}$

$\qquad\quad = \sqrt{0.4(1 - 0.4)/100}$

$\qquad\quad = 0.049$

Conclusion: There is a 64.43% probability that the study based on a sample size of $n = 100$ eggs would detect that lowered incubation temperature reduced the proportion of female hatchlings below 0.5 if the true proportion at this lower temperature was $P = 0.4$. Increasing the sample size from 50 to 100 reduced random sampling variation ($\sigma_{\hat{p}}$), so that a smaller MSD was required, increasing the power of the study. However, 0.6443 would still be considered modest power.   ▪

My assessments of the power for Examples 7.6 and 7.7 beg the question, "How much power is enough?" There is no widely accepted standard, similar to the standard for acceptable Type I error rate of $\alpha \leq 0.05$. However, consider the following argument. Suppose I asked you, "Does lowering incubation temperature reduce the number of female turtles hatched?", but you were unable to do any experiment. Your only option would be to guess, "yes" or "no." You would have a 50% probability of guessing the correct conclusion. Hence, the study in Example 7.6 (with $n = 50$ eggs) has a lower probability of correctly concluding lower temperatures reduce the proportion of female hatchlings than you would have if you simply guessed "yes." Hence, a study with power less than 50% would be totally inadequate. Even though there is not a widely accepted standard for minimum acceptable power, some have suggested that power should be at least 0.80 to 0.85, or it is not worth the effort of doing the study.

As I indicated earlier in the discussion of one- and two-tailed tests, the scientist studying gender determination in loggerhead turtles did not have a sufficiently strong basis for predicting how the proportion of female hatchlings would differ from $P = 0.5$. Hence, the appropriate test for his experiment is a two-tailed test. The only difference in the computation of power for one-tailed vs. two-tailed tests is in the selection of the critical Z-value used in the first step. For a one-tailed test, the entire $\alpha = 0.05$ area for the acceptable Type I error rate is in one tail, and the critical Z-value is $Z_{0.05} = 1.654$. However, for a two-tailed test this area corresponding to the acceptable Type I error rate must be divided into two tails, and the appropriate critical Z-value is $Z_{0.025} = 1.96$. (See Example 7.8.)

▪ **EXAMPLE 7.8**

Computing power for a two-tailed test of significance.   Suppose the scientist studying gender determination in loggerhead turtles applied a two-tailed test to the data derived from the study based on a sample size of $n = 100$ eggs. What is the power of this study to detect that the true proportion of female hatchlings produced by eggs incubated at a temperature 3°C lower than in natural nests differs from $P = 0.5$ by ± 0.1, with the allowable Type I error rate set at $\alpha = 0.05$? This is the same absolute effect size as the one-tailed power calculations in example 7.7, but now the investigator cannot predict the sign ($+$ or $-$) of the difference. Note that in two-tailed test situations, the minimum important difference is stated, rather than a specific value of the parameter (e.g., $P_a = 0.4$ in Examples 7.6 and 7.7).

1. Compute the value of $\hat{p}^*$ that corresponds to the minimum significant difference for a two-tailed test, with $\sigma_{\hat{p}} = \sqrt{0.5(1 - 0.5)/100} = 0.05$. (See Figure 7.8, Step 1.)

$$\hat{p}^* = Z_{0.025}\,\sigma_{\hat{p}} + P_0 = 1.96(0.05) + 0.5 = 0.598$$

The critical value of the sample proportion needed to conclude that the proportion of eggs incubated at the lower temperature that produced female hatchlings differed from $P = 0.5$ is $\hat{p}^* = 0.598$, based on a two-tailed test with $n = 100$. Another way of stating this is to say that the minimum significant difference $(\hat{p}^* - P_0) = 0.598 - 0.5 = 0.098$ (+ or −).

Note: The value of the minimum significant difference for a two-tailed test $[(\hat{p}^* - P_0) = 0.098)]$ is larger than for the one-tailed test in Example 7.6. $|(\hat{p}^* - P_0)| = 0.418 - 0.5 = 0.082$. A bigger difference and stronger evidence is required for two-tailed tests.

NOTE The sign of the critical Z-value used in Step 1 to compute power for a *two-tailed test* could be either + or −. In situations where a two-tailed test is used, the investigator cannot predict the direction of the difference (+ or −). However, if a real difference exists, it can be in only one direction. To compute power, you simply pick one direction and then compute power the same way as for a one-tailed test. The only difference is that the $Z_{critical}$ value used in Step 1 of the calculation is for a two-tailed probability $= \alpha/2$. I chose the positive Z-value here because Examples 7.5 and 7.6 had negative Z-values and I wanted to show you an example of power computations with a positive value.

Important: When you use a negative $Z_{critical}$ value, the inequality sign in the second step of the power calculation is ≤. When you use a positive $Z_{critical}$ value in the first step of the power calculation, you use the ≥ inequality sign in the second step.

2. Determine the probability of obtaining a sample proportion $\hat{p}^* \geq 0.598$ *if* the true proportion of female hatchlings produced by the lower incubation temperature is $P_a = 0.6$. (See Figure 7.8, Step 2.) The value 0.6 is determined by the investigator's specification of the minimum important difference $= \pm 0.1$. Given

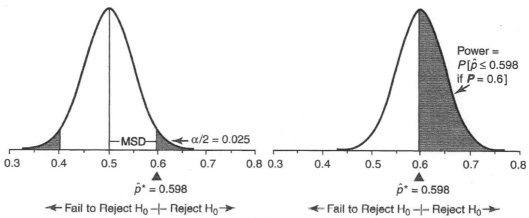

Step 1

Compute $\hat{p}^*$ that corresponds to a two-tailed minimum significant difference with $\alpha = 0.05$.

Step 2

Determine $P[\hat{p} \geq \hat{p}^*]$ if $P = 0.6$ (minimum important difference of ±0.1).

Power = $P[\hat{p} \leq 0.598$ if $P = 0.6]$

MSD $\leftarrow \alpha/2 = 0.025$

0.3   0.4   0.5   0.6   0.7   0.8
$\hat{p}^* = 0.598$
◄— Fail to Reject H$_0$ —|— Reject H$_0$ —►

0.3   0.4   0.5   0.6   0.7   0.8
$\hat{p}^* = 0.598$
◄— Fail to Reject H$_0$ —|— Reject H$_0$ —►

FIGURE 7.8 Diagram of power calculations for a two-tailed test of significance, corresponding to Example 7.8.

this, the P_a value could be either 0.4 or 0.6. I defined the "rejection region" as being in the right tail of the sampling distribution when I used a $+ Z$-value to compute $\hat{p}^*$ in Step 1. Therefore, I added the MID $+0.1$ to P_0 to determine $P_a = 0.6$. When determining power of two-tailed tests, you can choose to do the calculations in Steps 1 and 2 for either the left or right tail of the sampling distribution. Whichever tail you choose in Step 1, you must stay consistent with this choice in Step 2.

$$P[\hat{p} \geq 0.598 \text{ if } P_a = 0.6] = P[Z \geq (0.598 - 0.6)/0.049]$$
$$= P[Z \geq -0.04]$$
$$= 0.5160$$

where: $\sigma_{\hat{p}} = \sqrt{P_a (1 - P_a)/n}$
$$= \sqrt{0.6 (1 - 0.6)/100}$$
$$= 0.049$$

Conclusion: There is a 51.6% probability that the two-tailed test with a sample size of $n = 100$ would detect that lower incubation temperature increased the proportion of female hatchlings to $P > 0.5$, if the true value was $P = 0.6$. Likewise, there is a 51.6% probability that this test would detect that lowered incubation temperature *reduced* the proportion of female hatchlings, if the true value was $P = 0.4$. The point of a two-tailed test is that the investigator was unable to predict the direction of the deviation from $P = 0.5$. Nonetheless, the actual difference can only be in one direction. The power estimate of 51.6% applies regardless of the actual direction of this difference.

Note: Using a two-tailed test instead of a one-tailed test decreased the power of the study from 0.64 to 0.52. Two-tailed tests require a larger effect size than one-tailed tests for a difference to be significant. Hence, two-tailed tests are less likely to reject the Null hypothesis, and so, always have less power than one-tailed tests. ■

7.2

Assessing Precision of Statistical Estimates: Confidence Intervals

The popular media feed us a daily stream of statistics about the world around us. What percentage of the population thinks the president is doing a good job? What proportion of the U.S. population is HIV positive? Are people who are exposed to secondhand smoke more likely to develop lung cancer than people who live in smoke-free environments? We know that all of these studies were based on samples taken from some larger population (at least I know nobody asked me if I liked the president or if I'm HIV positive). As smart consumers, it is up to each of us to decide how much credence to give this information. By what criteria should a smart consumer evaluate statistics they encounter in everyday life? Of course, we should always look for possible biases in the way the data were collected. If an experiment to test whether or not a daily aspirin reduces heart attack risk included only males, women might reasonably wonder whether the results apply to them. If the survey question regarding approval of the president were worded, "Given that the president is a lying bum, what do you think of his job performance?", we might reasonably decide the resulting statistic was a biased estimate of true public opinion. However, even otherwise unbiased studies can produce statistical estimates of population parameters that vary in quality.

Random variation is a fact of life in all statistical analyses. We know that if we take a random, unbiased sample and compute the sample proportion for a categorical variable, the sample proportion $\hat{p}$ should be an unbiased estimate of the true, but unknown population proportion **P**. However, we also know that if we took independent samples from the same population and computed a proportion from each, the sample proportions would have a range of different values. Hence, we recognize that sample proportions are approximations of the true population proportion, and rarely are they actually equal to the population proportion.

If the sample size is large, the range of sample proportion values obtained from repeated samples will be small, and vice versa (see Figure 6.4). That is, random sampling variation decreases as sample size increases. Hence, repeated estimates of the sample proportion are more likely to fall close to the true population proportion at the center of the sampling distribution (assuming a randomized, unbiased study). Therefore, we expect the sample proportion to be a more precise estimator of the population proportion as sample size increases (remember the Law of Large Numbers). The bottom line is that the amount of "confidence" we have in a sample proportion as an estimate of the population proportion is determined by the sampling procedure (random, unbiased) and the sample size. It is always possible that the sample proportion derived from a random, unbiased sample will be quite different from the true population proportion. However, the probability of this error decreases as sample size increases. We quantify the precision of a sample statistic, and the associated confidence we have in this estimate of a population parameter, by computing a confidence interval.

Description of a Confidence Interval

A **confidence interval** has two parts: (1) an **interval** or range of values for the variable of interest, and (2) a **confidence level** that quantifies the probability the interval will include the true population parameter value. The general formula (for any population parameter) is:

$$\text{Confidence Interval} = \text{Sample Statistic} \pm (\text{Critical Value})(\sigma_{\text{statistic}})$$

The **critical value** in this formula comes from a probability distribution (e.g., the standard Normal distribution) and $\sigma_{\text{statistic}}$ is the standard deviation of the sample statistic (spread of the sampling distribution). The product $(\text{Critical Value})(\sigma_{\text{statistic}})$ is called the **margin of error**. This is a measure of how much random sampling variability we expect in the value of the statistic if we took repeated samples of size n from the same population. The margin of error is a measure of the precision of the sample statistic. The symbol $\pm$ indicates that the margin of error value is subtracted from the sample statistic value to obtain the lower bound of the confidence interval, and the margin of error is added to the sample statistic to obtain the upper bound of the interval.

Confidence intervals are computed based on the sampling distribution of the sample statistic. For example, if the sample size is sufficiently large, the sampling distribution of $\hat{p}$ is Normal, centered on $E(\hat{p}) = P$, and with standard deviation $\sigma_{\hat{p}} = \sqrt{P(1-P)/n}$. We know that 95% of the area under any Normal curve falls in the range $\mu \pm 1.96\,\sigma$, where μ represents the true center of the distribution and σ the spread. That is, the probability of obtaining a value for any Normal variable within the range $\mu \pm 1.96\,\sigma$ is 0.95. Therefore, 95% of $\hat{p}$ values from repeated samples from the same population should fall within the interval defined by $P \pm 1.96\ \sigma_{\hat{p}}$. The complement of this probability statement is that 5% of $\hat{p}$ values from repeated samples taken from exactly the same population would fall outside the range $P \pm 1.96$ $\sigma_{\hat{p}}$. (See part A of Figure 7.9).

The main point.

If 95% of all $\hat{p}$ values fall within the interval defined by $P \pm 1.96\ \sigma_{\hat{p}}$, then 95% of all confidence intervals computed as $\hat{p} \pm 1.96\ \sigma_{\hat{p}}$ from repeated samples of size n from the same population, *will include the true population proportion value* P. (See part B of Figure 7.9.)

An obvious problem with this logic is that we cannot compute $\sigma_{\hat{p}} = \sqrt{P(1 - P)/n}$ unless we know the true value of the population proportion P. If we knew this, we wouldn't need to compute a confidence interval for P. To compute the confidence interval for the value of P from sample data, we substitute the value $\hat{p}$ for the value P in the equation to compute the spread of the sampling distribution for $\hat{p}$, as shown below:

Standard error of p̂

$$S_{\hat{p}} = \sqrt{\frac{\hat{p}(1 - \hat{p})}{n}}$$

This value is called the **standard error of p̂** to distinguish it from the standard deviation of $\hat{p}$ ($\sigma_{\hat{p}}$), computed using an assumed value of P.

The 95% confidence interval for the true value of the proportion of a population that falls into some category can be computed as shown below.

95% confidence interval for the population proportion P

$$\hat{p} \pm 1.96 \sqrt{\frac{\hat{p}(1 - \hat{p})}{n}}$$

Interpretation of a 95% confidence interval.

When we compute a specific confidence interval based on a sample from the population of interest, this interval may or may not actually include the true value for the population parameter. The probability statement in the confidence level (e.g., 95% confidence) has the following interpretation: *The method used produces confidence intervals that include the true value of the population parameter value 95% of the time.* The "method" includes both random, unbiased sampling and the specific calculations used to obtain the confidence interval.

Error rate of confidence intervals (α).

If we are 95% confident that a confidence interval includes the true population parameter value, there is a probability of 0.05 (5%) that the interval does *not* include the parameter value. That is, there is a probability 0.05 that the confidence interval is in error. The relationship between error rate α and the confidence level is:

$$\text{Confidence Level} = 100(1 - \alpha)\%$$

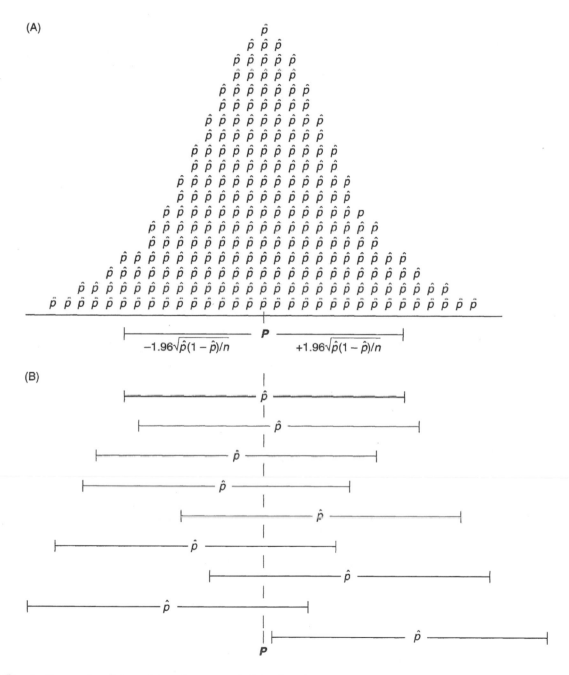

FIGURE 7.9 **$\hat{p}$ values and confidence intervals.**   95% of all $\hat{p}$ values fall within ± 1.96 $S_{\hat{p}}$ of P (part A). Therefore, 95% of all conficence intervals $\hat{p} \pm 1.96$ $\sqrt{\hat{p}(1-\hat{p})/n}$ include P (part B).

where the probability α is expressed as a proportion (e.g., 0.05). The confidence level and associated error rate are determined when the scientist chooses a critical value to calculate the confidence interval, and they are not influenced by the sample size.

The critical value used in the formula for a confidence interval is determined by the investigator's judgment about the desired level of confidence and corresponding error rate. To determine the critical value, you divide the desired error rate (α) by 2 and look for this probability value in the body of the standard Normal table to determine the Z-value that corresponds to this probability ($Z_{\alpha/2}$). Table 7.2 lists some confidence levels, error rates, and corresponding critical Z-values.

> **NOTE** Higher confidence levels have lower error rates and larger critical Z-values. If we want to be more confident that the interval $\hat{p} \pm Z_{\alpha/2}S_{\hat{p}}$ will include the true value of **P**, the interval must be wider. However, this makes the estimate of the parameter value less precise. ■

TABLE 7.2   Critical Z-values for commonly used confidence levels.

Confidence Level	Error Rate (α)	Critical Z-Value ($Z_{\alpha/2}$)
90%	0.10	1.645
95%	0.05	1.96
99%	0.01	2.575

Important relationships for confidence intervals.

- For a given confidence level, increasing sample size n decreases the width of the confidence interval, since the margin of error $= Z_{\alpha/2}\sqrt{\hat{p}(1-\hat{p})/n}$. (See Figure 7.10.)

- For a given sample size n, higher confidence levels require a larger critical Z-value, and the width of the confidence interval must be greater.

- The population proportion **P** and the sample estimate $\hat{p}$ also influence the width of the confidence interval. The product $\hat{p}(1-\hat{p})$ in the calculation of $S_{\hat{p}} = \sqrt{\hat{p}(1-\hat{p})/n}$ has a maximum value of 0.25 when $\hat{p} = 0.5$. This product progressively decreases as the value of $\hat{p}$ increases or decreases from 0.5. That is, the values of sample proportions obtained from repeated samples from the same population will be more variable if **P** and $\hat{p}$ are close to 0.5, and less variable if **P** and $\hat{p}$ are much bigger or smaller than 0.5.

> **NOTE** The sample statistic (e.g., $\hat{p}$) is *always* in the middle of any confidence interval. Hence, our "95% confidence" does *not* pertain to whether or not the confidence interval includes the sample statistic $\hat{p}$. Rather, this expression of confidence applies to whether or not the interval includes the unknown population parameter value (e.g., **P**). Therefore, the correct way to refer to this confidence interval is "95% confidence interval for **P**." (See Example 7.9.) ■

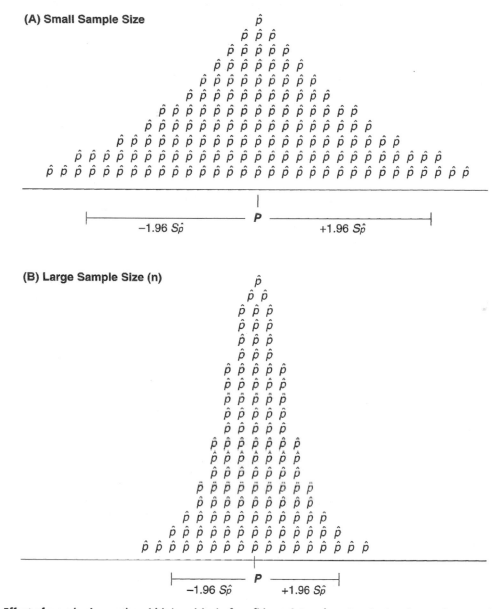

FIGURE 7.10   Effect of sample size on the width (precision) of confidence intervals. *Conclusion:* Increasing sample size *n* causes sample proportions from repeated independent samples from the same population to be less variable and more tightly clumped around the true population proportion. The result is that the confidence interval is narrower (more precise), yet has the same error rate α.

■ **EXAMPLE 7.9**

Computing confidence intervals. Suppose a biology student in Italy takes a random sample of $n = 100$ students to determine the proportion of blue-eyed people in the student body. She finds $\hat{p} = 0.1$ blue-eyed students.

a. Compute the 95% confidence interval for **P**, the true proportion of blue-eyed students, as follows.

95% confidence interval for **P** (with $n = 100$) $= 0.1 \pm 1.96 \sqrt{0.1(1 - 0.1)/100}$

$$= 0.1 \pm 1.96(0.03)$$

$$= 0.1 \pm 0.0588 = 0.0412, 0.1588$$

Because 95% of all confidence intervals computed by this method will include the true value of the population proportion **P**, the student is 95% confident that the true proportion of blue-eyed students at this university lies within the interval 0.0412 to 0.1588.

b. Compute the 99% confidence interval, using the critical value $Z_{\alpha/2 = 0.005}$:

99% confidence interval for **P** (with $n = 100$) $= 0.1 \pm 2.575 \sqrt{0.1(1 - 0.1)/100}$

$$= 0.1 \pm 2.575(0.03)$$

$$= 0.1 \pm 0.0773 = 0.0228, 0.1773$$

The student is 99% confident that the true value for **P** lies between 0.0228 and 0.1773. To be more confident the interval includes **P**, the interval must be wider (and less precise).

c. Compute the 95% confidence interval if $\hat{p} = 0.1$ was based on a sample size of $n = 200$:

95% confidence interval for **P** (with $n = 200$) $= 0.1 \pm 1.96 \sqrt{0.1(1 - 0.1)/200}$

$$= 0.1 \pm 1.96(0.0212)$$

$$= 0.1 \pm 0.0416 = 0.058, 0.142$$

Because statistics computed from larger samples exhibit less random sampling variation, the student can be 95% confident the true value for **P** lies within the narrower range 0.058 to 0.142. The narrower 95% confidence interval indicates a more precise estimate.

d. Suppose the biology student does her study at a university in Norway and the proportion of blue-eyed students in her sample is ($\hat{p} = 0.4$). Compute the 95% confidence interval for **P** in this student population, with $n = 100$.

$$= 0.4 \pm 1.96 \sqrt{0.4(1 - 0.4)/100}$$

$$= 0.4 \pm 1.96(0.049) = 0.4 \pm 0.096$$

Because the value of $\hat{p}$ is closer to 0.5, the standard error ($S_{\hat{p}}$) would be larger and therefore the confidence interval would be wider (less precise). ■

Computing the Required Sample Size for a Desired Confidence Interval

In some situations, the maximum allowable margin of error for a statistic is stipulated by standards set by a general consensus of scientists in a discipline. For example, most public opinion surveys presented by the popular media stipulate that the margin of error is ± 3% (usually in fine print at the bottom of the TV screen). The only determiners of the margin of error that a researcher has control over are the choice of confidence level and the sample size. The researcher has no control over the population parameter *P*, which will influence the value of $\hat{p}$ and therefore the value of $S_{\hat{p}} = \sqrt{\hat{p}(1 - \hat{p})/n}$. In most scientific studies, however, the required confidence level is set by standards of the scientific discipline or by the specific needs of the study (usually 95%). Thus, the only variable that can be adjusted by the scientist to attain the desired margin of error is the sample size (*n*). Given a specified confidence level and an estimate of the population proportion $\hat{p}$, the sample size (n^*) needed to obtain a margin of error equal to some specific value is computed as follows:

Sample Size Needed (n^*) for a Specific Margin of Error

$$n^* = (Z_{\alpha/2}/m^*)^2 \, \hat{p}(1 - \hat{p})$$

where $Z_{\alpha/2}$ is the Z-value corresponding to the desired confidence level $100(1 - \alpha)$, m^* is the desired margin of error, and n^* is the sample size required to attain this margin of error.

Sometimes this calculation is done *before* any sample data are collected (the whole purpose is to determine how big a sample to take). But how do we know the value of $\hat{p}$ needed for this calculation before we collect the data? To address this question, we must first go back to how the standard for desired margin of error is stated. In general, the standard stipulates that the margin of error will be *no larger than* some specific value. Of course, a margin of error smaller than this required value would indicate an even more precise estimate of the population parameter value, and would thus be entirely acceptable. With this in mind, we can now stipulate that the value we use in place of the unknown $\hat{p}$ value should ensure that the margin of error will be no larger than the specified value m^*. A $\hat{p}$ value of 0.5 will result in a maximum estimate for $S_{\hat{p}}$, so using the value 0.5 in the equation for desired sample size will produce an n^* value that will always have a margin of error less than or equal to the desired m^*. With $\hat{p} = 0.5$, the formula for desired sample size simplifies to:

Sample size formula

$$n^* = (Z_{\alpha/2}/m^*)^2 \, \hat{p}(1 - \hat{p}) = (Z_{\alpha/2}/m^*)^2 (0.25)$$

Examples 7.10 and 7.11 show how this formula is applied by the student doing the eye-color experiment.

■ **EXAMPLE 7.10**

Using the sample size formula. Suppose the student studying the prevalence of blue eye color in Example 7.9 wanted to know how much she should increase her sample size to attain a 95% confidence interval with a margin of error of ± 0.05. Based on the data she already has ($n = 100$, $\hat{p} = 0.1$), she could estimate the necessary sample size as:

$$n^* = (Z_{\alpha/2}/m^*)^2 \, \hat{p}(1 - \hat{p}) = (1.96/0.05)^2 (0.1)(1 - 0.1) = 138.3$$

To ensure that the margin of error would be *no larger than* $m^* = 0.05$, fractional sample sizes are always *rounded up* (to $n^* = 139$ students in this case).

Suppose this student did this calculation prior to collecting any data. She would have to estimate the necessary sample size using the value $\hat{p} = 0.5$, as shown below:

$$n^* = (Z_{\alpha/2}/m^*)^2 \, (0.25) = (1.96/0.05)^2(0.25) = 385$$

> **NOTE** When the student had some data to estimate $\hat{p} = 0.1$, she was able to use this value to compute a more precise (and smaller) estimate of the necessary sample size than when she used $\hat{p} = 0.5$ to compute necessary sample size. Notice that when the value $\hat{p} = 0.5$ is used to compute necessary sample size, the resulting value n^* will produce confidence intervals with a margin of error that is always less than or equal to the specified value m^*. (See Example 7.11.) However, in many cases the computed required sample size will be *larger than necessary* (involve more time, effort, resources than is required). Hence, many investigators will collect a preliminary sample to obtain a value for $\hat{p}$, and then use this value to compute how much the sample size should be increased to obtain the desired margin of error. ▪

▪ **EXAMPLE 7.11**

A student studying the prevalence of blue-eyed people at an Italian university used the sample size $n^* = 385$ to arrive at her estimate of $\hat{p} = 0.1$. The 95% confidence interval is:

$$0.1 \pm 1.96 \; \sqrt{0.1(1 - 0.1)/385} = 0.1 \pm 1.96(0.015) = 0.1 \pm 0.03$$

A student studying prevalence of blue-eyed people at a Norwegian university used the sample size $n^* = 385$ to arrive at her estimate of $\hat{p} = 0.4$. Her 95% confidence interval would be:

$$0.4 \pm 1.96 \; \sqrt{0.4(1 - 0.4)/385} = 0.4 \pm 1.96(0.025) = 0.4 \pm 0.049$$

In both cases, the margin of error of the 95% confidence interval for P was less than the stipulated value $m^* \leq 0.05$ in Example 7.10. Hence, these confidence intervals are more precise than required. Although better precision is always good, to attain this higher than necessary precision, larger sample sizes (and more effort) were expended than were necessary. Notice that when the true population proportion is closer to 0.5, the margin of error is closer to the stipulated $m^* = 0.05$. ▪

Requirements for a Valid Confidence Interval

The fundamental requirement for a confidence interval to be valid is that the sampling distribution for the sample statistic be Normal and centered over the true population parameter value. One way to verify that these assumptions are met is to take thousands of repeated samples of size n from the population, compute the sample statistic for each sample, and look at the distribution. Of course, this is unrealistic. Alternatively, probability theory allows us to assume the sampling distribution of the sample statistic is Normal and centered over the true parameter value if the following conditions are met:

1. Data were acquired from an unbiased, random sample, so you can assume the expected value of the statistic is the true parameter value. For example, $E(\hat{p}) = P$.

2. The sample size n is sufficiently large that the Central Limit Theorem applies. For example, the sampling distribution for $\hat{p}$ is approximately Normal if $n\hat{p} \geq 10$ *and* $n(1 - \hat{p}) \geq 10$.

NOTE For a summary of concepts and applications described in this chapter, see the description of the one-sample Z-test for proportions in Chapter 8. ∎

CHAPTER SUMMARY

1. **Test of significance:** A test of significance determines the probability that the observed difference between a sample statistic value and the parameter value stipulated by the Null hypothesis (for example, $\hat{p} - P_0$) is due only to random sampling variation.

   a. A test of significance is based on the sampling distribution for the statistic of interest, with the center (expected value) of that distribution specified by the Null hypothesis of "no difference." For a test of significance regarding a single sample proportion:
   - The **center of the sampling distribution**, $E(\hat{p})$, is assumed to be equal to the Null hypothesis value for the true population proportion P_0.
   - The **spread of the sampling distribution** is also computed using the Null hypothesis value P_0: $\sigma_{\hat{p}} = \sqrt{P_0(1 - P_0)/n}$.
   - If we assume that the Null hypothesis is true, the **shape of the sampling distribution for** $\hat{p}$ will be approximately Normal if the following conditions are both met:

   $$nP_0 \geq 10 \quad and \quad n(1 - P_0) \geq 10$$

   b. The output from a test of significance is the **p-value**, which is the probability that the observed difference between the sample statistic value and the value of the population parameter stipulated by the Null hypothesis (e.g., $\hat{p} - P_0$) is *due only to random sampling variation.*

   c. **One-tailed tests of significance:** These tests evaluate sample data as evidence for a difference in a specific direction. These tests are appropriate only if there is a theory-based reason, well documented in the scientific literature, for expecting a specific kind of difference. **Two-tailed tests of significance** should be used when there is insufficient basis to predict the direction of the difference prior to the study.

   d. **Significance:** A sample statistic is *significant* if it represents strong evidence that the true value of a population parameter differs from the value stated in the Null hypothesis. If the observed difference between the statistic value and the H_0

   value has a low probability of occurring due only to sampling variability (small p-value), the more likely explanation for the observed difference is that true population parameter value differs from the Null hypothesis value.

2. **Type I error rate (α):** The Type I error rate gives the probability that a conclusion to reject the Null hypothesis value of the population parameter is *incorrect*, quantified by the p-value. The p-value has two interpretations: (1) It indicates the probability of getting the observed sample statistic value due only to random sampling variability (when the Null hypothesis is true) and (2) the probability you are mistaken if, based on the sample statistic, you reject the Null hypothesis. The maximum allowable Type I error rate (α) is fixed when the investigator decides how small the p-value must be before he or she will conclude there is sufficient evidence to reject the Null hypothesis value for the population parameter.

3. **Power of a test of significance ($1 - \beta$):** The power is the probability that a test of significance will correctly reject the Null hypothesis value for the population parameter when there actually *is* a specified difference between the Null hypothesis value and the true population parameter value. The "specified difference" is based on the investigator's judgment about what would constitute the minimum "important" difference.

4. **Type II error rate (β):** The Type II error rate is the probability that a test of significance will fail to reject the Null hypothesis value when there really *is* a difference. (The test fails to "detect" a real difference.) When a test of significance fails to reject H_0, there are two possible reasons for this outcome: (1) there really is no difference ($P = P_0$) or (2) there *is* a difference, but the test of significance had insufficient power to detect it, usually due to inadequate sample size.

5. **Absolute effect size:** The absolute effect size is the actual difference between the observed sample statistic value and the value for the population parameter specified by the Null hypothesis (e.g., $\hat{p} - P_0$). **Relative effect size** expresses this difference as a percent of P_0 (i.e., $100(\hat{p} - P_0)/P_0\%$).

6. **Significant vs. important effect size:** A difference ($\hat{p} - P_0$) is *significant* merely if it is unlikely that the difference is due only to random sampling variation. A difference is *important* if it is sufficiently large and "interesting" to prompt a change in theory or standard procedures.

7. **Confidence interval:** The confidence interval specifies a range of values, within which the true value of the population parameter will occur with a specified probability.

   a. **Confidence level:** The confidence level is the probability that confidence intervals produced by the specified method will include the true population parameter value.

   b. **Error rate of a confidence interval (α):** The error rate is the probability that a confidence interval produced by the specified method will *not* include the true population parameter value. Confidence level 100(1 − α)%, and the error rate α, are fixed by the investigator's decision regarding how "confident" that person wants to be that the interval will include the true parameter value. The investigator's choice of error rate determines the critical value used to calculate the interval. *Confidence level and error rate are not affected by sample size or by the amount of random sampling variation in the statistic.*

   c. The 100 (1 − α)% confidence interval for the true value of a population proportion **P** is computed as follows:
   - Compute the standard error of $\hat{p}$ ($S_{\hat{p}}$) = $\sqrt{\hat{p}(1 - \hat{p})/n}$.
   - Compute the margin of error = $Z_{\alpha/2}(S_{\hat{p}})$.
   - Compute the confidence interval for **P** = $\hat{p} \pm Z_{\alpha/2}(S_{\hat{p}})$.

   d. The sample size required to obtain a confidence interval with a desired precision, as quantified by the desired margin of error, can be computed as follows:
   - If a prior estimate $\hat{p}$ for the population proportion **P** is available:

   $$n^* = (Z_{\alpha/2}/m^*)^2 \hat{p}(1 - \hat{p})$$

   - If no prior estimate $\hat{p}$ for the population proportion **P** is available:

   $$n^* = (Z_{\alpha/2}/m^*)^2(0.25)$$

8. **Important relationships**
   a. *p-Values*
   - Increasing the effect size (e.g., $\hat{p} - P_0$) decreases the p-value from a test of significance. Large effects are unlikely to be due to random sampling variation only.
   - Increasing the sample size decreases random sampling variation (e.g., $\sigma_{\hat{p}}$) and so decreases the p-value (all other things being held constant).
   - One-tailed tests of significance always produce lower p-values than a two-tailed test applied to the same data.

   b. *Power*
   - Increasing sample size reduces random sampling variation (e.g., $\sigma_{\hat{p}}$) and increases the power of a test of significance.
   - One-tailed tests of significance have a higher probability of rejecting the Null hypothesis than two-tailed tests, and so always have greater power.
   - Specifying a smaller minimum important difference decreases the power of a test of significance. (Small differences are less likely to be detected.)
   - Decreasing the allowable Type I error rate (α) sets a more rigorous standard of evidence for rejecting the Null hypothesis. By making it more difficult to reject the Null hypothesis, the power of a test of significance is reduced and the Type II error rate (β) is increased.

   c. *Confidence intervals*
   - Increasing the sample size decreases $S_{\hat{p}}$ and so decreases the width of confidence intervals.
   - Choosing a higher confidence level [100(1 − α)%] requires a larger Z_{critical} value and increases the width of confidence intervals. Hence, higher confidence is obtained by reducing precision.
   - Because the confidence level [100(1 − α)%] and the error rate (α) are chosen by the investigator, these probabilities are *not affected* by either sample size or random sampling variation of the statistic.
   - Increasing the desired confidence level [100(1 − α)%] decreases the error rate (α) of confidence intervals.

SECTION 2

Applications

SECTION 2   Applications

The remainder of this book describes applications of the fundamental concepts presented in Section 1 to specific types of research questions, under various circumstances defined by study design and the nature of the data. At this point, many students will be concerned that they do not sufficiently understand the fundamental concepts to be able to apply them. However, if you heed the suggestions that follow here, you will be able to use the various applications presented in Section 2 as opportunities to further develop your understanding of the fundamentals. The key to understanding is to look for the commonalities shared by all the different procedures presented in Section 2. To help you better see these commonalities, each application will be presented in the following standardized format:

Application: Describes the circumstances under which the test is appropriate, as determined by the nature of the question, the study design, and the data. Many statistical tests have *assumptions* that must be fulfilled for the test to be valid. I will describe these assumptions and how you can determine whether they are fulfilled.

Statement of Hypotheses: Describes the specific format of the Null and Alternative hypotheses that are tested by the statistical test under discussion. These hypotheses will state the population parameter of interest, as determined by the scientific question, study design, and nature of the data.

Sampling Distribution: Describes the *center* (expected value), *spread* (standard deviation or standard error), and *shape* of the sampling distribution under the assumption that the population parameter value is equal to that stated by the Null hypothesis. This description of the sampling distribution will be valid only if the assumptions described in the Application section are fulfilled.

Test Statistic: Describes how the test statistic is computed. This test statistic measures how much the observed sample statistic value deviates from the expected value defined by the Null hypothesis. The test statistic will have a known probability distribution that is used to obtain a p-value for the test of significance.

p-Value: Describes how the p-value is determined from the test statistic and how it is interpreted with regard to the original scientific question.

Confidence Interval: This section describes how the $100(1 - \alpha)\%$ confidence interval for the population parameter is computed, and the interpretation of this interval for each specific application. ∎

You have already done each of these steps in the context of analyses for a single sample proportion for a categorical variable that has only two classes. By repeating these steps in the context of a variety of different analyses, your understanding of the fundamental concepts will be strengthened. To get the full advantage of this "understanding through repetition," focus on the commonalities shared by the various tests rather than the confusing diversity of details and equations. Most of these details pertain to calculations that are usually done by computer statistics programs. The modern scientist needs to know how to determine the best analysis for a particular study and how to interpret the statistical computer output to make an appropriate inference regarding the scientific question being investigated. Hence, hand calculations will be kept to a minimum in this latter part of the course.

For each specific application presented in Section 2, I will reiterate the following fundamental concepts of statistical analyses:

1. Sample statistics exhibit *random sampling variation*.

2. The *sampling distribution of a sample statistic* displays which values are likely and unlikely, based on assumptions or hypotheses about the true value of the population parameter.

3. A *test of significance* determines the probability of obtaining the observed value of the sample statistic if the Null hypothesis ("no difference" or "no association") value of the population parameter is true. The sampling distribution for the test is centered over the Null hypothesis value, with a spread determined by the characteristics of the population and the sample size.

4. If an appropriate study design has been correctly implemented, there should be *only two possible explanations* for why a sample statistic value deviates from the exact value stated in the Null hypothesis: (1) The deviation from the Null hypothesis value is due only to random sampling variation, or (2) the deviation reflects the fact that the true population parameter value does not equal the Null hypothesis value. The p-value is the probability associated with the random-variation explanation. If this explanation is sufficiently unlikely, the evidence (data) supports the real-difference explanation.

5. *Confidence intervals* stipulate a range of values that will contain the true value of the population parameter with a specified probability. Confidence intervals provide information about the precision of estimates of populations parameter values.

Even though the reiteration of concepts in each application may sometimes seem repetitive, I believe it will help you to fully develop your understanding of these difficult and important concepts. Do not skim over these parts. Take advantage of these opportunities to see how the same fundamental concepts are applied in many different circumstances.

Choosing an appropriate statistical test—A flowchart.

Many "scientists-in-training" find it difficult to determine which of the many tests available is most appropriate for their particular data analysis situation. The *flowchart* beginning on the following page is intended to help the beginner choose the most appropriate statistical test from those presented in this text. The chart is divided into two sections, part A for analyses of quantitative data (such as data for mass, concentration, metabolic rate), part B for analyses of categorical data (such as data for gender, blood type, color). For each of these types of data a series of descriptions about the nature of the study question, study design, and data distribution are given. At each step, choose the option that best describes your specific situation and then follow the arrows to the next appropriate step in the chart. This listing of statistical tests covers most of those presented in this text, but is not an exhaustive listing of all statistical tests available. When in doubt, it is always a good idea to consult with an experienced scientist or applied statistician.

Flowchart Key to Tests of Significance

Variable of interest observed on each subject is <u>categorical</u>: Go to Part B

Or

Variable of interest measured on each subject is <u>quantitative</u>: Go to Part A

Part A: Tests for Quantitative Variables

Research question about an <u>association</u> between variables

    Or

Research question about a <u>difference</u> between groups

      Data for <u>more than one variable</u> or data from <u>more than two samples</u> of subjects

         Or

      Data for <u>one variable</u> obtained from <u>two samples</u> of subjects

         Or

      Data for <u>one variable</u> obtained from <u>one sample</u> of subjects

      Objective is to compare one sample mean to a hypothetical value.

         One measurement made on each subject. Sampling distribution of the mean can be assumed Normal. **One sample *t* test (Ch. 9)**

           Or

         Data from Before-After or Matched-Pairs study design. Sample statistic of interest is the mean difference. Sampling distribution of mean difference can be assumed Normal. **Paired *t*-test (Ch. 9)**

           Or

         Sampling distribution of mean difference cannot be assumed normal. **Sign test (Ch. 9)**

      Objective is to compare <u>two sample variances</u> to each other. *F*-test or **Levene test (Ch. 10)**

         Or

      Objective is to compare <u>two sample means</u> to each other.

         Sampling distributions for one or both sample means cannot be assumed Normal. **Mann-Whitney *U*-test (Ch. 9)**

           Or

         Sampling distributions for both sample means can be assumed Normal.

           Population variances can be assumed equal. **Two-sample *t*-test (pooled variance) (Ch. 9)**

              Or

           Population variances cannot be assumed equal. **Two-sample *t*-test (separate variance) (Ch. 9)**

Objective to compare sample means for <u>multiple variables</u> among two or more groups; separate test performed for each variable. **Bonferroni procedure (Ch. 11)**

    Or

Objective is to compare sample means for <u>one variable</u> among <u>more than two groups</u>.

    Population distributions for all groups can be assumed Normal and population variances for all groups can be assumed equal. **Analysis of variance (Ch. 11)**

      Or

    Population distributions for all groups *cannot* be assumed Normal or population variances for all groups *cannot* be assumed equal. **Kruskal-Wallis test (Ch. 11)**

Objective is to use linear association to predict values for one variable based on knowledge of the value for the other variable. **Linear regression (Ch. 13)**

    Or

Objective is to simply determine the strength and direction (+ or −) of the association.

    Association is <u>linear</u>; both variables are Normally distributed. **Pearson's correlation coefficient (Ch. 13)**

      Or

    Association is monotonic but <u>nonlinear</u> or variables are <u>not</u> Normally distributed. **Spearman's rank correlation (Ch. 13)**

Part B: Tests for Categorical Variables

Research question about an <u>association</u> or <u>independence</u> between two or more variables

> Or

> Research question about a <u>difference</u> between groups

> > Data obtained from <u>two or more samples</u> of subjects

> > > Or

> > **Data obtained from <u>one sample</u> of subjects**

> > Compare the sample proportion for <u>one class of one categorical variable</u> to a single hypothetical value. **One-sample Z-test (for proportions) (Ch. 8)**

> > Or

> > Compare sample proportions for <u>two or more classes of one categorical variable</u> to their corresponding hypothetical values. χ^2 **Goodness-of-fit test (Ch. 12)**

> Compare sample proportions for <u>one class of one categorical variable</u> between two independent samples. **Two-sample Z-test (for proportions) (Ch. 8)**

> > Or

> Compare sample proportions for <u>two or more classes</u> of one categorical variable among <u>two or more independent samples</u>. χ^2 **Test of homogeneity (Ch. 12)**

Determine if the proportion of individuals that fall into various classes of one categorical variable is independent of (or related to) whether or not the individual belongs to various classes of another categorical variable. χ^2 **test of independence (Ch. 14)**

8 One- and Two-Sample Tests for Proportions

INTRODUCTION Many variables in biology represent counts of individuals or events that fall into classes of a categorical variable (e.g., taxonomic groups, gender, blood types, and genetic phenotypes and genotypes). In most cases, the count of individuals or events (X) is divided by the total number of individuals or events in the sample (n) and converted to a proportion ($\hat{p}$). By converting the count to a proportion, the data are more easily interpreted without referring back to the sample size.

One-sample tests of significance for a sample proportion test Null hypotheses regarding the value of the true population proportion [H_0: $P = P_0$]. **Two-sample tests** evaluate the difference between two sample proportions ($\hat{p}_1 - \hat{p}_2$) as evidence to test the Null hypothesis that the two population proportions are not different [H_0: ($P_1 - P_2$) = 0].

A critical consideration for determining if the tests in this chapter are appropriate for a given situation is whether or not it is valid to assume that the sampling distributions of $\hat{p}$ and ($\hat{p} - \hat{p}_2$) are Normal. Either of these sampling distributions will be approximately Normal if the sample size(s) is sufficiently large (Chapter 6.2). However, if the sample size(s) is *not* sufficiently large, then the appropriate probability distribution will be Binomial, and the

p-value associated with an observed difference between the sample statistic and the Null hypothesis value should be obtained from the Binomial distribution. ■

CHAPTER GOALS AND OBJECTIVES

In this chapter you will learn how and when to use tests of significance for evaluating the difference between one sample proportion and a hypothetical population proportion, and how and when to use the difference between two sample proportions as evidence of differences between their corresponding population proportions.

Given the study design and a description of the data, by the end of the chapter you will be able to:

1. Identify the appropriate test of significance for sample proportions, implement the test, and draw appropriate conclusions with regard to the scientific question of interest. You will learn about the following tests: (a) the one-sample Z-test (for proportions) and (b) the two-sample Z-test (for proportions).
2. Compute the confidence intervals for the population proportion P and the difference between two population proportions ($P_1 - P_2$).

8.1
One-Sample Z-Test

1. **Application:** The one-sample Z-test for proportions is used to evaluate the difference between a single sample proportion and a hypothesized value of the population proportion ($\hat{p} - P_0$). This difference is used to test hypotheses about the true nature of the population. **Assumptions:**

   1. The observations used to compute $\hat{p}$ are random, unbiased, and independent.
   2. The population size N is greater than 100× the sample size n, so the value of the population proportion P can be assumed constant.

3. The sample size n is fixed.

4. The sample size is sufficiently large that $nP_0 \geq 10$ *and* $n(P_0 - 1) \geq 10$. (That is, the Normal approximation for sampling distribution of $\hat{p}$ is valid.)

2. **Statement of Hypotheses**
   - One-tailed test
     - H_0: $P = P_0$
     - H_a: $P < P_0$ *or* H_a: $P > P_0$
   - Two-tailed test
     - H_0: $P = P_0$
     - H_a: $P \neq P_0$

The sample proportion $\hat{p}$ will almost never exactly equal the Null hypothesis value P_0. In those rare cases where it does, there is no need to perform a test of significance; the data do not provide evidence to reject the Null hypothesis value. When $\hat{p}$ does differ from P_0, there should be only two possible explanations: (1) *random variation:* The difference is due merely to random sampling variation and the true value of P really does equal P_0, or (2) *real difference:* The value of $\hat{p}$ provides evidence that the true value of P differs from the Null hypothesis value P_0 in the manner stated by the alternative hypothesis.

3. **Sampling Distribution for $\hat{p}$**, Assuming H_0 is true, $P = P_0$

   Center: $E(\hat{p}) = P_0$

   Spread: $\sigma_{\hat{p}} = \sqrt{\dfrac{P_0(1 - P_0)}{n}}$

   Shape: Normal

   The sampling distribution displays those values of $\hat{p}$ that have a high probability of occurring if the Null hypothesis is true. If the observed value of $\hat{p}$ falls far from the expected value under the Null hypothesis, this would constitute evidence that the true population proportion value P is not equal to the Null hypothesis value P_0.

4. **Test Statistic**

   $$Z_{\text{test}} = \frac{\hat{p} - P_0}{\sigma_{\hat{p}}}$$

   The Z_{test} statistic represents the difference between the sample proportion $\hat{p}$ and the Null hypothesis value for the population proportion P_0 in "standard deviation units." You can determine the exact probability of getting the observed difference ($\hat{p} - P_0$) by referring to the standard Normal table.

5. *p*-Value
   a. *One-tailed test.* The probability $P[Z \geq +Z_{\text{test}}]$ or $P[Z \leq -Z_{\text{test}}]$, obtained from the standard Normal table, is the *p*-value for the one-tailed test of significance. This is the probability of obtaining values for the sample proportion that are greater than or equal to (or $\leq$) the observed $\hat{p}$ if the true population proportion value is equal to the Null hypothesis value P_0.
   b. *Two-tailed test.* The probability $2 \times P[Z \geq |Z_{\text{test}}|]$ is the *p*-value for the two-tailed test of significance. The term $|Z_{\text{test}}|$ is the absolute value of Z_{test} (i.e., any $-$ sign is deleted). In a two-tailed test, there was no prior ex-

pectation that the difference between $\hat{p}$ and $\boldsymbol{P}_0$ would be in a particular direction. The p-value is the probability of getting an absolute difference $|(\hat{p} - \boldsymbol{P}_0)|$ greater than or equal to the observed difference, if the true value of $\boldsymbol{P}$ is equal to the Null hypothesis value.

c. Stating conclusions:

If $p \leq 0.05$: The probability associated with the random-variation explanation is sufficiently small to reject this explanation. If an appropriate study design has been correctly implemented, the only remaining likely explanation supported by the data is that the true value of the population proportion differs from the Null hypothesis value. Regardless of whether your test of significance was one-tailed or two-tailed, your conclusion should indicate a specific direction of the difference, as indicated by the sample data.

■ **EXAMPLE**   "The proportion of individuals who exhibited the recessive trait for both characteristics was greater than expected if the traits were on different chromosomes ($p = 0.0025$). Therefore, the genes for these two traits are linked on the same chromosome."   ■

If $0.05 < p \leq 0.10$: The probability associated with the "random-variation" explanation is sufficiently high that we cannot confidently exclude this explanation. However, the probability is sufficiently small that we might suspect there is a real difference. Some scientists would conclude that the data "suggest" $\boldsymbol{P}$ is less than (or greater than) $\boldsymbol{P}_0$.

■ **EXAMPLE**   "The proportion of individuals who exhibited the recessive trait for both characteristics was somewhat greater than expected, suggesting that the genes for these traits are on the same chromosome ($p = 0.073$), but more data are needed to verify this."

If $p > 0.10$: The probability associated with the "random-variation" explanation is sufficiently high that this explanation cannot be excluded. Therefore, there is insufficient evidence to conclude that $\boldsymbol{P} \neq \boldsymbol{P}_0$.

■ **EXAMPLE**   "The observed proportion of individuals who exhibited the recessive trait for both characteristics did not significantly differ from 0.25 ($p = 0.48$). Therefore, the two genes for these characteristics are on separate chromosomes."   ■

6. **$100(1 - \alpha)\%$ Confidence Interval for P**

Confidence Interval for $\boldsymbol{P} = \hat{p} \pm Z_{\alpha/2} \sqrt{\dfrac{\hat{p}(1 - \hat{p})}{n}}$

NOTE: The confidence interval for $\boldsymbol{P}$ is not based on any assumption or hypothesis regarding the true value of $\boldsymbol{P}$. Therefore, you must rely entirely on the sample data, estimating the spread of the sampling distribution with the standard error, $S_{\hat{p}} = \sqrt{\hat{p}(1 - \hat{p})/n}$.

Interpretation of the confidence interval: Because $100(1 - \alpha)\%$ of all confidence intervals computed by this method include the true value of $\boldsymbol{P}$, we can be $100(1 - \alpha)\%$ "confident" that our specific confidence interval includes $\boldsymbol{P}$. However, there is a probability α (e.g., 0.05 for a 95% confidence interval) that the interval does not include the true population proportion value.

Example 8.1 shows how these six steps are applied to a genetics experiment that is analyzed using the one-sample Z-test.

▓ **EXAMPLE 8.1**

One-sample Z-test for proportions.

1. **Application:**

   The PTC taste test is a common lab exercise in human genetics. Some people can taste the chemical PTC while others cannot. This trait is controlled by a single gene. For over 25 years a biology teacher did the PTC taste test as part of his class. He kept records of the results, and for the thousands of students he tested in this class over this time, $P = 0.7$ were PTC "tasters." After retiring, he joined the Peace Corps and taught biology in Nepal. In his class of $n = 60$ Nepali students, 36 students were PTC tasters ($\hat{p} = 36/60 = 0.6$). Do these data provide evidence that the prevalence of the "PTC taster" gene differs between his American and Nepali students?

   Assessing Assumptions

   The teacher wants to compare the sample proportion of PTC tasters in his class of Nepali students to the proportion of PTC tasters in the population of American students he tested over many years. The 60 Nepali students in his class were not randomly selected individuals from the larger population of all Nepali students. Hence, we cannot be certain that the 60 students constitute independent observations. For this example, we will assume that they are independent. The population of all Nepali students numbers in the tens of thousands, so $N > 100n$. Because $60(0.7) = 42$ and $60(0.3) = 18$ are both greater than 10, the Normality assumption is fulfilled.

2. **Statement of Hypotheses:**

   - H_0: $P_{\text{Nepal}} = P_{\text{US}} = 0.7$
   - H_a: $P_{\text{Nepal}} \neq 0.7$

3. **Sampling Distribution of $\hat{p}$,** assuming H_0 is true

   Center: $E(\hat{p}) = P_0 = 0.7$

   Spread: $\sigma_{\hat{p}} = \sqrt{0.7(1 - 0.7)/60}$

   $\phantom{Spread: \sigma_{\hat{p}}} = 0.0592$

   Shape: Normal

   Note: Shaded area corresponds to the two-tailed p-value.

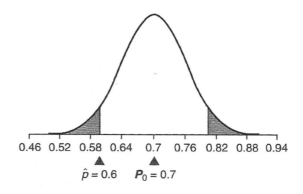

4., 5. Test Statistic and *p*-Value

$$Z_{test} = \frac{\hat{p} - P_0}{\sigma_{\hat{p}}} = \frac{0.6 - 0.7}{0.0592} = -1.69$$

$$P[\hat{p} \leq 0.6 \text{ if } P = 0.7] = P[Z \leq -1.69] = 0.0455$$

Because this is a two-tailed test, the *p*-value is $2 \times 0.0455 = 0.0910$.

Conclusion: The results of this study suggest ($p = 0.091$) that the proportion of PTC tasters among Nepali students is less than among the teacher's American students, but additional data are needed to verify this conclusion.

6. 95% Confidence Interval for P_{Nepali}

$$\hat{p} \pm Z_{\alpha/2} \sqrt{\hat{p}(1 - \hat{p})/n} = 0.6 \pm 1.96 \sqrt{0.6(1 - 0.6)/60} = 0.6 \pm 0.124$$

Interpretation: Because confidence intervals computed by this method include the true value of the population proportion (*P*) 95% of the time, the teacher is 95% confident that the true proportion of PTC tasters among Nepali students is within the range 0.6 ± 0.124 (between 0.476 and 0.724).

8.2

Two-Sample
Z-Test

1. **Application:** The two-sample *Z*-test for proportion is used to evaluate the difference between two sample proportions ($\hat{p}_1 - \hat{p}_2$) as evidence to test the Null hypothesis that there is no difference between the two population proportions [$H_0: (P_1 - P_2) = 0$]. The two sample proportions are based on independent samples of size n_1 and n_2 from the two populations. The sample statistic of interest is the value of the difference ($\hat{p}_1 - \hat{p}_2$). **Assumptions:**

   1. The data are obtained in a randomized, unbiased manner.
   2. The values for n_1 and n_2 are fixed.
   3. Population sizes N_1 and N_2 are greater than 100× sample sizes n_1 and n_2, respectively. Hence, the values for P_1 and P_2 can be assumed constant.
   4. The sample sizes are sufficiently large that the Normal approximation for the sampling distribution is valid. All four of the following products $n_1 \hat{p}_1$, $n_1 (1 - \hat{p}_1)$, $n_2 \hat{p}_2$, *and* $n_2 (1 - \hat{p}_2)$ must be greater than or equal to 5 to fulfill the Normality assumption.

2. **Statement of Hypotheses**
   - One-tailed test: $H_0: (P_1 - P_2) = 0$

     $\qquad\qquad\qquad H_a: (P_1 - P_2) < 0$ *or* $H_a: (P_1 - P_2) > 0$
   - Two-tailed test: $H_0: (P_1 - P_2) = 0$

     $\qquad\qquad\qquad H_a: (P_1 - P_2) \neq 0$

   The Null hypothesis always predicts "no difference." In this context, the Null expectation is that the difference between the two population proportions is zero. However, two proportions computed from independent samples will rarely be exactly equal, even if the two samples were taken from the same population. In rare cases when the two sample proportions are equal, no test of significance is required to conclude there is "no difference." When the two sample proportions are different, there are two possible explanations: (1) Random variation or (2) the true proportions for the two populations sampled really are different.

3. **Sampling Distribution of** $(\hat{p}_1 - \hat{p}_2)$**, assuming H$_0$ is true**

   Center: $E\,(\hat{p}_1 - \hat{p}_2)\ =\ 0$

   Spread: $S_{\hat{p}1-\hat{p}2}\ =\ \sqrt{\hat{p}_{\text{pooled}}(1 - \hat{p}_{\text{pooled}})\left[\dfrac{1}{n_1}\ +\ \dfrac{1}{n_2}\right]}$

   where: $\hat{p}_{\text{pooled}}\ =\ \dfrac{n_1\hat{p}_1\ +\ n_2\hat{p}_2}{n_1\ +\ n_2}$

   > In the two-sample situation, there are no Null hypothesis values for the true population proportions **P$_1$** and **P$_2$**. Rather, the Null hypothesis simply states that **P$_1$** = **P$_2$**. This does not provide the values for **P$_1$** and **P$_2$** that are required to compute the standard deviation of the sampling distribution of $(\hat{p}_1 - \hat{p}_2)$, $\sigma_{\hat{p}1-\hat{p}2}$. Therefore, the spread of the sampling distribution must be estimated by the standard error of the difference, $S_{\hat{p}1-\hat{p}2}$, computed from the data as shown above.

   Shape: The distribution of the difference between two Normally distributed variables will also be Normal. If sample sizes are sufficiently large for the Normal approximation to be valid, the sampling distribution for $(\hat{p}_1 - \hat{p}_2)$ will be Normal.

   The sampling distribution described above displays values for the difference $(\hat{p}_1 - \hat{p}_2)$ that have a high probability of occurring if the Null hypothesis is true (i.e., **P$_1$** = **P$_2$**). If the observed difference $(\hat{p}_1 - \hat{p}_2)$ falls far from the expected value of $(\mathbf{P_1} - \mathbf{P_2})$ = 0, this would constitute evidence that the two population proportions really are different.

4. **Test Statistic**

   $$Z_{\text{test}}\ =\ \frac{(\hat{p}_1 - \hat{p}_2) - (\mathbf{P_1} - \mathbf{P_2})_0}{S_{\hat{p}1-\hat{p}2}}\ =\ \frac{(\hat{p}_1 - \hat{p}_2)}{S_{\hat{p}1-\hat{p}2}}$$

   The Z_{test} statistic measures the deviation of the observed $(\hat{p}_1 - \hat{p}_2)$ from the expected value of zero, relative to the amount of random sampling variation, $S_{\hat{p}1-\hat{p}2}$. Look up this Z_{test} value in the standard Normal distribution to determine the exact probability that the observed difference $(\hat{p}_1 - \hat{p}_2)$ is due only to random variation.

5. ***p*-Value**

   a. *One-tailed test.* The probability $P[Z \geq\ +\ Z_{\text{test}}]$ or $P\,[Z \leq\ -\ Z_{\text{test}}]$, obtained from the standard Normal table, is the *p*-value for the one-tailed test of significance. This is the probability of obtaining a difference$(\hat{p}_1 - \hat{p}_2)$ that is greater than or equal to (or $\leq$) the observed value, due only to random variation, if the true population proportions are equal (i.e., **P$_1$** = **P$_2$**).

   b. *Two-tailed test.* The probability $2\ \times\ P[Z \geq\ |Z_{\text{test}}|]$ is the *p*-value for the two-tailed test of significance. The term $|Z_{\text{test}}|$ is the absolute value of Z_{test} (i.e., any $-$ sign is deleted). In a two-tailed test, there was no prior expectation that the difference between **P$_1$** and **P$_2$** would be in a particular direction. The *p*-value is the probability of getting an absolute difference

$|(\hat{p}_1 - \hat{p}_2)|$ greater than or equal to the observed difference, due only to random variation, if the two population proportions P_1 and P_2 are equal.

c. Stating conclusions:

If $p \leq 0.05$: The probability associated with the random-variation explanation is sufficiently small to reject this explanation. If an appropriate study design has been correctly implemented, the only remaining likely explanation supported by the data is that the true values of the two population proportions are different. Regardless of whether your test of significance was one-tailed or two-tailed, your conclusion should indicate the specific direction of the difference as indicated by the sample data.

■ **EXAMPLE** "The proportion of children who develop asthma was significantly greater in 'smoking' households than in 'nonsmoking' households ($p = 0.001$)." ▨

If $0.05 < p \leq 0.10$: The probability associated with the random-variation explanation is sufficiently high that we cannot confidently exclude this explanation. However the probability is sufficiently low that we might suspect there is a real difference. Some scientists would conclude that the data "suggest" P_1 is less than (or greater than) P_2.

■ **EXAMPLE** "The data suggest a higher proportion of children in 'smoking' households develop asthma than children in 'nonsmoking' households ($p = 0.085$), but more data are needed to verify this." ▨

If $p > 0.10$: The probability associated with the random-variation explanation is so high that this explanation cannot be excluded. This result provides insufficient evidence to conclude that there is a difference between P_1 and P_2.

■ **EXAMPLE** "The proportion of children with asthma was not significantly different between 'smoking' versus 'nonsmoking' households ($p = 0.45$)." ▨

6. $100(1 - \alpha)\%$ Confidence Interval for $(P_1 - P_2)$

$$(\hat{p}_1 - \hat{p}_2) \pm Z_{\alpha/2} \sqrt{\frac{\hat{p}_1(1 - \hat{p}_1)}{n_1} + \frac{\hat{p}_2(1 - \hat{p}_2)}{n_2}} = (\hat{p}_1 - \hat{p}_2) \pm Z_{\alpha/2}\, S_{(\hat{p}1 - \hat{p}2)}$$

NOTE Because the confidence interval is *not* based on a hypothesis that $P_1 = P_2$, we must compute the standard error using the separate sample proportions $\hat{p}_1$ and $\hat{p}_2$. This standard error is used *only for computing the confidence interval.*

Interpretation of the confidence interval: Because $100(1 - \alpha)\%$ of all confidence intervals computed by this method include the true value for the difference between the two population proportions, we are $100(1 - \alpha)\%$ confident that our specific interval includes the true difference ($P_1 - P_2$) for the variable and populations of interest.

In Example 8.2 we show the use of the two-sample Z-test in a medical study.

■ **EXAMPLE 8.2**

Two-sample Z-test for proportions.

1. Application:

There is a serious risk that women infected with HIV will pass the virus to their baby during the birth process, via exposure to infected blood; HIV is not usually passed to the fetus while in the uterus. For many years, the best treatment to reduce this risk was to treat the mother with daily doses of AZT for five months prior to giving birth. However, the cost of this preventive treatment is high (approximately $1,000), which makes it all but unavailable to poor women in underdeveloped countries that have high HIV infection rates. A new treatment proposed to prevent HIV transmission from mothers to infants involves giving a single dose of the antiviral drug Neviraprine to the mother at the beginning of labor, and a single dose to the infant within three days after birth. The cost for this treatment is $4. If it is effective, this treatment could be widely used, even in the poorest countries. A clinical trial was required to determine if the new treatment was as effective as the more expensive AZT treatment. In an experiment conducted in Uganda, 300 pregnant women were given the more expensive AZT treatment and 300 women were given the Neviraprine treatment. Women were randomly assigned to the two treatment groups. Of the women who received the Neviraprine treatment, 40 transmitted the HIV virus to their babies ($\hat{p}_N = 0.133$), compared to 77 women who received the more expensive AZT treatment ($\hat{p}_A = 0.257$). Is this sufficient evidence to conclude that Neviraprine treatment works as well as the AZT treatment?

Assessing Assumptions

Given random assignment of the 600 women to two treatment groups, we can assume an unbiased study design and independent observations. Sample sizes $n_N = n_A = 300$ were fixed at the beginning of the study. With hundreds of thousands of HIV-infected women in Africa, we can assume $N > 100n$. With sample sizes of 300 in each treatment group, and $\hat{p}_N = 0.133$ and $\hat{p}_A = 0.257$, $n_N\hat{p}_N$, $n_N(1 - \hat{p}_N)$, $n_A\hat{p}_A$, and $n_A(1 - \hat{p}_A)$ are all greater than 5. Hence, the Normality assumption is fulfilled.

2. **Statement of Hypotheses**

   - H_0: $(P_N - P_A) = 0$
   - H_a: $(P_N - P_A) \neq 0$

   NOTE Because there was insufficient basis to predict how the effectiveness of the Neviraprine treatment would compare to the AZT treatment, a two-tailed test was required.

3. **Sampling Distribution of $(\hat{p}_N - \hat{p}_A)$**, assuming H_0 is true

   - First compute $\hat{p}_{pool}$, an estimate of $P = P_N = P_A$ under the assumption that H_0 is true.

$$\hat{p}_{pool} = \frac{(300)(0.133) + (300)(0.257)}{300 + 300}$$

$$= (40 + 77)/(600)$$

$$= \mathbf{0.195}$$

   - Then compute $S_{\hat{p}N-\hat{p}A}$, using $\hat{p}_{pool}$.

$$S_{\hat{p}1-\hat{p}2} = \sqrt{\hat{p}_{pool}(1 - \hat{p}_{pool})\left[\frac{1}{n_1} + \frac{1}{n_2}\right]} = \sqrt{0.195(0.805)\left[\frac{1}{300} + \frac{1}{300}\right]}$$

$$= \mathbf{0.03235}$$

Center: $E(\hat{p}_N - \hat{p}_A) = 0$

Spread: $S_{(\hat{p}N-\hat{p}A)} = 0.03235$ rounded to 0.03 to scale the X-axis of the sampling distribution.

Shape: Normal

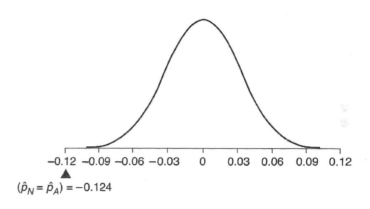

$(\hat{p}_N = \hat{p}_A) = -0.124$

Shaded area corresponding to the p-value is too small to see.

4., 5. Compute the Z_{test} Statistic and Determine the p-Value

$$Z_{test} = \frac{(\hat{p}_N - \hat{p}_A) - (P_N - P_A)_0}{S_{\hat{p}N-\hat{p}A}} = \frac{(0.133 - 0.257) - 0}{0.03235} \qquad = -3.83$$

$P[(\hat{p}_N - \hat{p}_A) < -0.124$ if $(P_N - P_A) = 0] = P[Z \le -3.83] \quad = p < 0.0002$

Because this is a two-tailed test, the p-value is $< 0.0002 \times 2 = \mathbf{p < 0.0004}$.

Conclusion: The new Neviraprine treatment resulted in a lower HIV transmission rate from mothers to newborn infants than did the more expensive AZT treatment ($p < 0.0004$). Having drawn this conclusion, the probability that this conclusion is incorrect is less than 0.0004.

NOTE Given that we had the question "Is Neviraprine *as effective* as AZT?", some might have believed that the test of significance should have addressed the one-tailed question "Is the transmission rate for Neviraprine higher than the transmission rate for AZT?" No one predicted or expected that a $4 treatment would reduce HIV transmission *better* than the $1,000 treatment. However, there was also no strong basis in the literature or past experience for predicting that Neviraprine *wouldn't* be better than AZT. Hence, the two-tailed test was the most appropriate.

6. **95% Confidence Interval for $(P_N - P_A)$**

- First compute the standard error of the difference $S_{\hat{p}_N - \hat{p}_A}$.

$$S_{\hat{p}_N - \hat{p}_A} = \sqrt{\frac{\hat{p}_N(1 - \hat{p}_N)}{n_N} + \frac{\hat{p}_A(1 - \hat{p}_A)}{n_A}}$$

$$= \sqrt{\frac{0.133(1 - 0.133)}{300} + \frac{0.257(1 - 0.257)}{300}}$$

$$= 0.0320$$

- Then compute the 95% confidence interval.

$$(\hat{p}_N - \hat{p}_A) \pm Z_{0.025}\, S_{\hat{p}_N - \hat{p}_A} = (0.133 - 0.257) \pm 1.96\,(0.032)$$

$$= -0.124 \pm 0.0626$$

$$= -0.1866, -0.0614$$

Interpretation of the confidence interval: Because 95% of confidence intervals computed by this method will include the true difference between two population proportions, the investigators are 95% confident that using Neviraprine instead of AZT will reduce HIV infection of newborns by 614 to 1,866 babies per 10,000 births to HIV-positive mothers. This is a statement of the "effect size" of using Neviraprine instead of AZT. However, this does not take into account that the much less expensive Neviraprine can be made more available to women in poor countries. The researchers who did this study estimated that the combination of higher effectiveness and wider availability of Neviraprine would reduce HIV infection of newborns in Africa alone by 1,000 per day. Any assessment of the practical importance of this new treatment would balance the social and financial costs of this reduction in the number of HIV-infected infants versus the cost of the new drug. In this case, the practical importance of this new treatment is obvious.   ▪

CHAPTER SUMMARY

1. **The one-sample Z-test (for proportions)** is used to evaluate the significance of a difference between one sample proportion and a Null hypothesis value for the true population proportion ($\hat{p} - P_0$). This test requires that the sample size is sufficiently large that the Normal approximation for Binomial variables applies [$nP_0 \geq 10$ and $n(1 - P_0) \geq 10$].

   The p-value from this test of significance represents the probability of getting the observed difference ($\hat{p} - P_0$) due only to random sampling variation. If this probability is sufficiently small (< 0.05), the sample proportion provides sufficient evidence to conclude that the population proportion is not equal to P_0, and to reject the Null hypothesis of "no difference."

2. The **two-sample Z-test (for proportions)** is used to evaluate the significance of a difference between two sample proportions ($\hat{p}_1 - \hat{p}_2$) from independent samples as evidence to test the Null hypothesis that two population proportions are equal ($P_1 - P_2 = 0$). This test requires that the sample sizes for both samples are sufficiently large that the Normal approximation for Binomial variables applies [$n_1\hat{p}_1 \geq 5$, $n_1(1 - \hat{p}_1) \geq 5$, $n_2\hat{p}_2 \geq 5$, $n_2(1 - \hat{p}_2) \geq 5$].

   The p-value from this test represents the probability of getting the observed difference between the sample proportions ($\hat{p}_1 - \hat{p}_2$) due only to random sampling variation. If this probability is sufficiently small (< 0.05), the difference between the sample proportions provides sufficient evidence to conclude that the population proportions are not equal ($P_1 \neq P_2$).

3. The **100(1 − α)% confidence interval for a single population proportion P** (where α is the probability that the interval will *not* include the true value of *P*) is:

$$\hat{p} \pm Z_{\alpha/2} \sqrt{\hat{p}(1 - \hat{p})/n}$$

Assuming randomized unbiased sampling, 100(1 − α)% of all intervals computed by this method will include the true value for the population proportion. Therefore, we can be 100(1 − α)% confident that any specific confidence interval will include **P**.

4. The **100(1 − α)% confidence interval for the difference between two population proportions** (where α is the probability that the interval will *not* include the true difference of $P_1 - P_2$) is:

$$(\hat{p}_1 - \hat{p}_2) \pm Z_{\alpha/2} \sqrt{\frac{\hat{p}_1(1 - \hat{p}_1)}{n_1} + \frac{\hat{p}_2(1 - \hat{p}_2)}{n_2}}$$

Assuming randomized unbiased sampling, 100(1 − α)% of all intervals computed by this method will include the true difference between two population proportions. Therefore, we can be 100(1 − α)% confident that any specific confidence interval will include $P_1 - P_2$.

9 One- and Two-Sample Tests for Sample Means

INTRODUCTION   Tests of significance presented in this chapter evaluate the difference between a single sample mean and a hypothesized value for the population mean (e.g., $\bar{x} - \mu_0$), or the difference between two sample means (e.g., $\bar{x}_1 - \bar{x}_2$). **One-sample tests** evaluate the hypothesis that the true population mean is equal to some Null hypothesis value ($\mu = \mu_0$). **Two-sample tests** evaluate the Null hypothesis that two populations means are equal, stated as ($\mu_1 - \mu_2$) = 0.

A critical consideration for determining which of the tests in this chapter is appropriate for a given situation is whether or not it is valid to assume that the sampling distribution of the mean is Normal. The **Central Limit Theorem (CLT)** states that the sampling distribution of the sample mean will be Normal if the population distribution is Normal *or* if the sample size *n* is sufficiently large (see the "rules of thumb" for applying the CLT described in Chapter 6). When the data distribution indicates the population distribution is not Normal and sample sizes are too small for the CLT to apply, **nonparametric tests** may be the only appropriate way to analyze the data. This class of significance tests does *not* require a Normal sampling distribution. Sometimes, **transformations** can be applied to the data values to make the shape of the data distribution approximately Normal. In this case, the transformed data values can be used to perform a parametric test that requires that the sampling distribution of the mean is Normal. When the transformations are correctly done, conclusions based on tests of significance applied to transformed data values are equally applicable to the original data.

Regardless of which test of significance you use, the format and interpretation of test results are the same as described in Chapters 7 and 8. For all tests of significance, you will (1) describe the sampling distribution of the sam-

ple statistic if the Null hypothesis of "no difference" is true, (2) use this sampling distribution to determine the probability of obtaining the observed value of the sample statistic due only to random sampling variation (the *p*-value), and (3) based on this *p*-value, with due consideration given to the power of the study, draw an inference as to whether or not the data support the Null hypothesis (with your conclusion always stated in terms of the original scientific question).   ▨

CHAPTER GOALS AND OBJECTIVES

In this chapter you will learn how and when to use tests of significance for evaluating the difference between a single sample mean and a hypothetical mean, and for evaluating the difference between two sample means.

Given the study design and the characteristics of the data, by the end of this chapter you will be able to:

1. Identify the appropriate test of significance from among those listed below, implement the test, and draw appropriate conclusions with regard to the scientific question of interest.

   One-sample Z-test (for means)

   One-sample *t*-test

   Paired *t*-test

   Sign test

   Two-sample Z-test (for means)

   Two-sample *t*-test (pooled variance)

   Two-sample *t*-test (separate variance)

   Mann-Whitney *U*-test

2. Compute confidence intervals for the population mean μ and the difference between two population means ($\mu_1 - \mu_2$).

9.1

One-Sample Tests for Means

One-Sample Z-Test

1. **Application:** The **one-sample Z-test** for the sample mean evaluates the significance of a difference between a single sample mean and a Null hypothesis value for the true population mean ($\bar{x} - \mu_0$). The population standard deviation σ must be known or estimated by a sample standard deviation S computed from a large sample ($n \geq 100$). *Assumptions:* (1) *The data were obtained in a randomized, unbiased manner,* so that the expected value of the sample mean is the true population mean, $E(\bar{x}) = \mu$. (2) *The sampling distribution of $\bar{x}$ is Normal.* This assumption is fulfilled if one of two conditions are met: (a) The population distribution is Normal (as indicated by a boxplot or Normal quantile plot of the data distribution) *or* (b) the sample size is sufficiently large that the Central Limit Theorem applies (see "rules of thumb" in Chapter 6).

   In actual practice this test is rarely used, as there are very few circumstances in which the population mean μ is unknown, but the population standard deviation σ is known. When sample size is large, the outcome of this test is similar to that obtained from a one-sample t-test (described in the next section, One-Sample t-Test). The one-sample Z-test is usually presented in statistics courses to help students better understand the more commonly used t-test.

2. **Statement of Hypotheses**

   - One-tailed test:

     H_0: $\mu = \mu_0$

     H_a: $\mu < \mu_0$ or $\mu > \mu_0$

   - Two-tailed test:

     H_0: $\mu = \mu_0$

     H_a: $\mu \neq \mu_0$

   The sample mean $\bar{x}$ rarely exactly equals the Null hypothesis value for the population mean μ_0. In those rare cases when $\bar{x} = \mu_0$, a test of significance is not required to conclude there is no difference. When $\bar{x}$ does differ from μ_0, there should be only two possible explanations for this difference: (1) random-sampling variation or (2) the true population mean really is different from the Null hypothesis value. These two explanations correspond to the Null and Alternative hypotheses.

3. **Sampling Distribution of $\bar{x}$**, assuming H_0 is true

   a. Center: $E(\bar{x}) = \mu_0$

   b. Spread: $\sigma_{\bar{x}} = \sigma/\sqrt{n}$

   c. Shape: Normal

   This sampling distribution displays those values of $\bar{x}$ that have a high probability of occurring if the Null hypothesis value for μ is correct. If the observed value for $\bar{x}$ falls far from the expected value under the Null hypothesis, this would constitute evidence that the Null hypothesis should be rejected.

4. **Test Statistic**

   $$Z_{\text{test}} = \frac{\bar{x} - \mu_0}{\sigma_{\bar{x}}}$$

In this relationship, the standard deviation of the mean, $\sigma_{\bar{x}}$, quantifies the expected random sampling variation in the value of $\bar{x}$, given the amount of variation in the population σ and the sample size n (Chapter 6). The Z_{test} statistic expresses the difference between $\bar{x}$ and μ_0 relative to the expected amount of random variation. This test statistic has a standard Normal distribution. Therefore, you can determine the exact probability of getting the observed difference $(\bar{x} - \mu_0)$ by referring to the standard Normal table.

5. **p-Value**

   a. *One-tailed test:* The probability $P[Z \geq +Z_{test}]$ or $P[Z \leq -Z_{test}]$ is the p-value for the one-tailed test of significance. This is the probability of obtaining values for the sample mean greater than or equal to (or $\leq$) the observed $\bar{x}$ due only to random sampling variation, when the true population mean value is equal to the Null hypothesis value μ_0.

   b. *Two-tailed test:* The probability $2 \times P[Z \geq |Z_{test}|]$ is the p-value for the two-tailed test of significance. The term $|Z_{test}|$ is the absolute value of Z_{test} (i.e., any $-$ sign is deleted). In a two-tailed test, there was no prior expectation that the difference between $\bar{x}$ and μ_0 would be in one direction or the other. The p-value is the probability of getting an absolute difference $|\bar{x} - \mu_0|$ greater than or equal to the observed difference due only to random sampling variation, when the true population mean is equal to the Null hypothesis value.

   c. *Stating conclusions:*

   If $p \leq 0.05$: The probability associated with the random-variation explanation is sufficiently low to reject this explanation. If an appropriate study design has been correctly implemented, the only remaining likely explanation supported by the data is that the true value of the population mean differs from the Null hypothesis value. Regardless of whether the test was one-tailed or two-tailed, your conclusion should state a specific direction of the difference, as indicated by the sample data.

   ■ **EXAMPLE** "The mean of 100 repeated measurements of a standard solution was less than the known value ($p = 0.011$). Therefore, the instrument is out of calibration." ▦

   If $0.05 < p \leq 0.10$: The probability associated with the random-variation explanation is sufficiently high that we cannot confidently exclude this explanation. However the probability is sufficiently low that we might suspect there is a real difference. Some scientists would conclude that the data "suggest" μ is less than (or greater than) μ_0.

   ■ **EXAMPLE** "The difference between the mean for 100 repeated measurements of a standard solution and the known value suggests that the instrument is out of calibration ($p = 0.069$)." ▦

   If $p > 0.10$: The probability associated with the random-variation explanation is so high that this explanation cannot be excluded. Therefore, there is insufficient evidence to conclude that $\mu \neq \mu_0$.

   ■ **EXAMPLE** "The mean for 100 repeated analyses of a standard solution was not different from the known value ($p = 0.77$). The instrument is correctly calibrated." ▦

6. $100(1 - \alpha)\%$ **Confidence Interval for** $\mu = \bar{x} \pm Z_{\sigma/2} \dfrac{\sigma}{\sqrt{n}}$

where α is the error rate (probability that the confidence interval will *not* include the true value of the population mean) chosen by the investigator (e.g., $\alpha = 0.05$ for a 95% confidence interval). Confidence intervals are always two-tailed (i.e., $\bar{x}$ could be greater than or less than μ). Hence, the critical Z-value corresponds to the probability $\alpha/2$ (e.g., $\alpha/2 = 0.025$ for a 95% confidence interval).

Interpretation: Because $100(1 - \alpha)\%$ of all confidence intervals computed by this method include the true value of the population mean, we can be $100(1 - \alpha)\%$ confident that our specific confidence interval includes the true value μ for the variable of interest.

In Example 9.1 we show the use of the one-sample Z-test in a lab application.

■ **EXAMPLE 9.1**

The one-sample Z-test for means.

1. **Application:**

   A company that performs chemical analyses has a period of training and assessment for all new employees. During this period new employees are trained in all standard operating procedures by experienced personnel. At the end of this training, new employees are evaluated based on their ability to obtain the same analytical results as experienced employees. For each analytical procedure, the new employee is given a single standard sample and required to perform the analytical procedure on $n = 9$ subsamples. Based on a large number of repeated analyses of this standard sample by experienced employees, the mean value for the sample is $\mu = 25$.

   Even experienced employees do not get exactly the same value when the same sample is repeatedly analyzed. Suppose the standard deviation of values produced by experienced employees is $\sigma = 0.75$ and that the distribution of the nine data values is Normal. To pass the evaluation, the mean of the 9 values produced by the new employee should not be significantly different ($p \leq 0.05$) from the mean for analyses by experienced employees. Suppose the mean for the 9 repeated analyses performed by a new employee is $\bar{x} = 24.6$. Has the new employee passed the evaluation?

   Assessing Assumptions

   We will assume that the $n = 9$ repeated measurements the new employee made on the standard sample constitute a representative sample of his or her ability to implement the measurement protocol. We are told that $\sigma = 0.75$ and that the data distribution is Normal. A Normal data distribution is required, given the small sample size (CLT would not apply in this case).

2. **Statement of Hypotheses**

   $H_0: \mu = 25$ Mean for the new employee equals that for experienced employees.

   $H_a: \mu \neq 25$ Mean for the new employee differs from that for experienced employees.

3. Sampling Distribution of $\bar{x}$, assuming H_0 is true

Center: $E(\bar{x})$ = 25
Spread: $\sigma_{\bar{x}}$ = $\sigma/\sqrt{n}$
                    = $0.75/\sqrt{9}$
                    = 0.25
Shape: Normal

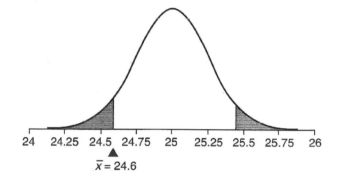

$\bar{x} = 24.6$

This sampling distribution displays which values of $\bar{x}$ have a high probability of occurring if the Null hypothesis value for μ is correct. If the observed value for $\bar{x}$ falls far from the expected value under the Null hypothesis, this would constitute evidence that the Null hypothesis should be rejected.

4. Test Statistic

$$Z_{\text{test}} = \frac{(\bar{x} - \mu_0)}{\sigma_{\bar{x}}} = \frac{24.6 - 25}{0.25} = -1.60$$

5. p-value

The probability of getting the observed difference between the new employee's mean and the mean for experienced employees due only to random sampling variation is equal to $P[Z_{\text{test}} \leq -1.60] = 0.0548$. There was no prior expectation that the difference would be positive or negative, so a two-tailed test is appropriate and the p-value is $0.0548 \times 2 = 0.1096$.

Conclusion: The mean of the new employee's 9 analyses is not significantly different from the mean obtained by experienced employees ($p = 0.1096$). Therefore, the new employee "passes" this evaluation.

6. 95% Confidence Interval for μ

Here μ is the true mean of the new employees' measurements.

$$\bar{x} \pm Z_{\alpha/2} \frac{\sigma}{\sqrt{n}} = 24.6 \pm 1.96 \frac{0.75}{\sqrt{9}} = 24.6 \pm 1.96(0.25) = 24.6 \pm 0.49$$

Interpretation: Because 95% of all confidence intervals computed by this method include the true value of the population mean, the evaluator is 95% confident that this confidence interval includes the true mean of the new employee's measurements for this standard sample.

NOTE The 95% confidence interval for the true mean of the new employee's measurements includes the value 25, the mean for measurements obtained by experienced employees. This indicates that the new employee can begin working. ■

One-Sample *t*-Test

1. **Application:** The **one-sample *t*-test** is used to evaluate the difference between a single sample mean and a Null hypothesis value for the population mean $(\bar{x} - \mu_0)$ as evidence that the true population mean does not equal the Null hypothesis value.

   NOTE The one-sample *t*-test is used under the same circumstances as the one-sample *Z*-test for the sample mean, except that the *t*-test does *not* require knowledge of the population standard deviation σ or very large sample sizes ($n > 100$). ■

   Assumptions: (1) *The data were obtained in a randomized, unbiased manner,* so $E(\bar{x})$ is the true population mean μ, and (2) *the sampling distribution of $\bar{x}$ is Normal.* This latter assumption is fulfilled if either of the following two conditions is met: (a) the population distribution is Normal (as indicated by a box-plot or Normal quantile plot of the data distribution), or (b) the sample size is sufficiently large that the Central Limit Theorem applies (according to "rules of thumb" in Chapter 6).

2. **Statement of Hypotheses**
   - One-tailed test:

     H_0: $\mu = \mu_0$

     H_a: $\mu < \mu_0$ or $\mu > \mu_0$
   - Two-tailed test:

     H_0: $\mu = \mu_0$

     H_a: $\mu \neq \mu_0$

3. **Sampling Distribution of $\bar{x}$,** assuming H_0 is true

   Center: $E(\bar{x}) = \mu_0$

   Spread: $S_{\bar{x}} = S/\sqrt{n}$

   In this formula, S is the standard deviation of the data values, and $S_{\bar{x}}$ is called the **standard error of the mean** to distinguish this from the standard deviation of the mean ($\sigma_{\bar{x}}$), which can only be computed if σ is known or $n > 100$. *Important:* When the spread of the sampling distribution of the mean is estimated by the standard error, $S_{\bar{x}}$, random variation among individuals in the population is no longer the only source of variability. *Measurement error* caused by sloppy technique or poor training increases the variability in data values, increasing S and $S_{\bar{x}}$. Increasing the spread of the sampling distribution will increase the width of confidence intervals (reducing precision) and reduce the power of tests of significance. Hence, it is critical that measurement error always be minimized through proper training.

   Shape: *t*-distribution [with degrees of freedom (df) $= n - 1$]

   Degrees of freedom refers to the number of data values that are free to vary after the value of the sample mean has been determined. Once the value of the sample mean is stipulated, all the data values save the last one can vary independently without changing the value of the mean. However, once you stipulate the sample mean and all but one data value, the last remaining data value can only have one value. Hence, it is not "free to vary."

The *t*-distribution looks similar to a Normal distribution (symmetric and bell-shaped), but has a wider spread. When a statistics computer program is used to perform the test of significance, this difference can be ignored. If you must perform this test by hand, you will need to use the Student's *t*-table (Appendix Table 4) instead of the standard Normal table. Read the next section, The *t*-Distribution, for a description of the differences between *t*- and *Z*-distributions and their respective tables.

4. **Test Statistic**

   Calculate t_{test}:

   $$t_{test} = \frac{(\bar{x} - \mu_0)}{S/\sqrt{n}} = \frac{(\bar{x} - \mu_0)}{S_{\bar{x}}}$$

   The t_{test} statistic represents the difference between the sample mean and the Null hypothesis value in "standard error units." This test statistic has a *t*-distribution with degrees of freedom $= n - 1$. Statistics computer programs will provide an exact *p*-value associated with the t_{test}. If you must do this test by hand, you can determine approximate *p*-values using the *t*-table, as described in the next section.

5. ***p*-Value**

   a. *One-tailed test:* The probability $P[t \geq t_{test}]$ or $P[t \leq -t_{test}]$, obtained from the *t*-distribution, is the *p*-value for the one-tailed test of significance. This is the probability of obtaining values for the sample mean less than or equal to (or $\geq$) the observed $\bar{x}$ if the true population mean value is equal to the Null hypothesis value μ_0.

   b. *Two-tailed test:* The probability $2 \times P[t \geq |t_{test}|]$ is the *p*-value for the two-tailed test of significance. The term $|t_{test}|$ is the absolute value of t_{test} (i.e., any − sign is deleted). In a two-tailed test, there was no prior expectation that the difference between $\bar{x}$ and μ_0 would be in one direction or the other. The *p*-value is the probability of getting an absolute difference $|\bar{x} - \mu_0|$ greater than or equal to the observed difference if the true population mean is equal to the Null hypothesis value.

   c. *Stating conclusions:*

   If $p \leq 0.05$: The probability associated with the random-variation explanation for the difference $(\bar{x} - \mu_0)$ is sufficiently low to reject this explanation. If an appropriate study design has been correctly implemented, the only remaining likely explanation supported by the data is that the true value of the population mean differs from the Null hypothesis value. Regardless of whether the test was one-tailed or two-tailed, your conclusion should state a specific direction of the difference, as indicated by the sample data.

   ■ **EXAMPLE**   "The mean of 10 repeated measurements of a standard solution was less than the known value ($p = 0.0012$). Therefore, the instrument is out of calibration."   ▨

   If $0.05 < p \leq 0.10$: The probability associated with the random-variation explanation is sufficiently high that we cannot confidently exclude this explanation. However, the probability is sufficiently low that we might suspect there really is a difference. Some scientists would conclude that the data "suggest" the true value of μ is less than (or >) μ_0.

▨ **EXAMPLE**   "The difference between the mean for 10 repeated analyses of a standard solution and the known value suggests that the instrument is out of calibration ($p = 0.069$), but more data are needed to verify this."   ▨

If $p > 0.10$: The probability associated with the random-variation explanation for the observed difference ($\bar{x} - \mu_0$) is so high that this explanation cannot be excluded. Hence, there is insufficient evidence to conclude that $\mu \neq \mu_0$.

▨ **EXAMPLE**   "The mean for 10 analyses of a standard solution did not significantly differ from the known value ($p > 0.15$). Therefore, the instrument is calibrated."   ▨

6. **$100(1 - \alpha)$% Confidence Interval for $\mu = \bar{x} \pm t_{df,\,\alpha/2}\,S/\sqrt{n}$**

NOTE When the spread of the sampling distribution is estimated using the standard error $= S/\sqrt{n}$, the critical value for the confidence interval must be obtained from the t-table rather than from the standard Normal table ($Z_{critical}$ is replaced by $t_{critical}$). The next section (The t-Distribution) gives directions on how to obtain a critical value from the t-table. ▪

Interpretation: Because $100(1 - \alpha)$% of all confidence intervals computed by this method include the true value of the population mean, we are $100(1 - \alpha)$% confident that our specific confidence interval includes the value of μ for the variable of interest.   ▨

In Example 9.2 we show the one-sample t-test used in a laboratory application.

▨ **EXAMPLE 9.2**

The one-sample t-test.

1. **Application:** A scientist wants to test his spectrophotometer to ensure it is properly calibrated. He uses a standard solution from the National Bureau of Standards (NBS) that has a known absorption of 50 at a specified wavelength of light. He uses subsamples from this solution to make $n = 16$ measurements with the spectrophotometer. If the mean of these 16 measurements is significantly ($p \leq 0.05$) different from the known value of 50, he will recalibrate the instrument. Suppose that he obtains a sample mean $\bar{x} = 51$, the standard deviation of the 16 data values is $S = 1.6$. A boxplot of the data values is symmetric, with no outliers. Do these results provide sufficient evidence that the spectrophotometer is out of calibration?

Assessing Assumptions

We will assume that the $n = 16$ subsamples are a representative sample from the NBS standard solution. The boxplot of data values is symmetric and has no outliers. With $n = 16$, the Central Limit Theorem rules of thumb allow us to assume the sampling distribution of $\bar{x}$ is Normal.

2. **Statement of Hypotheses**
   - H_0: $\mu = 50$
   - H_a: $\mu \neq 50$

3. **Sampling Distribution of $\bar{x}$**, assuming H_0 is true

Center: $E(\bar{x}) = 50$

Spread: $S_{\bar{x}} = S/\sqrt{n}$
$$= 1.6/\sqrt{16}$$
$$= 0.4$$

Shape: t-distribution with degrees of freedom df = 15, the shaded area corresponds to the two-tailed p-values.

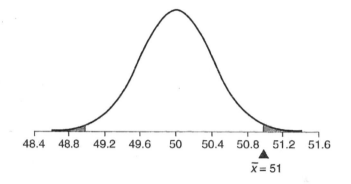

$\bar{x} = 51$

4.,5. **Test Statistic and p-Value**

$$t_{test} = \frac{(\bar{x} - \mu_0)}{S_{\bar{x}}} = \frac{51 - 50}{0.4} = 2.5$$

p-value $= 2 \times P[t_{df=15} \geq 2.5]$

With degrees of freedom df = 15, this t-value falls between the t-table values of 2.249 and 2.602, corresponding to a p-value $0.01 < p < 0.02$. The question was framed such that a two-tailed test of significance is required, so the appropriate p-value is $0.02 < p < 0.04$.

Conclusion: Based on these results, the scientist concludes that the mean of the 16 analyses was greater than the known standard value ($p < 0.04$). Therefore, the spectrophotometer is out of calibration. There is a probability < 0.04 that this conclusion is incorrect due to a Type I error. Before the scientist decides to go through the effort of recalibration, he wants to document just how much his instrument is out of calibration (the "effect size"). This is usually done by computing the confidence interval.

6. **95% Confidence Interval for μ**

Confidence Interval for $\mu = \bar{x} \pm t_{df=15,\ \alpha/2=0.025}\ S/\sqrt{n}$
$$= 51 \pm 2.131\ (1.6/\sqrt{16})$$
$$= 51 \pm 0.85$$
$$= 50.15,\ 51.85$$

Interpretation: Because 95% of all confidence intervals computed by this method include the true value of the population mean, the scientist is 95% confident that the true mean absorption for the standard sample, as measured by the spectrophotometer in its current calibration, is within the range of 50.15 to 51.85. Because this 95 % confidence interval does not include the known value for the standard (50), there is strong evidence that $\mu \neq 50$ for the spectrophotometer in its current condition. Hence, it should be recalibrated.   ■

The *t*-Distribution

Because the t_{test} statistic is computed based on two random variables ($\bar{x}$ and S), the **t-distribution** has a greater spread than the Normal distribution for Z_{test}. Although the probability distribution of $(\bar{x} - \mu)/S_{\bar{x}}$ is still symmetric and bell-shaped like the Normal distribution, more of the area under the probability density curve is located in the tails, and less around the center, than in a Normal distribution (see Figure 9.1). The probability distribution for $(\bar{x} - \mu)/S_{\bar{x}}$ is called the **Student's t-distribution** (Appendix Table 4) or simply the **t-distribution**.

One difference between the standard Normal distribution and the *t*-distribution is that the shape of the *t*-distribution changes as sample size changes. Increasing the sample size reduces random sampling variation of the sample standard deviation S. Hence, as sample size increases, the value of S approaches the population parameter value σ and the shape of the *t*-distribution approaches the shape of the Normal distribution. In Figure 9.1 below, even with df = 32 (n = 33), the Normal and *t*-distributions are almost indistinguishable.

A complete listing of probabilities for all possible *t*-values similar to the standard Normal table would require a separate two-page *t*-table for each possible sample size! Because this is not practical, the standard *t*-table (Appendix Table 4) lists only one row of 12 *t*-values and their associated probabilities for each *t*-distribution. Only positive *t*-values are listed, and all probabilities reflect the area in the right-hand tail (i.e., $P[t \geq t_{test}]$).

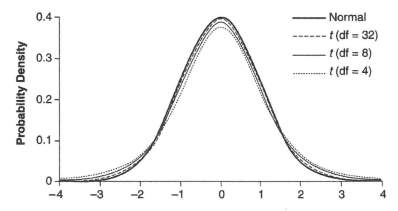

FIGURE 9.1   Comparison of the Normal distribution and *t*-distributions for a range of degrees of freedom.

Using the *t*-table.

The following steps and examples show how to use the *t*-table to obtain approximate probabilities for a range of *t*-values.

1. Compute the degrees of freedom (df $= n - 1$) and locate the row in the *t*-table that corresponds to this df value.

2. Read across the columns on the row with the appropriate df until you find the two *t*-values in the table that bracket your t_{test} value.

3. The *p*-value that corresponds to your t_{test} value is between the two probabilities at the top of the two columns for the *t*-values that bracket the t_{test} value.

Example 9.3 shows these steps as they are used in determining a specific *t*-value.

■ **EXAMPLE 9.3**

Determine the *p*-value for $t_{test} \geq 1.98$, with a sample size $n = 20$. On the row for df $= 19$, $t_{test} = 1.98$ is bracketed by the table values 1.729 and 2.093. The probabilities associated with these two *t*-values are 0.05 and 0.025, respectively. Hence, $P[t_{test} \geq 1.98]$ is $0.025 < p < 0.05$.

NOTE   By convention, the *p*-value is reported only as *p* less than the larger of the two *t*-table probabilities ($p < 0.05$). ■

				Tail Probability						
df	.25	.20	.15	.10	.05	.025	.02	.01	.005	. . . .
1	1.000	1.376	1.963	3.078	6.314	12.71	15.89	31.82	63.66	. . . .
2	.816	1.061	1.386	1.886	2.920	4.303	4.849	6.965	9.925	. . . .
.	.	.	.			.	.	.	.	
18	.688	.862	1.067	1.330	1.734	2.101	2.214	2.552	2.878	. . . .
19	.688	.861	1.066	1.328	1.729	2.093	2.205	2.539	2.861	. . . .
20	.687	.860	1.064	1.325	1.725	2.086	2.197	2.528	2.845	. . . .
.	.	.	.	. .	.	.	.	.	.	

Determining the *p*-value for a negative *t*-value.

First, the *t*-table does not list negative *t*-values, so ignore the negative sign and look for tabled *t*-values that bracket the positive value of t_{test}. The *t*-distribution is symmetric, so the probability $P[t \geq +t_{test}]$ from the t-table is equal to the probability $P[t \leq -t_{test}]$. Example 9.4 shows how negative *t*-values are handled.

■ **EXAMPLE 9.4**

Determine the *p*-value for $t_{test} \leq -1.98$, with a sample size $n = 20$. In most cases when t_{test} is negative, you will be interested in the probability $P[t \leq -t_{test}]$, that is, the area under the curve in the left-hand tail. Because the *t*-distribution is symmetric, $P[t \leq -1.98] = P[t \geq +1.98]$, which was $p < 0.05$ in Example 9.3. See the accompanying figure below.

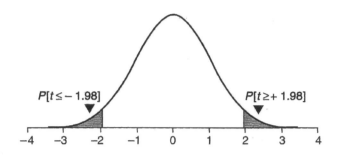

Determining a *p*-value for *t*-values outside the table range.

When t_{test} is larger than the largest value on a row in the *t*-table, the *p*-value is reported as $p < 0.0005$ (smallest *p*-value in table) for a one-tailed test and $p < 0.001$ for a two-tailed test.

■ **EXAMPLE** $P[t \geq 5.58]$ with df = 15 would be stated $p < 0.0005$. ■

Determining *p*-values for *t*-values that are smaller than the smallest tabled value in a row is a bit more difficult. First, the *t*-value at the center of the bell-shaped *t*-distribution is zero. Therefore, $P[t \geq 0] = 0.5$. The *p*-value associated with the smallest *t*-value in a row is 0.25. Hence, $P[0 \leq +t_{test} \leq$ smallest *t*-value in a row) is 0.25 $< p < 0.5$. Likewise, $P[$smallest $-t$-value $\leq -t_{test} \leq 0]$ is also $0.25 < p < 0.5$. The probability associated with *t*-values greater than or equal to a negative *t*-value between 0 and the smallest value on a row would be $0.5 < p < 0.75$. See Figure 9.2 for diagrams of these probabilities.

Determining critical values $t_{\alpha/2,df}$ for confidence intervals.

Using the *t*-table, look across the column titles (where probability values are listed) to locate the column that corresponds to the value $\alpha/2$ ($= 0.025$ for a 95% confidence interval). Read down that column to the row that corresponds to the degrees of freedom for your data ($n - 1$). The tabled *t*-value is the critical value $t_{critical}$ required to compute the $100(1 - \alpha)$% confidence interval.

(A) Probability $P[t \geq +0.42]$ with $n = 15$ corresponds to the shaded area, which is $0.25 < p < 0.5$.

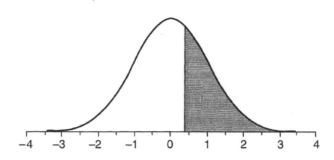

(B) Probability $P[t \leq -0.42]$ with $n = 15$ is equal to $P[t_{test} \geq +0.42]$, and corresponds to the shaded area that is $0.25 < p < 0.5$.

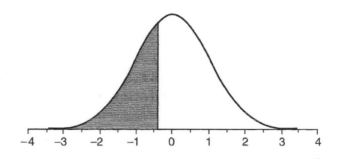

(C) Probability $P[t \geq -0.42]$ with $n = 15$ is $1 - (0.25 < p < 0.5)$, which is $0.5 < p < 0.75$.

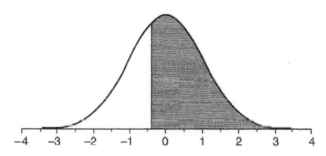

FIGURE 9.2   Determining probabilities for t-values less than the smallest t-value in a row of the t-table.

Using a computer spreadsheet to obtain t-test p-values and critical t-values.

1. You can use a computer spreadsheet to obtain p-values based on computed values for t_{test}, df, and one- or two-tailed tests.

   Excel: = TDIST($+t_{test}$, df, **tails** < 1 *or* $2 >$)
   For example: = TDIST(2.21,18,1) = 0.02015 (one-tailed p-value)

   QuattroPro: @TDIST(t_{test}, df, **tails** < 1 *or* $2 >$)
   For example: @TDIST(2.21,18,2) = 0.0403 (two-tailed p-values)

2. You can also obtain critical t-values for a $100(1 - \alpha)\%$ confidence interval.

   Excel: = TINV($\alpha/2$,df)
   For example: = TINV(0.025,20) = 2.423

   QuattroPro: @TINV($\alpha/2$,df)
   For example: @TINV(0.025,14) = 2.51

Example 9.5 gives some additional examples of the use of the t-table.

■ **EXAMPLE 9.5**

Using the t-table to obtain p-values and critical t-values for confidence intervals. Try to determine these probabilities and critical t-values yourself and compare your answers to mine.

$P[t_{test} \geq +0.55]$ with 10 df $= 0.25 < p < 0.50$

$P[t_{test} \leq -0.55]$ with 10 df $= 0.25 < p < 0.50$

$P[t_{test} \geq -0.55]$ with 10 df $= 0.5 < p < 0.75$

$P[t_{test} \geq +1.05]$ with 25 df $= 0.15 < p < 0.20$

$P[t_{test} \leq +1.05]$ with 25 df $= 0.85 > p > 0.80$

$P[t_{test} \geq +3.05]$ with 45 df $= 0.001 < p < 0.0025$

$P[t_{test} \geq -3.05]$ with 45 df $= 0.999 > p > 0.9975$

Critical t-value ($\alpha = 0.05$, 95% confidence interval, $n = 20$) $= 2.093$

Critical t-value ($\alpha = 0.01$, 99% confidence interval, $n = 30$) $= 2.756$

Critical t-value ($\alpha = 0.05$, 95% confidence interval, $n = 45$) $= 2.021$ ▤

NOTE When the exact number for the degrees of freedom is not listed in the t-table, always round your df value *down* to the closest value in the table. For example, df $= 45$ was rounded down to df $= 40$, the closest value. ■

Paired t-Test

1. **Application:**

The **paired t-test** is a special case of the one-sample t-test used for before–after or matched-pairs experimental designs. In before–after designs, study subjects are each measured twice, once before and once after a treatment. In matched-pairs designs, study subjects are paired based on their similarity with regard to the variable of interest. In each matched pair, one subject is randomly assigned to receive the treatment and the other does not, serving as a control. In both types of study designs, the difference (D) between the two measurements is computed for each subject or matched pair of subjects. The sample statistics used for the test of significance are the mean $(\overline{D})$ and standard deviation (S_D) of these D-values, and sample size is $n =$ number of D-values.

The before-after and matched-pairs experimental designs reduce random sampling variation by eliminating variation among individual experimental units from the calculation of the sample standard deviation S. That is, S_D reflects the variation in the difference values computed for each individual or matched pair, but *not* the variation among individuals. For example, two individuals could have very different blood pressures, but both could experience the same reduction in blood pressure when treated with a drug. If the individual measurements were used to assess for the effect of the drug, the large differences among individuals would be included in the sample standard deviation, resulting in a sampling distribution with a large spread. However, if these very different individuals experience similar responses to the treatment, there will be relatively small variations in the difference values ($D =$ pre $-$ post) and the sampling distribution of $\overline{D}$ will have a relatively small spread. Reducing the spread of the sampling distribution increases the power of the matched-pairs study designs to detect small differences, compared to randomized two-sample designs, with independent control and treatment groups.

Assumptions: (1) *The data were obtained by a randomized, unbiased study design,* so the expected value $E(\overline{D})$ is the true mean difference μ_D, and (2) *the sampling distribution of $\overline{D}$ is Normal.* This Normality assumption is fulfilled if either one of the following conditions is met: (a) The population distribution of the differences (D-values) is Normal or (b) the sample size n = the number of D-values is sufficiently large for the Central Limit Theorem to apply (as determined by the "rules of thumb" presented in Chapter 6). The Normality assumption is assessed using a boxplot or Normal quantile plot of the D-values, *not* the original paired data values.

2. **Statement of Hypotheses**
   - One-tailed test:

     H_0: $\mu_D = 0$

     H_a: $\mu_D < 0$ or $\mu_D > 0$
   - Two-tailed test:

     H_0: $\mu_D = 0$

     H_a: $\mu_D \neq 0$

   The Null hypothesis is always associated with the "no difference" outcome. In the context of a before-after or matched-pairs study design, this translates to no difference between the before and after measurements, or between the measurements on matched pairs of subjects. Even though random variation will usually result in small differences between these pairs of measurements, the average difference μ_D is expected to be zero.

3. **Sampling Distribution of $\overline{D}$**, assuming H_0 is true

   Center: $E(\overline{D}) = 0$

   Spread: $S_{\overline{D}} = S_D/\sqrt{n}$ where: S_D is standard deviation of D-values

   Shape: t-Distribution with df = number of D-values − 1

   The sampling distribution displays the likely values of $\overline{D}$ if the Null hypothesis is true (and there really is no difference). The more the observed mean difference $\overline{D}$ deviates from zero, the less likely this is due only to random variation.

4. **Test Statistic**

   $$t_{test} = \frac{(\overline{D} - 0)}{S_{\overline{D}}} = \frac{\overline{D}}{S_{\overline{D}}}$$

   The t_{test} statistic represents how far $\overline{D}$ deviates from the expected value of zero under the Null hypothesis, measured in standard error units. An approximate p-value for this test statistic can be determined from the t-table with $n - 1$ degrees of freedom. Statistics computer programs provide exact p-values.

5. **p-Value**
   a. *One-tailed test:* The probability $P[t \geq +t_{test}]$ or $P[t \leq -t_{test}]$, is the p-value for the one-tailed test of significance. This is the probability of obtaining values for the *mean difference* greater than (or less than) the observed $\overline{D}$ if the true mean difference μ_D is equal to zero.
   b. *Two-tailed test:* The probability $2 \times P[t \geq |t_{test}|]$ is the p-value for the two-tailed test of significance. The term $|t_{test}|$ is the absolute value of t_{test}

(i.e., any − sign is deleted). In a two-tailed test, there was no prior expectation that the mean difference $\overline{D}$ would be positive or negative. The p-value is the probability of getting an absolute mean difference $|\overline{D}|$ greater than or equal to the observed value if the true mean difference μ_D is equal to zero.

c. *Stating conclusions:*

If $p \leq 0.05$: The probability associated with the random-variation explanation for the observed mean difference $\overline{D}$ is sufficiently low to reject this explanation. If an appropriate study design has been correctly implemented, the only remaining likely explanation is that the *true mean difference* is greater than (or less than) zero. Regardless of whether your test of significance was one-tailed or two-tailed, your conclusion should state how the mean difference differs from zero, as indicated by the sample data.

▓ **EXAMPLE**   "On average, blood pressure decreased 50 points after a week on the new drug treatment ($p = 0.0012$), indicating that the new drug was effective for reducing high blood pressure." ▓

If $0.05 < p \leq 0.10$: The probability associated with the random-variation explanation is sufficiently high that we cannot confidently exclude this explanation. However, the probability is sufficiently low that we might suspect there is a difference. Some scientists would conclude that the data "suggest" there is a difference.

▓ **EXAMPLE**   The mean difference between paired subjects (-25 points) suggests that the treatment may reduce blood pressure ($p < 0.1$), but more data are needed to verify this. ▓

If $p > 0.10$: The probability associated with the random-variation explanation is so high that this explanation cannot be excluded. Hence, there is insufficient evidence to conclude that there is a difference.

▓ **EXAMPLE**   "The mean difference between pre- and post-treatment measurements was not significant ($p > 0.25$). The treatment was not effective for reducing blood pressure." ▓

6. $100(1 - \alpha)\%$ **Confidence Interval for μ_D** $= \overline{D} \pm t_{df=n-1,\alpha/2}\, S_D/\sqrt{n}$

Interpretation: Because confidence intervals computed by this method include the *true mean difference* (μ_D) $100(1 - \alpha)\%$ of the time, we can be $100(1 - \alpha)\%$ confident that the interval we computed from our sample includes the true mean difference. ▓

Example 9.6 illustrates how the paired t-test is used in a study of the effectiveness of a new drug to lower cholesterol levels.

▓ **EXAMPLE 9.6**

The paired t-test.

1. **Application:** Fifteen people with high cholesterol take part in a study of a new drug for reducing blood cholesterol levels. Cholesterol levels above 200 are associated with increased risk of vascular disease. Levels above 300 might cause

doctors to prescribe drugs to reduce blood cholesterol. Blood cholesterol levels among individuals in the study varied from 350 to 1200. Because of this high variability among individuals, a before-after study was done. The subjects had their blood cholesterol measured prior to the experiment, they took the drug for a month, and then their cholesterol was remeasured. The difference (D = After-Before) between these two measurements represents the effect of the drug. The blood cholesterol data and summary statistics from this study are presented below.

Before	360	364	505	578	623	684	724	732	787	823	877	901	933	1114	1200
After	401	430	629	530	387	687	603	611	698	519	631	712	777	768	911
D-values	+41	+66	+124	−48	−236	+3	−121	−121	−89	−307	−246	−189	−156	−346	−289

Exploratory data analysis for difference values

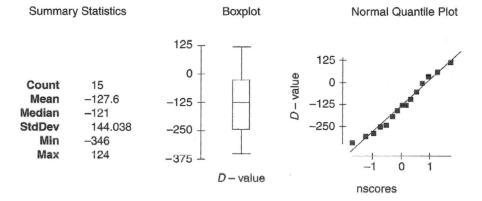

Summary Statistics		Boxplot	Normal Quantile Plot
Count	15		
Mean	−127.6		
Median	−121		
StdDev	144.038		
Min	−346		
Max	124		

Assessing Assumptions

As in all human research, the study subjects are volunteers, not a random sample from the population of interest. There is no ethical way around this problem, and researchers generally assume their samples of human volunteers constitute a representative sample. The symmetric boxplot and straight-line Normal quantile plot of the D-values indicate that the data distribution is approximately Normal. hence, the sampling distribution of $\overline{D}$ can be assumed Normal, regardless of the sample size.

The mean difference between pre- and post-tests is $\overline{D} = -127.6$, with $S_D = 144$. Do these results provide strong evidence that the drug reduced cholesterol?

2. **Statement of Hypotheses**

H_0: $\mu_D = 0$

H_a: $\mu_D < 0$

Drug is expected to *reduce* cholesterol.

3. **Sampling Distribution of $\overline{D}$,** assuming H_0 is true

Center: $E(\overline{D}) = 0$
Spread: $S_{\overline{D}} = S_D/\sqrt{n}$
$= 144/\sqrt{15}$
$= 37.2$
Shape: t-Distribution, with df $= 14$

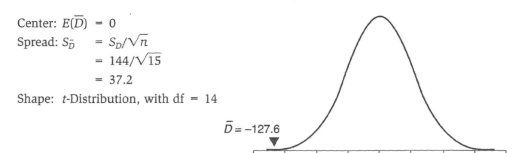

$\overline{D} = -127.6$

$-148 \quad -111 \quad -74 \quad -37 \quad 0 \quad 37 \quad 74 \quad 111 \quad 148$

4. **Test Statistic**

$$t_{\text{test}} = \frac{\overline{D} - 0}{S_{\overline{D}}} = \frac{-127.6 - 0}{37.2} = -3.43$$

5. **p-value**

$$P[\overline{D} \leq -127.6 \text{ if } \mu_D = 0] = P[t_{\text{df}=14} \leq -3.43]$$
$$= P[t_{\text{df}=14} \geq +3.43]$$

$0.001 < p < 0.0025$

Conclusion: Because the investigator wanted to determine only if the new drug *reduces* blood cholesterol levels, this was a one-tailed test. The investigator concludes that the new drug reduced blood cholesterol ($p < 0.0025$). Having drawn this conclusion, the p-value now is the probability that this conclusion is a Type I error (the drug had no effect and $\overline{D}$ differed from zero due only random sampling variation).

6. **95% Confidence Interval for μ_D:** Now that the investigator has concluded that the drug reduces blood cholesterol, he wants to document the effect size (i.e., the average change in blood cholesterol after taking the new drug for one month). The 95% confidence interval is an effective way to convey the effect size of the drug.

$$\overline{D} \pm t_{0.025,\,14}\, S_D/\sqrt{n} = -127.6 \pm 2.145\,(144/\sqrt{15})$$
$$= -127.6 \pm 79.8$$
$$= -207.4 \text{ to } -47.8$$

Interpretation: Because 95% of all confidence intervals computed by this method include the true mean difference μ_D, the investigator is 95% confident that this drug will reduce blood cholesterol between 47.8 and 207.4 points.

Computing the Power of a Paired t-Test

The calculation of power for the paired t-test is done using the same procedure as described in Chapter 7 for the one-sample Z-test for proportions. The procedure is done in two steps: 1. Determine the critical value of the sample statistic that defines the "rejection region" of values that would provide sufficient evidence to reject the Null hypothesis for a specified Type I error rate (usually set at $\alpha = 0.05$). 2. Determine the probability of obtaining a value for the sample statistic within this rejection region if the true value of the population parameter is a specified value that would constitute some "minimum important difference" from the Null hypothesis value. The only differences between computing power for the paired t-test and the Z-test for proportions are related to the description of the sampling distribution (*t*-distribution vs. *Z*-distribution, calculation of the spread). Once the sampling distributions have been described, the procedure is the same. See Example 9.7 for an example of computing power for a paired *t*-test.

■ **EXAMPLE 9.7**

Compute the Power of the Study in Example 9.6 to "Detect" a 100 Point Drop in Cholesterol if α is Set at 0.05. *Note:* The "100 Point Drop" stipulation is a statement of "Minimum Important Difference" that would indicate the drug can lower blood cholesterol enough to be considered effective for treating patients with high cholesterol. By setting $\alpha = 0.05$, we are trying to minimize the probability that we would conclude the drug is effective when it actually does not lower cholesterol. Computing power is a two-step process:

Step 1: Compute the *critical value* $\overline{D}^*$ that represents the minimum significant difference required to reject H_0. Since this was a one-tailed test with $\alpha = 0.05$ and since $\overline{D}$ values less than 0 indicate the drug *reduces* cholesterol, the appropriate critical *t*-value is $-t_{df=14, p=0.05} = -1.761$.

Sampling Distribution of $\overline{D}$ Assuming $\mu_D = 0$

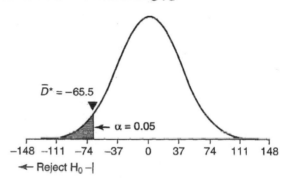

$$D^* = -t_{df=14,\, \alpha=0.05}\ (S_{\overline{D}})$$
$$= -1.761\ (37.2)$$
$$= -65.5$$

Reject H_0 if $\overline{D} \le -65.5$

Step 2: Compute power. Power equals the probability of getting values of $\overline{D} \le -65.5$ if the drug actually reduces cholesterol an average of 100 points (i.e., assume $\mu_D = -100$).

Sampling Distribution of $\overline{D}$ Assuming $\mu_D = -100$

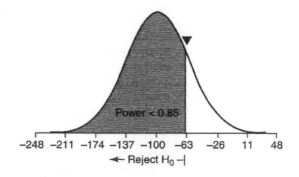

$$\begin{aligned}
P[\overline{D} \le -65.5 \text{ if } \mu_D = -100] &= P[t_{\text{df}=14} \le (-65.5 - (-100))/37.2] \\
&= P[t_{\text{df}=14} \le +0.927] \\
&= 1 - P[t_{\text{df}=14} \ge +0.927] \\
&= (1 - 0.2) < \text{Power} < (1 - 0.15) \\
&\quad\; 0.8 < \text{Power} < 0.85
\end{aligned}$$

Conclusion: The probability that this study would be able to detect a 100-point drop in blood cholesterol after the subjects took the drug for one month is between 0.8 and 0.85. This would generally be considered adequate power. If the study failed to detect a drop in cholesterol after the drug treatment, the investigator would have some basis for concluding that the drug probably does not have the desired effect.   ■

9.2

Two-Sample Tests for Means

In many studies, we want to determine if the difference between two means obtained by sampling two populations ($\bar{x}_1 - \bar{x}_2$), or the difference between two means from an experimental control group and a treatment group ($\bar{x}_C - \bar{x}_T$), is statistically significant. That is, does the difference between the two sample means provide strong evidence of a real difference between the population means ($\mu_1 - \mu_2$)? This differs from the paired t-test in that the sampling of subjects from the two populations, or the assignment of subjects to the control and treatment groups, is completely randomized. Hence, the resulting data values in the two samples are independent of each other. In this situation, the sample statistic of interest is the difference between two sample means ($\bar{x}_1 - \bar{x}_2$), each of which is computed from an independent random sample. If the Null hypothesis is true, this difference will be zero. The Alternative hypothesis states that this differences is less than, greater than, or not equal to 0. The sample statistic ($\bar{x}_1 - \bar{x}_2$) is obtained by combining two independent sample statistics, each of which is a random variable. To do a test of significance, we must define the mean, standard deviation or standard error, and the shape of the sampling distribution of ($\bar{x}_1 - \bar{x}_2$).

Two-Sample Z-Test

1. **Application:** The **two-sample Z-test** is used to evaluate a difference between two sample means ($\bar{x}_1 - \bar{x}_2$) as evidence to test the Null hypothesis that the means of the two population from which the samples were taken are equal (H_0: $\mu_1 - \mu_2 = 0$). This test is used when the population standard deviations σ_1 and σ_2 are known *or* estimated with both sample sizes n_1 and $n_2 \ge 100$. *Assumptions:* (1) *Independent samples were obtained from the two populations in a randomized, unbiased manner.* Hence, we can assume that the expected value for the difference between the two sample means is the actual difference between the two population means, $E(\bar{x}_1 - \bar{x}_2) = (\mu_1 - \mu_2)$. (2) *Sampling distributions for both $\bar{x}_1$ and $\bar{x}_2$ are Normal, which means that the sampling distribution for ($\bar{x}_1 - \bar{x}_2$) will also be Normal.* This assumption is fulfilled if either one of the following conditions is fulfilled: (a) Both population distributions are Normal (as indicated by separate Normal quantile plots or boxplots for the data distributions of each group), or (b) both sample sizes n_1 and n_2 are sufficiently large that the Central Limit Theorem applies to both of the sampling distributions of $\bar{x}_1$ and $\bar{x}_2$.

In actual practice this test is rarely used. There are very few circumstances in which the two population means would be unknown (hence the need for a test of significance), but the two population standard deviations would be known. Also, when sample sizes are large, the outcome of this test is virtually identical to that obtained from a two-sample t-test (described in the next section). The two-sample Z-test for means is usually presented in statistics courses to help students better understand the two-sample t-test.

2. **Statement of Hypotheses**

   - One-tailed test:

     H_0: $\mu_1 - \mu_2 = 0$

     H_a: $\mu_1 - \mu_2 < 0$ or $\mu_1 - \mu_2 > 0$

   - Two-tailed test:

     H_0: $\mu_1 - \mu_2 = 0$

     H_a: $\mu_1 - \mu_2 \neq 0$

3. **Sampling Distribution of** $(\bar{x}_1 - \bar{x}_2)$, assuming H_0 is true

   Center: $E(\bar{x}_1 - \bar{x}_2) = 0$

   Spread: $\sigma_{\bar{x}1-\bar{x}2} = \sqrt{\dfrac{\sigma_1^2}{n_1} + \dfrac{\sigma_2^2}{n_2}}$

NOTE   The value of the sample statistic $(\bar{x}_1 - \bar{x}_2)$ is obtained by combining two sample statistics, each of which is a random variable. The variance of the sum (or difference) of two random variables is equal to the sum of their variances. To compute the standard deviation for the sampling distribution of $(\bar{x}_1 - \bar{x}_2)$ you must add the variances for $\bar{x}_1$ and $\bar{x}_2$ and then take the square root. ■

Shape: Normal

This sampling distribution displays the values of the difference $(\bar{x}_1 - \bar{x}_2)$ that have a high probability of occurring if the Null hypothesis is true (H_0: $\mu_1 - \mu_2 = 0$). The more the observed difference deviates from the expected value of zero, the stronger the evidence that the two population means really are different.

4. **Test Statistic**

$$Z_{\text{test}} = \frac{(\bar{x}_1 - \bar{x}_2) - (\mu_1 - \mu_2)_0}{\sqrt{\dfrac{\sigma_1^2}{n_1} + \dfrac{\sigma_2^2}{n_2}}} = \frac{(\bar{x}_1 - \bar{x}_2) - 0}{\sigma_{\bar{x}_1 - \bar{x}_2}} = \frac{(\bar{x}_1 - \bar{x}_2)}{\sigma_{\bar{x}_1 - \bar{x}_2}}$$

NOTE   The standard deviation of the difference $\sigma_{\bar{x}_1 - \bar{x}_2}$ quantifies the expected amount of random sampling variation for the value of $(\bar{x}_1 - \bar{x}_2)$, given the amount of variation among individuals in the two populations and the two sample sizes. The Z_{test} statistic expresses the size of the observed difference $(\bar{x}_1 - \bar{x}_2)$ relative to the expected amount of random variation. This test statistic has a standard Normal distribution. You can determine the exact probability of getting the observed difference $(\bar{x}_1 - \bar{x}_2)$ by referring to the standard Normal table. ■

5. *p*-Value

   a. *One-tailed test:* The probability $P[Z \geq +Z_{test}]$ or $P[Z \leq -Z_{test}]$ is the *p*-value for the one-tailed test of significance. This is the probability of obtaining the observed difference between sample means $(\bar{x}_1 - \bar{x}_2)$ if the two population means were equal, i.e., $(\mu_1 - \mu_2) = 0$.

   b. *Two-tailed test:* The probability $2 \times P[Z \geq |Z_{test}|]$ is the *p*-value for the two-tailed test of significance. The term $|Z_{test}|$ is the absolute value of Z_{test} (i.e., any − sign is deleted). In a two-tailed test, there was no prior expectation that the difference $(\bar{x}_1 - \bar{x}_2)$ would be positive or negative. The *p*-value is the probability of getting an absolute difference $|(\bar{x}_1 - \bar{x}_2)|$ greater than or equal to the observed value if the true difference between population means is $(\mu_1 - \mu_2) = 0$.

   c. *Stating conclusions:*

   If $p \leq 0.05$: The probability associated with the random-variation explanation for the observed difference $(\bar{x}_1 - \bar{x}_2)$ is sufficiently low to reject this explanation. If an appropriate study design has been correctly implemented, the only remaining likely explanation is that the true difference between the two population means is greater than (or less than) zero. Regardless of whether the test of significance was one-tailed or two-tailed, your conclusion should state how $(\bar{x}_1 - \bar{x}_2)$ differs from zero, as indicated by the sample data.

   ■ **EXAMPLE** "Mean blood pressure for the treatment group was 50 points less than the mean for the control group ($p = 0.012$), indicating that the new treatment was effective for reducing high blood pressure." ■

   If $0.05 < p \leq 0.10$: The probability associated with the random-variation explanation is sufficiently high that we cannot confidently exclude this explanation. However, the probability is sufficiently low that we might suspect there is a difference. Some scientists would conclude that the data "suggest" there is a difference.

   ■ **EXAMPLE** "The 25-point difference in mean blood pressure between control and treatment groups suggests that the treatment may reduce blood pressure ($p = 0.064$), but more data are needed to verify this." ■

   If $p > 0.10$: The probability associated with the random-variation explanation is so high that this explanation cannot be excluded. Hence, there is insufficient evidence to conclude that there is a difference.

   ■ **EXAMPLE** "Mean blood pressure was not significantly different between control and treatment groups ($p > 0.25$). The new drug was not effective for reducing blood pressure." ■

6. **$100(1 - \alpha)\%$ Confidence Interval for $\mu_1 - \mu_2$**

$$(\bar{x}_1 - \bar{x}_2) \pm Z_{\alpha/2}\sqrt{\frac{\sigma_1^2}{n_1} + \frac{\sigma_2^2}{n_2}} = (\bar{x}_1 - \bar{x}_2) \pm Z_{\alpha/2}\,\sigma_{\bar{x}_1 - \bar{x}_2}$$

   where α is the error rate (probability that the confidence interval will *not* include the true value of the difference between the two population means). This probability is chosen by the investigator when selecting the desired confidence level (e.g., $\alpha = 0.05$ for a 95% confidence interval). Confidence intervals are always two-tailed, so the critical Z-value corresponds to the probability $\alpha/2$ (e.g., $\alpha/2 = 0.025$ for a 95% confidence interval).

Interpretation: Because $100(1 - \alpha)\%$ of all confidence intervals computed by this method will include the true difference between the two population means, we can be $100(1 - \alpha)\%$ confident that our specific confidence interval includes $(\mu_1 - \mu_2)$.

Example 9.8 applies the two-sample Z-test to a biological study of wolf body mass.

■ **EXAMPLE 9.8**

The two-sample *Z*-test for sample means.

1. **Application:**

   Bergmann's rule states that within widely distributed mammal species, individuals that live in colder climates have larger body size (body mass), on average, than individuals of the same age and gender that live in warmer climates. A wildlife biologist wants to determine if this general rule applies to the gray wolf, a species that lives in areas from Mexico to the Arctic, and throughout most of Eurasia. The biologist searches existing databases, mainly from wolf "control" programs, and obtains body mass measurements for 123 mature male wolves killed in Alaska and 218 wolves killed in the western United States south of Canada. The biologist restricted her sample to mature male wolves to control for extraneous variation due to gender or age so as to more clearly assess the effect of environmental temperature. Mean body mass for the "Northern" (n) wolves was $\bar{x}_n = 41.5$ kg, and it was $\bar{x}_s = 38.1$ kg for the "Southern" (s) wolves. Because the sample sizes are at least 100, the sample standard deviations can be used in place of the population standard deviations: $\sigma_n \approx S_n = 9.8$ kg and $\sigma_s \approx S_s = 10.1$ kg.

   Assessing Assumptions

   Given that past wolf control programs killed the animals indiscriminately, we can probably assume that the samples are representative of the larger populations of interest. With the very large sample sizes, the Central Limit Theorem allows us to assume the sampling distributions of $\bar{x}_n$ and $\bar{x}_s$ are Normal, and the sample standard deviations can be assumed to be approximately equal to the population standard deviations. Hence, the two-sample Z-test for means can be applied.

2. **Statement of Hypotheses**

   - H_0: $(\mu_n - \mu_s) = 0$
   - H_a: $(\mu_n - \mu_s) > 0$

   Bergmann's rule indicates wolves in the colder northern area should be larger (have greater body mass) than wolves in the warmer southern area.

3. **Sampling Distribution of $(\bar{x}_n - \bar{x}_s)$, assuming H$_0$ is true**

Center: $E(\bar{x}_n - \bar{x}_s) = 0$

Spread: $\sigma_{\bar{x}_n - \bar{x}_s} = \sqrt{\dfrac{S_n^2}{n_n} + \dfrac{S_s^2}{n_s}}$

$= \sqrt{\dfrac{9.8^2}{123} + \dfrac{10.1^2}{218}}$

$= 1.12$

Shape: Normal

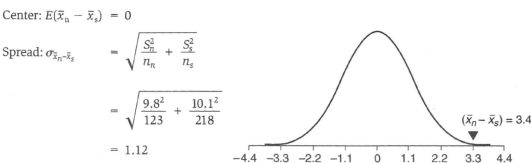

$(\bar{x}_n - \bar{x}_s) = 3.4$

4. **Test Statistic**

$$Z_{\text{test}} = \frac{(\bar{x}_n - \bar{x}_s)}{\sigma_{\bar{x}_n - \bar{x}_s}} = \frac{(41.5 - 38.1)}{1.12} = 3.04$$

5. **p-Value**

The probability of getting the observed difference in mean body mass between the northern and southern wolves due only to random sampling variation is equal to $P[Z_{\text{test}} > 3.04] = 0.0012$. Because Bergmann's rule states that body size (mass) should be *larger* in regions in cold climates, this is a one-tailed test.

Conclusion:

Male gray wolves in colder, northern regions had greater mean body mass than male gray wolves in warmer southern regions ($p = 0.0012$). The p-value is also the probability that this conclusion is incorrect due to a Type I error (the observed difference is really only due to random sampling variation).

6. **95% Confidence Interval for $(\mu_n - \mu_s)$**

$(\bar{x}_n - \bar{x}_s) \pm Z_{\alpha/2}\, \sigma_{\bar{x}_n - \bar{x}_s} = (41.5 - 38.1) \pm 1.96(1.12) = 3.4 \pm 2.2$

Effect size is often expressed in the form of a confidence interval. Because confidence intervals computed by this method include the true value of the difference between two population means 95% of the time, we can be 95% confident that the true difference in mean body mass of adult male gray wolves between northern and southern populations falls within the interval 3.4 ± 2.2 (1.2 to 5.6 kg).

Introduction to Two-Sample t-Tests

Like the one-sample t-test described earlier, **two-sample t-tests** for comparing sample means are used when the population standard deviations σ_1 and σ_2 are *unknown* and estimated by sample standard deviations S_1 and S_2, with sample sizes n_1 and $n_2 < 100$. In the discussion of the one-sample t-test, I described how replacing a fixed parameter σ with a sample statistic S increased the variability of the test statistic. The result was that the test statistic has a t-distribution rather than a Normal distribution. In the two-sample case, *two* fixed parameters σ_1 and σ_2 are replaced by sample statistics S_1 and S_2, both of which are random variables. The resulting t_{test} statistic is even more variable, and its probability distribution has a greater spread than a t-distribution. In fact, the theoretical probability distribution of the t-test statistic computed with two-sample standard deviations has not yet been determined by mathematical statisticians. While we wait for the mathematicians to resolve this problem, scientists still need to determine the significance of differences between sample means computed with small samples sizes.

Two versions of the two-sample t-test have been developed to deal with the problem described above:

1. **Pooled variance t-test:** If the two populations being compared can be assumed to have equal variances ($\sigma_1^2 = \sigma_2^2 = \sigma^2$), we pool the data from both samples to estimate this single value for σ^2. In this case, the test statistic is computed using only one sample standard deviation, S_{pooled}, and t_{test} has a t-distribution with ($n_1 + n_2 - 2$) degrees of freedom.

2. **Separate variance t-test:** If the two population standard deviations cannot be assumed equal, use the two separate sample standard deviations S_1 and S_2 to compute the t_{test} statistic. However, to determine the p-value for t_{test}, we use a t-distribution with an "adjusted degrees of freedom" that is less than ($n_1 + n_2 - 2$). With fewer degrees of freedom, the t-distribution will have increased spread, reflecting the increased random variation that results from using two-sample standard deviations to compute t_{test}. These two versions of the two-sample t-test are described in detail below.

Two-Sample t-Test (Pooled- and Separate-Variance Tests)

1. **Application:** The **two-sample t-test** is used to evaluate the significance of the difference between two sample means ($\bar{x}_1 - \bar{x}_2$) as evidence to test the Null hypothesis that there is no difference between the true population means [H_0: ($\mu_1 - \mu_2$) = 0]. This test is used instead of the two-sample Z-test when the population standard deviations σ_1 and σ_2 are *unknown* and estimated by S_1 and S_2, with sample sizes n_1 and $n_2 < 100$.

   Assumptions:

   a. For both the pooled- and separate-variance tests, *the data are obtained from the two populations in a randomized, unbiased manner, so $E(\bar{x}_1 - \bar{x}_2) =$ ($\mu_1 - \mu_2$).*

      NOTE   The data values in the two samples are completely independent of each other, in contrast to the paired t-test. ■

   b. For both the pooled- and separate-variance tests, *the sampling distribution of ($\bar{x}_1 - \bar{x}_2$) is Normal.* This assumption is fulfilled if the two population distributions can be assumed to be Normal. This is assessed using *separate* boxplots or Normal quantile plots of the data distributions for each of the two samples, to identify skewness and outliers. Both data distributions should be approximately Normal. Alternatively, the sampling distribution of ($\bar{x}_1 - \bar{x}_2$) can be assumed to be Normal if the sample sizes for *both* samples are sufficiently large that the Central Limit Theorem applies, as described in the "rules of thumb" in Chapter 6.

   c. **For the pooled-variance t-test only:** The population variances can be assumed equal ($\sigma_1^2 = \sigma_2^2$). The F_{max} test described below can be used to assess whether or not this assumption is valid, based on a comparison of the sample variances.

The F_{max} Test for Equality of Variances

- Compute F_{max}, the ratio of the two sample *variances*, placing the larger of the two variances in the numerator and the smaller in the denominator.

$$F_{max} = S^2_{larger}/S^2_{smaller}$$

The expected value for this test statistic if the two population variances are equal is 1.0. However, the statistic will generally be greater than this expected value due to random sampling variation.

- Look-up the critical F_{max} value for a *p*-value 0.05 in Appendix Table 7. The number of samples being compared is $k = 2$, and the degrees of freedom value is the smaller of $n_1 - 1$ or $n_2 - 1$. If the df value falls between two table df values, round df down to the lower tabled value.

- If the computed $F_{max} \geq$ critical F_{max} from Table 7, there is sufficient evidence to conclude that the two population variances are *not* equal, and the separate-variance *t*-test should be used.

If the computed $F_{max} <$ critical F_{max}, the assumption of equal population variances is valid, and the pooled-variance *t*-test can be used.

NOTE   The pooled variance *t*-test will always have degrees of freedom greater than or equal to the degrees of freedom for the separate-variance *t*-test. Hence, all other things being equal, the pooled-variance *t*-test has more statistical power to detect a difference between means than the separate-variance *t*-test. Therefore, *if the equal variances condition is met, the pooled-variance t-test is the most appropriate test.*

2. **Statement of Hypotheses**
   - One-tailed test:

     H_0: $\mu_1 - \mu_2 = 0$

     H_a: $\mu_1 - \mu_2 < 0$ or $\mu_1 - \mu_2 > 0$
   - Two-tailed test:

     H_0: $\mu_1 - \mu_2 = 0$

     H_a: $\mu_1 - \mu_2 \neq 0$

3. **Sampling Distribution of $(\bar{x}_1 - \bar{x}_2)$, assuming H_0 is true**

   Center: $E(\bar{x}_1 - \bar{x}_2) = 0$

   Spread

   - *For the pooled-variance t-test:* Computing the standard error of the difference $S_{\bar{x}1-\bar{x}2}$ for the pooled variance *t*-test is a two-step process: (1) Compute the pooled estimate of the standard deviation S_p and (2) use S_p in place of S_1 and S_2 to compute the standard error of the difference $S_{\bar{x}1-\bar{x}2}$. The calculations for each step are as follows:

   $$(1) \ S_p = \sqrt{\frac{(n_1 - 1) S_1^2 + (n_2 - 1)S_2^2}{n_1 + n_2 - 2}}$$

   If the two sample sizes are equal ($n_1 = n_2$), this formula for S_p simplifies to the square root of the average of the two sample variances:

   $$S_p = \sqrt{\frac{S_1^2 + S_2^2}{2}}$$

(2) $S_{\bar{x}1-\bar{x}2} = S_p \sqrt{\dfrac{1}{n_1} + \dfrac{1}{n_2}}$

- *For the separate-variance t-test:*

$$S_{\bar{x}1-\bar{x}2} = \sqrt{\dfrac{S_1^2}{n_1} + \dfrac{S_2^2}{n_2}}$$

Important: When the spread of the sampling distribution is estimated using the sample data, measurement error due to inconsistent technique will increase the sample standard deviation values and so increase the spread of the sampling distribution. This will reduce the power of the test of significance and increase the width of the confidence interval (reduce precision of the estimated difference).

Shape: *t*-distribution. Degrees of freedom for this *t*-distribution differ between the pooled-variance and separate-variance *t*-tests:

- *For the pooled-variance t-test:* df $= n_1 + n_2 - 2$
- *For the separate-variance t-test:*

$$df_{adj} = \dfrac{\left[\dfrac{S_1^2}{n_1} + \dfrac{S_2^2}{n_2}\right]^2}{\dfrac{(S_1^2/n_1)^2}{(n_1 - 1)} + \dfrac{(S_2^2/n_2)^2}{(n_2 - 1)}}$$

NOTE   The adjusted degrees of freedom (df_{adj}) computed using this formula will decrease as the difference between the two sample variances increases. In general, you will use a computer program to do two-sample *t*-tests, and you will rarely be required to use this formula to hand-calculate the adjusted degrees of freedom for the separate-variance *t*-test. ■

4. **Test Statistic**

$$t_{test} = \dfrac{(\bar{x}_1 - \bar{x}_2) - (\mu_1 - \mu_2)_0}{S_{\bar{x}_1-\bar{x}_2}} = \dfrac{(\bar{x}_1 - \bar{x}_2) - 0}{S_{\bar{x}_1-\bar{x}_2}} = \dfrac{(\bar{x}_1 - \bar{x}_2)}{S_{\bar{x}_1-\bar{x}_2}}$$

The t_{test} statistic represents the difference between the observed value of $(\bar{x}_1 - \bar{x}_2)$ and the expected value $(\mu_1 - \mu_2) = 0$ under the Null hypothesis relative to the expected amount of random sampling variation, as measured by $S_{\bar{x}1-\bar{x}2}$. This test statistic has a *t*-distribution with degrees of freedom computed as described above. The *t*-table can be used to obtain an approximate *p*-value for this test statistic. Statistics computer programs provide exact *p*-values.

5. ***p*-Value**

   a. *One-tailed test:* The probability $P[t \geq +t_{test}]$ or $P[t \leq -t_{test}]$ is the probability of obtaining the observed difference between sample means $(\bar{x}_1 - \bar{x}_2)$ if the two population means were equal, $(\mu_1 - \mu_2) = 0$.

   b. *Two-tailed test:* The probability $2 \times P[t \geq |t_{test}|]$ is the *p*-value for the two-tailed test of significance. The term $|t_{test}|$ is the absolute value of t_{test} (i.e., any $-$ sign is deleted). In a two-tailed test, there was no prior expectation that the difference $(\bar{x}_1 - \bar{x}_2)$ would be positive or negative. The *p*-value is the probability of getting an absolute difference $|(\bar{x}_1 - \bar{x}_2)|$ greater than or equal to the observed value if the true difference between population means is $(\mu_1 - \mu_2) = 0$.

c. *Stating conclusions:*

If $p \leq 0.05$: The probability associated with the random-variation explanation for the observed difference $(\bar{x}_1 - \bar{x}_2)$ is sufficiently low to reject this explanation. If an appropriate study design has been correctly implemented, the only remaining likely explanation is that the two population means actually are different. Regardless of whether your test of significance was one-tailed or two-tailed, your conclusion should state how the means are different (which is larger or smaller), as indicated by the sample data.

■ **EXAMPLE** "Mean blood pressure for the treatment group was 50 points lower than for the control group ($p = 0.0012$), indicating that the new treatment was effective for reducing high blood pressure." ■

If $0.05 < p \leq 0.10$: The probability associated with the random-variation explanation is sufficiently high that we cannot confidently exclude this explanation. However, the probability is sufficiently low that we might suspect there is a difference. Some scientists would conclude that the data "suggest" there is a difference.

■ **EXAMPLE** "The 25-point difference in mean blood pressure between control and treatment groups suggests that the treatment may reduce blood pressure ($p = 0.064$), but more data are needed to verify this." ■

If $p > 0.10$: The probability associated with the random-variation explanation is so high that this explanation cannot be excluded. Hence, there is insufficient evidence to conclude that there is a difference.

■ **EXAMPLE** Mean blood pressure was not significantly different between control and treatment groups ($p > 0.25$). The new drug was not effective for reducing blood pressure. ■

6. **$100(1 - \alpha)\%$ Confidence Interval for $(\mu_1 - \mu_2)$**

● *Pooled-variance:*

$$(\bar{x}_1 - \bar{x}_2) \pm t_{\text{df},\alpha/2}\, S_p \sqrt{\frac{1}{n_1} + \frac{1}{n_2}}$$

● *Separate variance:*

$$(\bar{x}_1 - \bar{x}_2) \pm t_{\text{adj-df},\alpha/2} \sqrt{\frac{S_1^2}{n_1} + \frac{S_2^2}{n_2}}$$

Interpretation: Because $100(1 - \alpha)\%$ of confidence intervals computed by this method include the true difference between population means $(\mu_1 - \mu_2)$, we can be $100(1 - \alpha)\%$ confident that our specific confidence interval includes the true difference. The value of α is the error rate (probability the interval does not include the true difference), chosen by the investigator. Because confidence intervals are always "two-tailed," the critical t-value is associated with the probability $\alpha/2$. The confidence interval is often presented as an indication of the effect size.

Example 9.9 applies the two-sample t-test to a study of cardiovascular fitness.

■ **EXAMPLE 9.9**

The two-sample t-test.

1. **Application:** An exercise physiologist wants to determine if several short bouts of exercise provide the same benefit for cardiovascular fitness as one long bout of exercise. She obtains 40 volunteers who are all overweight female college students. The volunteers are randomly assigned to two treatment groups. Treatment Group 1 does standardized aerobic exercise on a stationary bicycle for 30 minutes, 1 time per day, 5 days per week. Group 2 does the same standardized exercise for 10 minutes, 3 times per day, 5 days per week. Cardiovascular fitness was measured by VO2 max (maximum oxygen consumption while exercising) at the beginning of the experiment and after the subjects had been doing the exercise regime for 12 weeks. The variable of interest for this study was the change in VO2 max between beginning and end of the study (ΔVO2), computed from the two measurements for each subject in each of the two treatment groups. Because mean ΔVO2 max is compared between two experimental groups formed by random assignment, a two-sample t-test is appropriate.

Exploratory data analysis

Summary Statistics for ΔVO2 for Experimental Groups						
Group	**Count**	**Mean**	**Median**	**StdDev**	**Min**	**Max**
1 (30 × 1)	20	0.2056	0.1926	0.1663	−0.1095	0.5511
2 (10 × 3)	20	0.0696	0.0390	0.2223	−0.2374	0.5033

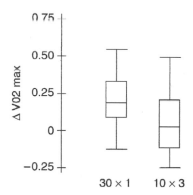

Assessment of Assumptions:

The study subjects were randomly assigned to treatment groups, and all subjects were given identical instructions, so we can assume the study is unbiased. The data distribution for both groups is symmetric, with no outliers. With sample sizes of 20 per group, we could assume the sampling distribution of

$(\bar{x}_1 - \bar{x}_2)$ is Normal, even if the data distributions were moderately skewed. Based on the following F_{max} test for equality of variances, we can assume the two population variances are equal.

$F_{max} = S^2_{larger}/S^2_{smaller} = (0.222)^2/(0.166)^2 = 1.79$
Critical F_{max} (for $\alpha = 0.05$, $k = 2$, df $= 19$) $= 2.86$
(from Appendix Table 7, df rounded down to 15)

Under these conditions, the pooled-variance two-sample t-test is most appropriate.

2. **Statement of Hypotheses**
   - H_0: $(\mu_1 - \mu_2) = 0$
   - H_a: $(\mu_1 - \mu_2) \neq 0$

3. **Sampling Distribution of $(\bar{x}_1 - \bar{x}_2)$**

   Center: $E(\bar{x}_1 - \bar{x}_2) = 0$

   $$\text{Spread: } S_p = \sqrt{\frac{(n_1 - 1)S_1^2 + (n_2 - 1)S_2^2}{n_1 + n_2 - 2}}$$

   $$= \sqrt{\frac{(20 - 1)(0.166)^2 + (20 - 1)(0.222)^2}{20 + 20 - 2}}$$

   $$= \sqrt{0.0384} = 0.196$$

   $$S_{\bar{x}_1 - \bar{x}_2} = S_p \sqrt{(1/n_1) + (1/n_2)} = 0.196 \sqrt{(1/20) + (1/20)} = 0.062$$

   Shape: t-distribution with df $= 38$

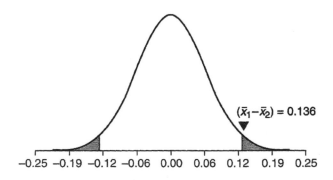

Shaded areas in the sampling distribution of $(\bar{x}_1 - \bar{x}_2)$ shown above correspond to the two-tailed p-value.

4. **Test of Significance and p-Value**

   The following print-out from a statistics computer program provides both the t_{test} statistic (here labeled t-statistic) and the p-value.

For pooled t-test of $\mu_1 - \mu_2$

$30 \times 1 - 10 \times 3$:

Test H_0: $\mu(30 \times 1) - \mu(10 \times 3) = 0$ vs. H_a: $\mu(30 \times 1) - \mu(10 \times 3) \neq 0$

Difference between Means = 0.13593435 t-statistic = 2.19 with 38 df

$p = 0.0347$

Interpretation of the t-test results: Based on the print-out from the statistics program, $t_{test} = 2.19$, indicating that the observed difference between the two sample means $(\bar{x}_1 - \bar{x}_2) = 0.136$ is 2.19 times larger than the expected amount of random variation. The probability associated with the random-variation explanation is $p = 0.0347$.

NOTE   Because the investigator couldn't predict which exercise regime would produce the greater change in cardiovascular fitness, the test was done as a two-tailed test. ■

Conclusion: Mean ΔVO2 of overweight college women after a 12-week exercise program was greater when daily exercise was done as a single long bout than when the same exercise time was divided into shorter bouts per day ($p = 0.0347$).

6. **95% Confidence Interval for $(\mu_1 - \mu_2)$: Expressing the Effect Size**

   Since the equal-variances assumption for the pooled-variance t-test was fulfilled, the confidence interval is computed using the pooled-variance estimate of the standard error of the difference $S_{\bar{x}1-\bar{x}2}$ and the degrees of freedom for the critical t-value are $n_1 + n_2 - 2 = 38$. The 95% confidence interval is calculated as follows:

   $(\bar{x}_1 \quad \bar{x}_2) \pm t_{38, 0.025} S_{\bar{x}1-\bar{x}2} - 0.136 \pm 2.042 (0.062) = 0.136 \pm 0.127$

   Because 95% of all confidence intervals computed by this method include the true difference between population means, the investigator is 95% confident that the difference in mean ΔVO2 max between the two exercise regimes is between 0.09 and 0.263 liters/min.

   NOTE   With the print-out from the two-sample t-test, we could have avoided the complex hand-calculations of $S_{\bar{x}1-\bar{x}2}$ described above. The t-test print-out gives us $(\bar{x}_1 - \bar{x}_2) = 0.136$ and $t_{test} = 2.19$. We can reverse the formula for computing t_{test} to calculate $S_{\bar{x}1-\bar{x}2}$: If

   $t_{test} = (\bar{x}_1 - \bar{x}_2)/S_{\bar{x}1-\bar{x}2}$,

   then

   $S_{\bar{x}1-\bar{x}2} = (\bar{x}_1 - \bar{x}_2)/t_{test} = 0.136/2.19 = 0.062$

Determining Power for a Two-Sample t-Test

The procedure for determining power for a two-sample t-test is the same as described earlier for one-sample Z- and t-tests. The only differences are related to specifications of the sampling distribution (calculation of the standard deviation or standard error, whether the shape is Normal or a t-distribution). Once you have specified the center, spread, and shape of the sampling distribution for the statistic of interest, the determination of power involves the same two steps, as follows:

1. Determine the critical value for $(\bar{x}_1 - \bar{x}_2)^*$ that defines the range of values that would provide sufficient evidence to reject the Null hypothesis (the "rejection region").

2. Determine the probability of obtaining a value for $(\bar{x}_1 - \bar{x}_2)$ somewhere in this rejection region if the true difference between the population means is some minimum important difference $(\mu_1 - \mu_2)_a$. Example 9.10 shows the application of this two-step procedure to the two-sample t-test presented in Example 9.9.

■ **EXAMPLE 9.10**

Computing the power of a two-sample t-test.
Let us suppose that the investigator who did the study described in Example 9.9 decided that a difference in mean ΔVO2 of 0.20 liters/min between the two exercise regimes would constitute the smallest "important" difference worthy of using to advise participants in exercise programs. She wants to hold the Type I error rate to α = 0.05 to avoid making exercise recommendations that would later be found inappropriate. We can determine the power of this study to detect this minimum important difference $(\mu_1 - \mu_2)_a$ = 0.20 following these two steps:

1. Determine the critical value of the difference $(\bar{x}_1 - \bar{x}_2)^*$ necessary to reject H_0 at the specified α level, using a sampling distribution of $(\bar{x}_1 - \bar{x}_2)$ as specified by H_0. The adjusted degrees of freedom = 38. Because the test of significance was a two-tailed test, the critical t-value is found in the t-table under the column labeled 0.025 = $\alpha/2$. Because this was a two-tailed test, power could be computed for either a positive or negative difference. Calculations are easier with positive numbers, so the power calculation is done to find a positive difference, and the critical t-value is positive.

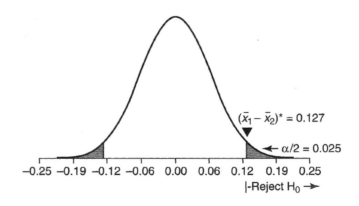

$$(\bar{x}_1 - \bar{x}_2)^* = t_{adj-df=38,\ .025} (S_{\bar{x}1-\bar{x}2})$$
$$= +2.042\ (0.062)$$
$$= 0.127$$

If $(\bar{x}_1 - \bar{x}_2) \geq 0.127$, the investigator will have sufficient evidence to conclude that ΔVO2 differs between the 30 × 1 and 10 × 3 exercise regimes, based on a two-tailed test.

2. Determine the probability of getting values of $(\bar{x}_1 - \bar{x}_2) \geq 0.127$ if $(\mu_1 - \mu_1) =$ MID $= 0.20$.

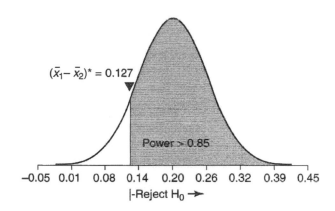

$(\bar{x}_1 - \bar{x}_2)^* = 0.127$

Power > 0.85

−0.05 0.01 0.08 0.14 0.20 0.26 0.32 0.39 0.45

|-Reject H_0 →

$$P[(\bar{x}_1 - \bar{x}_2) \geq 0.127 \text{ if } (\mu_1 - \mu_2) = 0.20] = P[(t_{df=38} \geq (0.127 - 0.20)/0.062)]$$
$$= P[t_{df=38} \geq -1.18)$$
$$= 1 - P[t_{df=38} \geq 1.18)$$

Therefore,

$$(1 - 0.15) < \text{Power} < (1 - 0.10)$$
$$0.85 < \text{Power} < 0.90$$

This would generally be considered adequate power for such an experiment. If the investigator had failed to detect a statistically significant difference, she would have been able to make the argument that the two exercise regimes really did not differ in ΔVO2. Otherwise, her study had a high probability of detecting such a difference. ■

9.3

One-Sample Tests for Medians When the Sampling Distribution of the Mean Is Not Normal

The one-sample Z- and t-tests for means are **parametric tests,** which require that the sampling distribution of the mean is Normal. If this assumption is not fulfilled, p-values from these tests provide inaccurate estimates for the probability of getting an observed difference from Null hypothesis value due simply to random sampling variation. Since the p-value provides the basis for drawing conclusions from the sample data, this is a highly undesirable situation. Because exploratory data analyses are not usually presented in the scientific literature, readers of research papers usually cannot determine that an otherwise valid-appearing analysis and conclusion actually present a false representation of the real world. Hence, you should not use these parametric tests of significance if the assumption of a Normal sampling distribution is violated.

However, the Central Limit Theorem (CLT) states the sampling distribution of the mean will be Normal if the sample size is "sufficiently large," regardless of the shape of the population distribution. When deciding whether or not it is valid to use a parametric test of significance, you should use the "rules of thumb" for applying the Central Limit Theorem (presented in Chapter 6):

$n < 15$ Use a parametric test only if the data distribution is approximately Normal (symmetric), with *no outliers*.

$15 \leq n < 40$ Use a parametric test even if the data distribution is moderately skewed, but not if it is extremely skewed or includes outliers.

$n \geq 40$ Use a parametric test in most cases. If data distribution has extreme outliers ($> 100\times$ the median), a sample size $n > 100$ may be necessary.

If the sample size is small and the data distribution is non-Normal, the test of significance must be modified. This modification can take one of two forms:

1. Apply a nonlinear data transformation (e.g., logarithmic, power, exponential, or root function) to the data values to change the shape of the data distribution to approximately Normal, then apply the appropriate parametric test to the transformed data. The application of a parametric test to transformed data will be covered in the next section and in Example 9.11.

2. Use a **nonparametric test** of significance that does *not* require that the sampling distribution is Normal. These tests of significance are sometimes called *distribution-free* tests because population parameters are not used in statements of hypotheses, and sample statistics are not used in the calculation of the test statistic. One-sample nonparametric tests also differ from parametric Z- and t-tests in that the nonparametric tests determine if the *median* (not the mean) of the data values differs from some Null hypothesis value. Hence, the wording of conclusions based on nonparametric tests must refer to the median rather than the mean.

If the assumptions for parametric tests are fulfilled, these tests often have slightly more statistical power than their nonparametric counterparts. Hence, if a transformation can make the data distribution sufficiently close to Normal that the CLT allows you to assume the sampling distribution is Normal, you should apply the parametric test to the transformed data. This will minimize both Type I and Type II errors. However, sometimes no transformation will be able to make the data distribution sufficiently close to Normal, and the nonparametric test must be used.

Applying a Parametric Test to Transformed Data Values

Transformations for a positively skewed data distribution.
To make the data distribution more symmetric, use a nonlinear function that has the effect of decreasing the size of large values more than it decreases the size of small values. Logarithmic and root functions (e.g., $\sqrt{\ }$, $\sqrt[3]{\ }$) have this effect. These functions "pull-in" the long right-hand tail of a positively ($+$) skewed distribution and make it more similar to the left tail of the distribution. (See Figure 9.3.)

Transformations for a negatively skewed data distribution.
To make the distribution more symmetric, use a nonlinear function that increases the value of large numbers more than it increases the value of small numbers. Power functions (x^2, x^3, etc.) and the exponential function (e^x) have this effect. These transformations spread out the right tail of the distribution, making it as large as the left tail. (See Figure 9.4.)

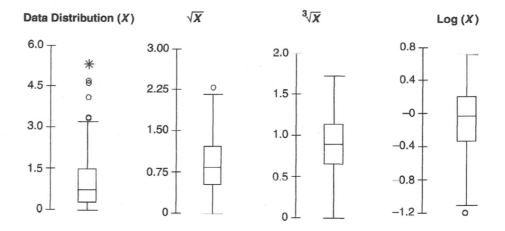

FIGURE 9.3   Effect of the log and root transformations on the distribution of positively skewed data values.   For this particular data distribution, the distribution of the cube root ($\sqrt[3]{\ }$) transformed data is closest to Normal, with no outliers.

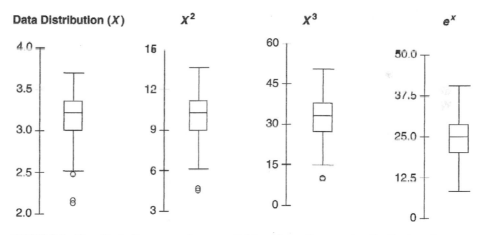

FIGURE 9.4   The effect of power and exponential transformations on the distribution of negatively skewed data values.   For the data distribution in the figure, the exponential function e^x, produced an approximately Normal distribution, whereas the square and cube functions were not quite strong enough.

Whether the distribution of data values is positively or negatively skewed, you will usually need to try a variety of progressively more powerful transformations until you find the transformation that modifies the data distribution to be approximately Normal. Use the least powerful transformation necessary to attain this goal.

Important: If the original data values include negative numbers, it is not possible to apply many nonlinear transformations (e.g., the log of a negative number is undefined). In this situation, you must add a constant to all data values that will make them all positive. As a rule of thumb, you should add the smallest constant that will convert the largest negative data value to a value greater than 1. However, when I apply the log transformation, I sometimes use a larger constant to convert the largest negative number to a value greater than 10. Once the data values have been adjusted to eliminate zeros and negative values, you can apply any nonlinear transformation.

Once you have found a data transformation that produces an approximately Normal (symmetric, "mound-shaped") data distribution, perform the parametric test appropriate for the study design using the *transformed* data values. *Important:* For one-sample tests of significance, you must apply the transformation applied to the data values to the population parameter value specified in the Null and Alternative hypotheses. If $\mu_0 = 0$ and the original data include negative values, you must add a constant to the μ_0 value and then apply the nonlinear transformation. Conclusions based on a test of significance applied to transformed data values are equally applicable to the untransformed data. Example 9.11 shows a paired t-test applied to the cholesterol study data (Example 9.6) that have been transformed using a square root function.

■ **EXAMPLE 9.11**

A paired t-test applied to transformed data.

1. **Application:**

Suppose the study of the new drug to reduce blood cholesterol described in Example 9.6 had included only $n = 10$ subjects, and that the resulting (after-before) difference values had a positively skewed distribution. The square root transformation is commonly applied in cases of positive skew, but cannot be applied to negative values. To use the square root ($\sqrt{\ }$) transformation, we would first add a constant to the D-values so that all are greater than or equal to 0 (e.g., +306), then take the square root of these values. The resulting transformed data values $\sqrt{D + 306}$ have an approximately Normal distribution, so a parametric paired t-test can be applied.

The data

D-Values	+66	−148	−43	−246	−221	−201	−179	−305	−266	−289
$\sqrt{(D + 306)}$	19.3	12.6	16.2	7.7	9.2	10.2	11.3	1	6.3	4.1

Exploratory data analysis of original D-values.

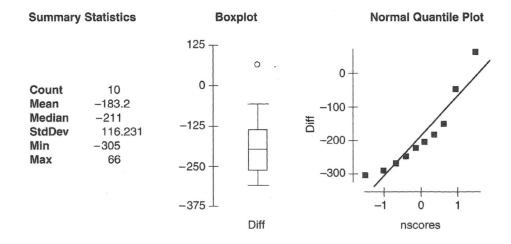

Summary Statistics		Boxplot	Normal Quantile Plot
Count	10		
Mean	−183.2		
Median	−211		
StdDev	116.231		
Min	−305		
Max	66		

Exploratory Data Analysis for Transformed $D-1$ Data Values: $\sqrt{D+306}$

Summary Statistics	**Boxplot**	**Normal Quantile Plot**

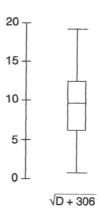
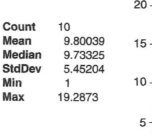
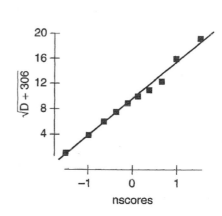

Count	10
Mean	9.80039
Median	9.73325
StdDev	5.45204
Min	1
Max	19.2873

$\sqrt{D+306}$

2. **Statement of Hypotheses**

   For the original D-values:

   - H_0: $\mu_D = 0$

   - H_a: $\mu_D < 0$ Drug is expected to *reduce* cholesterol.

   For the transformed D-values:

   - H_0: $\mu_D = \sqrt{0+306}$ Same transformation applied to the D-values
     $= 17.5$ was also applied to the Null hypothesis value.

   - H_a: $\mu_D < 17.5$

3. **Sampling Distribution of Transformed $\overline{D}_{Tr}$ assuming H_0 is true**

Center: $E(\overline{D}_{Tr}) = 17.5$

Spread: $S_{\overline{D}_{Tr}} = S_{D_{Tr}}/\sqrt{n}$
$= 5.45/\sqrt{10}$
$= 1.72$

Shape: t-distribution, with df $= 9$ $\overline{D}_{Tr} = 9.8$

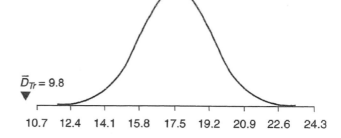

10.7 12.4 14.1 15.8 17.5 19.2 20.9 22.6 24.3

NOTE The shaded area under the curve that corresponds to the p-value is too small to see. ■

4. **Test Statistic**

$$t_{test} = \frac{\overline{D}_{Tr} - 0}{S_{\overline{D}_{Tr}}} = \frac{9.8 - 17.5}{1.72} = -4.48$$

5. **p-Value**

$P[\overline{D}_{Tr} \leq 9.8 \text{ if } \mu_{D_{Tr}} = 17.5] = P[t_{df=9} \leq -4.48]$

$p < 0.0005$

Conclusion: This was a one-tailed test because the investigator was interested only in whether or not the new drug *reduces* blood cholesterol levels. Therefore, the significance level (*p*-value) is reported as $p < 0.001$. The investigator concludes that the new drug reduced mean blood cholesterol ($p < 0.001$). ■

The Sign Test: A Nonparametric Test for Matched-Pairs Data

1. **Application:** The **sign test** is used when data were produced by a before-after or matched-pairs study design and the variable of interest is the difference D between each pair of values. The sign test is used when the data distribution of the D-values deviates from Normal, the sample size is too small for the Central Limit Theorem to apply, and nonlinear transformations cannot produce a data distribution sufficiently close to Normal. In a before-after experimental design, if the treatment has no effect, both the true mean difference between paired measurements (μ_D) and the true *median* difference (M_D) are expected to be zero. If $M_D = 0$, half the D-values will be negative and half will be positive. Hence, another approach for testing the Null hypothesis is to determine the proportion of D-values that are positive and negative. The more the proportion of D-values that are positive (or negative) deviates from 0.5, the stronger the evidence that the treatment had an effect. *Assumptions:* The difference values (D) represent independent observations obtained by an unbiased, randomized study design.

2. **Statement of Hypotheses:**
   - One-tailed test:

     $H_0: M_D = 0$

     $H_a: M_D <, \text{ or } > 0$
   - Two-tailed test:

     $H_0: M_D = 0$

     $H_a: M_D \neq 0$

   However, the test statistic is based on the proportion of D values that fall below (or above) zero. If $M_D = 0$, then $E(\hat{p})$ of D-values below (or above) 0 is $P = 0.5$.

3. **Sampling Distribution of $\hat{p}$**, assuming H_0 is true, $M_D = 0$, and $P = 0.5$

   Center: $P_0 = 0.5$

   Spread: $\sigma_{\hat{p}} = \sqrt{0.5(1 - 0.5)/n}$

   Shape: If $n \geq 20$, approximately Normal

   If $n < 20$, the sampling distribution of $\hat{p}$ is Binomial (n, $P = 0.5$)

4. **Test Statistic**

To calculate the test statistic, drop all D-values that have zero difference. Of the remaining nonzero D-values, determine the proportion $\hat{p}$ of D-values that support the Alternative hypothesis. In the cholesterol drug example, this would be the proportion of all subjects with D-value that showed a decrease in blood cholesterol (negative D = After-Before). If the statement of hypotheses is two-tailed, compute the proportion of either positive or negative differences.

$$Z_{\text{test}} = \frac{\hat{p} - P_0}{\sqrt{P_0(1 - P_0)/n}} = \frac{\hat{p} - 0.5}{\sqrt{0.5(1 - 0.5)/n}}$$

NOTE   Although this is a test of the hypothesis that the population median is zero, neither the Null hypothesis value nor the sample median is used in calculations for the test of significance. This is typical of 'distribution-free,' nonparametric tests. ■

5. *p*-**Value**

a. *One-tailed test:* The probability $P[Z \geq +Z_{\text{test}}]$ or $P[Z \leq -Z_{\text{test}}]$, obtained from the standard Normal table, is the p-value for the one-tailed test of significance. This is the probability of obtaining values for the *median difference* (m_D) greater than (or less than) the observed sample median if the true median difference M_D is zero.

b. *Two-tailed test:* The probability $2 \times P[Z \geq |Z_{\text{test}}|]$ is the p-value for the two-tailed test of significance. The term $|Z_{\text{test}}|$ is the absolute value of Z_{test} (i.e., any − sign is deleted). In a two-tailed test, there is no prior expectation that the median difference would be positive or negative. The p-value is the probability of getting an absolute median difference $|m_D|$ greater than or equal to the observed value if the true median difference M_D is equal to zero.

c. *Stating conclusions:*

If $p < 0.05$: The probability associated with the random-variation explanation for the observed *median difference* m_D is sufficiently low to reject this explanation. If an appropriate study design has been correctly implemented, the only remaining likely explanation is that the true median difference is greater than (or less than) zero. Regardless of whether your test of significance was one-tailed or two-tailed, your conclusion should state how the median difference differs from zero, as indicated by the sample data.

■ **EXAMPLE**   "The median change in blood pressure (−45 points) indicated that the new treatment was effective for reducing high blood pressure $(p = 0.0012)$." ■

If $0.05 < p \leq 0.10$: The probability associated with the random-variation explanation is sufficiently high that we cannot confidently exclude this explanation. However, the probability is sufficiently low that we might suspect there is a difference. Some scientists would conclude that the data "suggest" there is a difference.

■ **EXAMPLE**   "The 20-point median difference between paired-subjects, one given the treatment and the other given a placebo, suggests that the treatment may reduce blood pressure $(p = 0.087)$, but more data are needed to verify this." ■

If $p > 0.10$: The probability associated with the random-variation explanation is so high that this explanation cannot be excluded.

■ **EXAMPLE**   "The median difference between pre- and post-treatment measurements was not significant ($p = 0.39$). The treatment was not effective for reducing blood pressure."   ■

Example 9.12 illustrates the sign test and compares results from this test of results obtained from a paired t-test applied to the same data..

■ **EXAMPLE 9.12**

The sign test.

1. **Application:** This example will both demonstrate the sign test and illustrate the difference in power between the sign test and the paired t-test. The data for change in blood cholesterol after treatment with a new drug in Example 9.11 were positively skewed, with one outlier. The sample size was too small for the Central Limit Theorem to apply. In Example 9.11 the paired t-test was applied to transformed difference values. In this example the original data are analyzed using the sign test to evaluate the evidence against the Null hypothesis that the true median difference was equal to zero.

2. **Statement of Hypotheses**

   - H_0: $M_D = 0$
   - H_a: $M_D < 0$

   NOTE   Since the investigator was only interested in whether the new drug *reduced* blood cholesterol levels, the test of significance was implemented as a one-tailed test. Because the difference values were computed for each subject as (Diff $=$ Post $-$ Pre), negative difference values would provide evidence that the drug was effective. ■

3. **Sampling Distribution**

   In this case, we want to know the probability of getting 9 negative D-values out of 10 if the true median difference is $M_D = 0$. Because the sample size was less than 20, the appropriate sampling distribution to determine the p-value for the Null hypothesis is a Binomial distribution with $n = 10$ and $P = 0.5$. However, when you are using a computer statistics program, this is all done "behind the scenes." All that is required of the investigator is to choose the appropriate test and then interpret the p-value.

4. **Test of Significance**

   The results shown below are from a print-out produced by a statistics computer program. This print-out includes both the Z-test statistic and the p-value.

   Paired Sign Test

   Post $-$ Pre
   Test H_0: Median(Post $-$ Pre) $= 0$
   H_a: Median(Post $-$ Pre) < 0
   Total Observations: 10
   Observations Post $>$ Pre: 1
   Tied Values between Samples: 0
   $p = 0.0107$

Conclusion: Based on the print-out, we can conclude that the median difference is less than zero ($p = 0.0107$). Thus, the drug was effective for lowering blood cholesterol.

Comparison of sign test and t-test on transformed data: The *p*-value for the sign test ($p = 0.0107$) was much larger than the *p*-value from the matched-pairs *t*-test performed on transformed data values (Example 9.11, $p < 0.0005$). Although the results of both tests provide strong evidence ($p \leq 0.05$) that the drug reduced blood cholesterol, the *t*-test applied to transformed data provided stronger evidence. This is an example of the general difference in the power of the sign test versus the matched-pairs *t*-test. To minimize both Type I and Type II error rates, you should use the t-test instead of the sign test whenever the Normality assumption is valid. ▚

9.4
Two-Sample Tests for Medians When the Sampling Distributions of the Means Are Not Normal

To obtain accurate *p*-values from **two-sample *t*-tests**, the sampling distribution of the difference $(\bar{x}_1 - \bar{x}_2)$ must be Normal. The difference between two Normally distributed variables $(\bar{x}_1 - \bar{x}_2)$ is also Normally distributed. Hence, if the sampling distributions of $\bar{x}_1$ and $\bar{x}_2$ are Normal, so too will be the sampling distribution of the difference $(\bar{x}_1 - \bar{x}_2)$. If boxplots and Normal quantile plots of the data values for each of the two samples indicate either one of the distributions is asymmetric or has outliers, the Normality assumption may not be valid. When the data distributions are not Normal, the Central Limit Theorem states that the sampling distributions of the two means $\bar{x}_1$ and $\bar{x}_2$ will be approximately Normal if the sample sizes are sufficiently large, as described by the "rules of thumb" in Chapter 6. However, if the sample sizes are small and the data distribution for one or both samples is skewed or includes outliers, you cannot assume the sampling distribution of the difference $(\bar{x}_1 - \bar{x}_2)$ is Normal and you should *not* use the two-sample *t*-test.

There are two alternatives for comparing two sample means when the Normality assumption is violated:

1. Apply the *same* nonlinear transformation to the data values in each of the two samples so that the transformed data distributions are *both* approximately Normal; then apply the two-sample *t*-test to the transformed data values. (See Example 9.13.)

   NOTE It is not uncommon that a transformation that makes the two data distributions approximately Normal will also make the two sample variances more similar. If the original data values also violated the equal-variances assumption of the pooled-variance *t*-test, the transformed data values may meet this assumption. This would allow you to use the more powerful pooled-variance test instead of the separate-variance *t*-test. ■

2. Use the Mann-Whitney U-test, a nonparametric test for the difference between two sample *medians*, which does not require that the sampling distribution is Normal. The Mann-Whitney U-test is discussed in more detail later in this chapter.

Nonparametric tests often have less statistical power to detect a difference between two populations than their parametric counterparts. Hence, if you can find a transformation that makes the data distributions for both samples approximately Normal, you should apply the parametric test to the transformed data. This will allow you to minimize the Type II error rate (failure to find a real difference), while obtaining a valid estimate of the Type I error rate (the p-value). However, sometimes there is no one transformation that can make the data distributions for both samples approximately Normal. In these cases, the Mann-Whitney U-test is an appropriate test of significance.

∎ **EXAMPLE 9.13**

The two-sample t-test for log-transformed data.

1. **Application:** Highly active anti-retroviral therapy (HAART), based on daily doses of a multidrug "cocktail," has proven an effective treatment for reducing AIDS mortality. This therapy reduces active viral RNA in the bloodstream to near zero level, but does not eradicate HIV. Hence, lifelong therapy is required. Unfortunately, long-term HAART has a number of negative side effects and is very costly. One proposal for reducing both side effects and cost is to use "intermittent" drug therapy (IDT) that cycles patients through alternating time periods on and off the daily drug cocktail. One study compared two groups of $n = 10$ HIV patients each, one that took the drugs continuously (HAART), the other that took the drugs intermittently, alternating 7 days on then 7 days off the drug regime. At the beginning of the study all patients had been practicing HAART, and had near zero blood HIV RNA counts. After 52 weeks of drug therapy, blood viral RNA counts were obtained for all patients and mean plasma RNA counts (copies/ml) compared between the two groups. The data are shown below, along with exploratory data analysis.

Summary Statistics

Group	Sample Size	Mean	Median	StdDev	Min	Max
HAART	10	121.5	94.04	96.71	23.17	330.2
IDT	10	162.0	112.9	150.7	28.17	499.5

Assessment of Assumptions for t-test (for original data):

a. *Randomized, unbiased study design:* Study subjects were randomly assigned to treatment groups. All subjects received similar counseling and were required to keep diaries to document compliance with the drug regime. Assumption is valid.

b. *Sampling distribution of $(\bar{x}_I - \bar{x}_H)$ is Normal:* Data distributions for both HAART (H) and IDT (I) groups are positively skewed with large outliers. With sample sizes of only $n = 10$ in each group, this assumption is *not* valid.

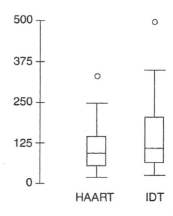

c. *For pooled-variance t-test: population variances are equal:*

Computed $F_{max} = 150.7^2/96.7^2 = 2.43$

Critical F_{max} (with $k = 2$ and df $= 9$) $= 4.03$

Population variances can be assumed equal.

Conclusion: Because the Normality assumption is violated, it would be invalid to use the parametric two-sample t-test to compare mean plasma viral loads between these two samples. Because both data distributions have similar shape, it is likely that the same data transformation could make both distributions approximately Normal.

Exploratory data analysis for log-transformed data

Summary Statistics

Variable	Count	Mean	Median	StdDev	Min	Max
HAART	10	1.967	1.972	0.3439	1.365	2.519
IDT	10	2.058	2.053	0.3810	1.450	2.699

Assessment of assumptions for t-test (for transformed data):

a. *Randomized, unbiased study design.*

b. *Sampling distribution of $(\bar{x}_I - \bar{x}_H)$ is Normal:* Transformed data distributions for both HAART and IDT groups are approximately Normal with no outliers. This assumption is valid.

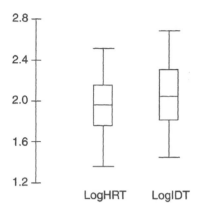

c. *For pooled t-test: population variances are equal:*

Computed $F_{max} = 0.381^2/0.344^2 = 1.23$

Critical F_{max} (with $k = 2$ and df $= 9$) $= 4.03$

Population variances can be assumed equal.

For both the original and log-transformed data values, the F_{max} test indicates that the sample variances are sufficiently similar to assume the population variances are equal. The student decides to use the pooled variance t-test, applied to the log-transformed data.

NOTE   The F_{max} value for the log-transformed data is smaller than that for the original data. This indicates that the sample variances of the transformed data are more similar than those of the original data. ■

2. **Statement of Hypotheses**
   - H_0: $(\log \mu_I - \log \mu_H) = 0$
   - H_a: $(\log \mu_I - \log \mu_H) \neq 0$

When one-sample tests are applied to transformed data, the Null hypothesis value for the population parameter was transformed exactly the same way as the data values. However, in the case of two-sample tests, the same transformation is applied to both samples. Hence, if the difference between the population means on the original data scale was equal to zero, the difference between the means on the transformed data scale would also be equal to zero.

3. **Sampling Distribution for $(\log \bar{x}_I - \log \bar{x}_H)$**

   Center: $E(\log \bar{x}_I - \log \bar{x}_H) = 0$

   Spread: $S_{\bar{x}_I - \bar{x}_H}$

   Pooled Estimate of the Standard Deviation S_p (with equal sample sizes)

   $$S_p = \sqrt{\frac{S_I^2 + S_H^2}{2}} = \sqrt{\frac{0.381^2 + 0.344^2}{2}} = \sqrt{0.132} = 0.363$$

   Standard Error of the Difference $(\log \bar{x}_I - \log \bar{x}_H)$

   $$S_{\bar{x}_I - \bar{x}_H} = S_p\sqrt{(1/n_I) + (1/n_H)} = 0.363\sqrt{(1/10) + (1/10)} = 0.162$$

   Shape: t-distribution with df $= 18$

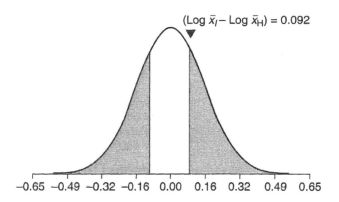

The shaded area under the curve corresponds to the two-tailed *p*-value.

4. **Test of Significance**

   The print-out from a statistics computer program that follows includes both the *t*-test statistic and the *p*-value.

   Pooled *t*-test of $\mu_1 - \mu_2$

   LogIDT − LogHRT:
   Test H_0: $\mu(\text{LogIDT}) - \mu(\text{LogHRT}) = 0$
   H_a: $\mu(\text{LogIDT}) - \mu(\text{LogHRT}) \neq 0$
   Difference between Means = 0.091509287   *t*-statistic = 0.5638 with 18 df
   p = .5798

   Conclusion: There is not a significant difference in plasma HIV RNA load between patients in the HAART and IDT groups (*p* = 0.58). Intermittent drug therapy is just as effective as continuous drug therapy for suppressing HIV.

   NOTE If the researcher had not bothered with exploratory data analysis and had simply used a two-sample *t*-test to analyze the original data values, the resulting *p*-value would have been *p* = 0.4835. This difference between the *p*-values from the valid test on the transformed data and the invalid test on the original data does not change the nature of the conclusion. However, in cases where the effect size is "borderline" (*p*-value close to 0.05), the difference between *p*-values from a valid test and an invalid test can lead to entirely different conclusions. Hence, it is essential that you always explore the data to verify that the assumptions of a test of significance are fulfilled prior to implementing the test. ■

Mann-Whitney *U*-Test

1. **Application:** The **Mann-Whitney *U*-test** is a nonparametric test of significance that examines whether the data values in one sample tend to be larger than the data values in a second sample. If the data distributions for the two samples (displayed by boxplots) have a similar shape, this test evaluates the evidence for the

Null hypothesis that the two populations have the same *median*. This test is typically used in place of a two-sample *t*-test when the Normality assumption is violated and transformations are not effective. *Assumption:* (1) The two samples were obtained by a randomized, unbiased study design, and (2) the data values in each sample are independent (i.e., *not* "matched-pairs" measurements).

NOTE   The Mann-Whitney *U*-test produces the same results as the *Wilcoxon two-sample U-test*. These different names refer to two different calculation procedures to obtain exactly the same result. Statistical computer programs may use one name or the other for this test. ■

2. **Statement of Hypotheses**

If the sample data values have distributions of similar, but non-Normal shape, the Null hypothesis is that the two populations have the same median:

- One-tailed tests:

   $H_0: M_1 = M_2$

   $H_a: M_1 < \text{ or } > M_2$

- Two-tailed tests:

   $H_0: M_1 = M_2$

   $H_a: M_1 \neq M_2$

If the sample data values have distributions of dissimilar shape, the Null hypothesis is that a randomly chosen value from one population is equally likely to be greater than or less than a randomly chosen value from the other population. That is, data values from the two populations are very similar.

3. **The U_{test} Statistic**

The Mann-Whitney *U*-statistic is a measure of how interspersed are the data values in two independent samples. If the data values in the samples are similar (high interspersion), U_{test} will be smaller. If one sample has predominantly larger data values, and the other predominantly smaller values (low interspersion), U_{test} will be larger. The U_{test} statistic is based on the "ranks" of the data values rather than the actual values, and is computed as follows:

a. The first step in computing the U_{test} statistic is to list the data values by rank order. First, all data values for both groups are combined. Then the smallest value for the combined data is assigned the rank 1, the second smallest data value is assigned the rank 2, and so on. The largest data value would have a rank $= n_1 + n_2$.

b. When there are two or more observations with the same data value, this situation is called a **tie**. The tied data values are assigned the average of sequential rank values that would otherwise be assigned to them (See Example 9.12).

c. The rank values for each of the two groups are summed to produce what are called **rank sums** (symbol ΣR). If one group tends to have larger data values (and so larger ranks), the two rank sums will be very different. If the data values for the two groups are highly interspersed, they will have similar rank sums.

d. U_{test} is the *larger* of the following two values computed for the two samples:

$$U_1 = n_1 n_2 + n_1(n_1 + 1)/2 - \Sigma R_1$$
$$U_2 = n_1 n_2 - U_1$$

e. This test statistic can be computed by hand calculation (See Example 9.14.), but you will usually use statistics computer programs to perform this test of significance. Nonetheless, the calculations below provide insight into how this test of significance works.

■ EXAMPLE 9.14

Calculation of Mann-Whitney U_{test}

Group 1 Data Values	11	15	22	29	36	
Rank Values	1	2.5	5	7	8	$\Sigma R = 23.5$ Rank total

Group 2 Data Values	15	19	25	38	40	
Rank Values	2.5	4	6	9	10	$\Sigma R = 31.5$ Rank total

$$U_1 = (5)(5) + (5)(5 + 1)/2 - 23.5 = 16.5$$
$$U_2 = (5)(5) - 16.5 = 8.5$$
$$U_{test} = 16.5$$

NOTE The second and third smallest data values are tied, with a value of 15. The ranks assigned to these tied data values are the average of the sequential ranks 2 and 3. ■

4. Sampling Distribution of U_{test}

Center: If the Null hypothesis is true, the rank sums for the two groups and the values U_1 and U_2 should be approximately equal. The Null hypothesis value for U_{test} is computed as:

$$U_{test} = n_1 n_2/2$$

In Example 9.14 above, this Null hypothesis value for U_{test} would be $(5)(5)/2 = 12.5$.

NOTE The Null hypothesis value for U_{test} is also equal to the average of U_1 and U_2, as computed above. If H_0 is false, the data value ranks will not be well interspersed, so that one group will have predominantly small ranks and the other larger ranks, and the larger of U_1 or U_2 will be much larger than the value $n_1 n_2/2$. ■

Spread: The standard deviation of the U_{test} statistic is computed as:

$$\sigma_{U\text{-test}} = \sqrt{n_1 n_2 (n_1 + n_2 + 1)/12}$$

For the two samples of five data values each in Example 9.14:

$$\sigma_{U\text{-test}} = \sqrt{(5)(5)(5 + 5 + 1)/12} = 4.79$$

Shape: When the smaller of the two groups has a sample size greater than 20, the U_{test} statistic has an approximately Normal distribution. When the smaller group has a sample size less than or equal to 20, U_{test} has a discrete probability distribution. However, you will typically use a statistics computer program to obtain the p-value.

5. *p*-**Value**

   a. *One-tailed test:* The *p*-value is the probability getting values of U_{test} greater than the observed value if the two populations have the same median, based on the area in one-tail of the Standard Normal distribution. You can determine the direction of the difference between sample medians from the rank totals (ΣR) or the mean ranks ($\Sigma R/n$) for the two groups that are provided in the computer print-out of the Mann-Whitney *U*-test.

   b. *Two-tailed test:* The *p*-value is the probability getting values of U_{test} greater than the observed value if the two populations have the same median, based on the area in both tails of the standard Normal distribution.

   c. *Stating Conclusions:*

   If $p \leq 0.05$: The probability associated with the random-variation explanation for the observed difference between the two sample medians is sufficiently low to reject this explanation. If an appropriate study design has been correctly implemented, the only remaining likely explanation is that the two populations do not have the same median. Regardless of whether your test was one-tailed or two-tailed, your conclusion should state which population has the larger median, as indicated by the data.

   ■ **EXAMPLE** "The median blood pressure for the treatment group was less than that for the control group ($p = 0.007$). Therefore, the new drug is effective for lowering blood pressure." ■

   If $0.05 < p \leq 0.1$: The probability associated with the random-variation explanation is sufficiently high that we cannot confidently exclude this explanation. However, the probability is sufficiently low that we have reason to suspect that there really is a difference. Some scientists would conclude that the data "suggest" there is a difference.

   ■ **EXAMPLE** "The difference in median blood pressure between treatment and control groups suggests the drug may be effective for reducing blood pressure ($p = 0.085$), but more data are needed to verify this." ■

   If $p > 0.10$: The probability associated with the random-variation explanation is so high that this explanation cannot be excluded.

   ■ **EXAMPLE** "Median blood pressure for the treatment group was not significantly different from that for the control group ($p = 0.37$)." ■

Examples 9.15 and 9.16 show the computation of the Mann-Whitney *U*-test and its application.

■ EXAMPLE 9.15

Example of computer print-out for the Mann-Whitney U-test. Italicized text was added to help you interpret the print-out. The data values listed in Example 9.14 were used in this example.

Group1 − Group2
- H_0: Median1 = Median2
- H_a: Median1 ≠ Median2

Ties Included *Tied ranks can pose problems for this test of significance. Most statistics programs perform the necessary adjustments to obtain a valid p-value when ties are included.*

	Rank Totals	Cases	Mean Rank (Rank total/Sample size)
Group 1	23.5	5	4.7
Group 2	31.5	5	6.3
Total	60	10	6
Ties between Groups	5	2	2.5

U-statistic:	16.5	*The rank totals and mean rank for Group 2 are greater than for Group 1, indicating the data values in Group 2 tend to be larger.*
U-prime:	8.5	
Sets of ties between included observations:	1	*When there are tied ranks, the spread of the sampling distribution must be adjusted.*
Variance:	22.917	
Adjustment to Variance for Ties:	−0.139	
Expected value:	12.5	*If the Null hypothesis were true, we would expect the U-values for both the two groups, and U_{test} to be equal to 12.5.*
Z-statistic:	0.838	*The observed value for U_{test} is 0.838 standard deviation units from the expected value under the Null hypothesis.*
$p = 0.4020$		*This is the two-tailed probability of obtaining the observed deviation from the Null hypothesis expected value due only to random variation.*

■ **EXAMPLE 9.16**

The Mann-Whitney U-test: When the researcher in Example 9.13 discovered that the Normality assumption for a two-sample t-test was violated, he could have chosen to use a nonparametric test of significance rather than transforming his data. Given the study question and experimental design, the Mann-Whitney U-test is the appropriate nonparametric test. The computer print-out for this test is provided below.

Mann-Whitney U-test

HAART − IDT:

H_0: Median(HAART) = Median(IDT)
H_a: Median(HAART) ≠ Median(IDT)

	Rank Totals	**Cases**	**Mean Rank**
HAART	97	10	9.7
IDT	113	10	11.3
Total	210	20	10.5
Ties between Groups	0	0	

U-statistic: 42

U-prime: 58

$p = 0.5789$

Because the investigator had no basis for predicting the nature of the difference in viral load between the IDT and HAART treatments, he specified a two-tailed test.

Conclusion: The results of this nonparametric test indicated that median plasma viral load was not different between the IDT and HAART groups. Hence, the intermittent drug therapy was equally effective as the continuous therapy ($p = 0.5789$). Given the reduced toxicity and cost associated with the intermittent therapy, this result has important implications for the long-term treatment of HIV. However, given the small sample size of this study, additional trials would be required before a recommendation to change from HAART to IDT would be issued to the public.

NOTE The two-sample t-test on the log-transformed data (Example 9.13) and Mann-Whitney U-test in this example are both valid in that no assumptions are violated. In this particular case, the p-values from these two tests were quite similar. However, the p-values will differ in some cases. Hence, it is important that you choose the best test under the specific circumstances of each study. Often, the t-test applied to transformed data will provide more power to detect a difference than the U-test. Hence, more powerful parametric tests should be used when their assumptions are fulfilled. ■

CHAPTER SUMMARY

Tests of significance for one or two sample means and medians were presented in this chapter.

1. The **one-sample Z-test (for means)** is used to evaluate the significance of a difference between a single sample mean and a Null hypothesis value for the true population mean $(\bar{x} - \mu_0)$. This test requires that the population standard deviation σ be known or estimated by a sample standard deviation S computed from a large sample $(n \geq 100)$.

2. The **one-sample t-test** is used to evaluate the significance of a difference between a single sample mean and a Null hypothesis value for the true population mean $(\bar{x} - \mu_0)$. This t-test does *not* require knowledge of the population standard deviation σ or very large sample sizes $(n > 100)$. Rather, the spread of the sampling distribution of the mean is estimated by the **standard error of the mean**, computed using the sample standard deviation $(= S/\sqrt{n})$.

3. The **paired t-test** is used to evaluate the significance of the mean difference between paired data values from before-after or matched-pairs experimental designs to test the Null hypothesis that the true mean difference is $\mu_D = 0$.

4. The **two-sample Z-test (for means)** is used to evaluate the significance of a difference between two sample means $(\bar{x}_1 - \bar{x}_2)$ as evidence to test the Null hypothesis that the corresponding means of two populations are equal $(H_0: \mu_1 - \mu_2 = 0)$. This test is used when the population standard deviations σ_1 and σ_2 are known *or* estimated with both sample sizes n_1 and $n_2 \geq 100$.

5. The **two-sample t-test** is used to evaluate the significance of the difference between two sample means $(\bar{x}_1 - \bar{x}_2)$ as evidence to test the Null hypothesis that the corresponding means of two populations are equal $(H_0: \mu_1 - \mu_2 = 0)$. This test is used instead of the two-sample Z-test for means when the population standard deviations σ_1 and σ_2 are *unknown* and estimated by sample standard deviations S_1 and S_2, with sample sizes n_1 and $n_2 < 100$.

6. Z-tests and t-tests for means both require that the sampling distribution of the mean is Normal. This requirement is fulfilled if either one of the following conditions are met: (a) The distribution of sample data values is approximately Normal or (b) the sample size is sufficiently large that the Central Limit Theorem applies.

7. If the Normality assumption for Z-tests and t-tests is violated, there are two alternative ways to analyze the data: (a) transform the data values so that the distribution becomes approximately Normal, then apply the Z- or t-test to the transformed data *or* (b) use a nonparametric test that does not require the sampling distribution is Normal.

8. The **sign test** is a nonparametric test used when the distribution of difference values from a before–after or matched-pairs study design deviates from Normal, the sample size is too small for the Central Limit Theorem to apply, and nonlinear transformations cannot produce a data distribution sufficiently close to Normal. This tests the Null hypothesis that the *median* difference is zero.

9. The **Mann-Whitney U-test** is a nonparametric test used in place of a two-sample t-test for means when the Normality assumption is violated and transformations are not effective. This test of significance evaluates if the data values from one population tend to be larger than the data values from the other. If the data distributions for the two samples have similar shape, this test evaluates the evidence for the Null hypothesis that the two populations have the same *median*.

10. A **100(1 − α)% confidence interval for a single population mean μ** (where α is the probability the confidence interval will *not* include the true value of the mean) is calculated based on the type of distribution.
    Based on the Normal Distribution (σ Known)

    $$\bar{x} \pm Z_{\alpha/2} \frac{\sigma}{\sqrt{n}}$$

    Based on the t-Distribution (σ estimated by S)

    $$\bar{x} \pm t_{\alpha/2, df=n-1} \frac{S}{\sqrt{n}}$$

    Interpretation: Because 100(1 − α)% of all confidence intervals computed by this method include the true value of the population mean, we can be 100(1 − α)% confident that any specific confidence interval includes the true value μ.

11. A **100(1 − α)% confidence interval for the difference between two population means** (where α is the probability that the confidence interval will *not* include the true value of the difference

between the two population means) is calculated based on the type of distribution.

Based on the Normal Distribution (σ_1 and σ_2 known)

$$(\bar{x}_1 - \bar{x}_2) \pm Z_{\alpha/2} \sqrt{(\sigma^2{}_1/n_1) + (\sigma^2{}_2/n_2)}$$

Based on the t-Distribution (σ_1^2 and σ_2^2 estimated by S_1^2 and S_2^2

$$(\bar{x}_1 - \bar{x}_2) \pm t_{adj-df,\ \alpha/2} \sqrt{(S_1^2/n_1) + (S_2^2/n_2)}$$

or

$$(\bar{x}_1 - \bar{x}_2) \pm t_{df,\ \alpha/2}\ S_p \sqrt{(1/n_1) + (1/n_2)}$$

Interpretation: Because $100(1 - \alpha)\%$ of all confidence intervals computed by this method include the true difference between the two population means, we can be $100(1 - \alpha)\%$ confident that any specific confidence interval includes $(\mu_1 - \mu_2)$.

10 Tests for Comparing Two Sample Variances

INTRODUCTION The most common comparisons among populations address questions about the *center* of the population distribution (mean, median), but there are circumstances in which we want to compare the *spread* of distributions. For example, are scores on standardized tests more variable among students at urban schools with higher ethnic, cultural, and racial diversity than among students at suburban schools with a more homogeneous student body? Some statistical tests have an equal-variances assumption that requires the investigator to compare the spread of distributions (e.g., the F_{max} test described in Chapter 9). Finally, some statistical tests used to compare more than two group means are based on a test statistic that compares a "within-group variance" to "between-group variance" (Chapter 11). In all these contexts, the investigator needs to assess the evidence for or against a Null hypothesis that two distributions have *equal spread*.

As I discussed in Chapter 2, there are several measures of spread for distributions that might be used to compare distribution spread, including the standard deviation, variance, and interquartile range. Although standard deviation and interquartile range are the more commonly reported measures of spread, the most commonly used tests of significance for comparing spreads are based on sample variances. You may wish to review Chapter 2, where the characteristics of these measures of spread are described.

The calculations and probability distributions for tests of significance to compare variances will be different from those described for comparing means, but the same fundamental concepts and general approach will apply:

1. Describe the sampling distribution for a test statistic under the assumption that a Null hypothesis of "no difference" is true.
2. Determine the probability of obtaining the observed value of a test statistic computed from sample data if the Null hypothesis is true.
3. Based on this *p*-value, draw an inference regarding whether or not the data support the Null hypothesis that there is no difference in the spread of two population distributions. ■

GOALS AND OBJECTIVES

In this chapter you will learn how and when to use tests of significance that evaluate sample data to determine if two population distributions have different spread. Given the study design and characteristics of the data, you will be able to select and perform the appropriate test for the Null hypothesis that two population variance values are equal.

In particular, by the end of the chapter you will be able to:

1. Evaluate sample data to determine whether or not the assumptions of the *F*-test are fulfilled.
2. Compute the F_{test} statistic and determine its degrees of freedom.
3. Use the *F*-distribution table to assign a *p*-value to F_{test}.
4. Draw appropriate conclusions from the *F*-test.
5. Implement the Levene test to compare the spread of two data distributions when the data indicate that the population distributions are not Normal.

10.1
The *F*-Test

1. **Application:** The *F*-test evaluates the ratio of two sample variances as evidence to test the Null hypothesis that two population variances are equal. *Assumptions:* (1) The data used to compute the two sample variances were obtained by a randomized, unbiased study design, so that the expected value of each sample variance $E(S^2)$ is the true population variance σ^2 and (2) both populations have a Normal distribution. *Important:* The *F*-test is very sensitive to violations of the Normality assumption. The Central Limit Theorem does *not* apply to sampling distributions of variances. You should use Normal quantile plots to assess whether or not the data indicate this assumption is violated. The Normality assumption of the *F*-test can be violated even when the data distribution is symmetric (e.g., a "flat" or "peaky" distribution, as described in Figure 2.4). Normal quantile plots provide a much better basis for assessing if the data distribution is "peaky" or "flat" than boxplots. Some statistics computer programs offer tests of significance for the Normality assumption. If the data indicate the Normality assumption is violated, the *F*-test is invalid. That is, the resulting *p*-value will be inaccurate, as will any conclusions based on that *p*-value.

2. **Statement of Hypotheses**

   - H_0: $\sigma^2_1 = \sigma^2_2$
   - H_a: $\sigma^2_1 \neq \sigma^2_2$

3. **The F_{test} Statistic**

   Comprison of means were based on the difference between the two sample means, whereas comparisons of variances are based on the ratio of the sample variances:

   $$F_{test} = S^2_1/S^2_2$$

   By convention, the larger of the two sample variance values is always placed in the numerator and the smaller variance value in the denominator. If the Null hypothesis is true, the expected value of this ratio is 1.0. However, random sampling variation will generally cause the ratio to be greater than 1. The larger the value of F_{test}, the stronger the evidence that the two population variances are unequal.

4. **Sampling Distribution of F_{test}**

   The probability distribution for the F_{test} statistic is an *F*-distribution. Like the *t*-distribution, the shape of the *F*-distribution depends of the number of degrees of freedom in the data used to compute the sample variances. However, the F_{test} statistic is computed from two variances, one in the numerator and one in the denominator. Hence, the *F*-distribution is defined by *two* degrees of freedom (one for the numerator variance and one for the denominator variance). Because variances are always positive, the range of values for F_{test} extends from zero to large positive numbers. Given that the center of the distribution is close to 1.0 under the Null hypothesis, the *F*-distribution is positively skewed. Figure 10.1 displays example probability density curves for the *F*-distribution.

5. **p-Value**

   a. Because the *F*-test procedure always puts the larger of the two sample variances in the numerator and the smaller in the denominator, the value for F_{test} will always be greater than the expected value stipulated by the Null hypothesis. Hence, the *p*-value will always be $P[F \geq F_{test}]$. Although

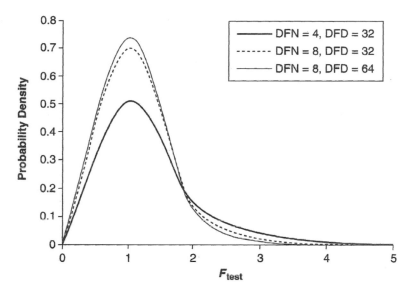

FIGURE 10.1   Sampling distribution of F_{test} for a range of degrees of freedom.   Under the Null hypothesis of equal variances, the expected value for $F_{test} = S_1^2/S_2^2 = 1.0$. Increasing degrees of freedom for either the numerator (DFN) or denominator (DFD) variance, reduces random sampling variation of the F_{test} statistic. This reduction is indicated by the reduced spread of the sampling distribution. Because F_{test} can have only positive values, the area in the right tail of the sampling distribution represents the *p*-value, even though the *F*-test is typically done as a two-tailed test.

this would appear to be a one-tailed test. the standard *F*-distribution always provides the two-tailed probability appropriate to the Alternative hypothesis $\sigma_1^2 \neq \sigma_2^2$. In most cases, you will use a statistics computer program to implement *F*-tests, and this program will provide *p*-values. However, it may sometimes be necessary for you to perform this test by hand. You can determine if the *p*-value is less than or equal to 0.05 using the *F*-distribution table (Appendix Table 5, see Section 10.2). Alternatively, you can obtain an exact *p*-value from many spreadsheet computer programs using the formula **FDIST(F_{test},DFN,DFD)**.

■ **EXAMPLE**   With $F_{test} = 4.4$, DFN = 3, and DFD = 18, in MS Excel type in = FDIST(4.4,3,18) and the program will return the *p*-value 0.0173. In Correl Quattropro, type in @FDIST(4.4,3,18) and the program gives the same *p*-value.   ▨

b. *Stating conclusions:*

If $p \leq 0.05$: The probability associated with the random-variation explanation for the observed difference between the two sample variances is sufficiently low to reject this explanation. If an appropriate study design has been correctly implemented, the only remaining likely explanation is that the two population variances are actually different. Even though the *F*-test is always two-tailed, your conclusion should indicate which population has the larger or smaller variance.

■ **EXAMPLE** "The variance of standardized test scores for students in urban schools is greater than that for students in suburban schools (p = 0.002)." ▨

If $0.05 < p \leq 0.10$: The probability associated with the random-variation explanation for the observed difference between the two sample variances is sufficiently high that we cannot confidently exclude this explanation. However, the probability is sufficiently low that we might suspect there is a difference among the population variances. Some scientists would conclude that the results "suggest" that the population variances differ, but more data are needed to verify this.

■ **EXAMPLE** "The results suggest that the variance of test scores for students in urban schools is greater than for students in suburban schools (p = 0.083), but more data are needed to verify this." ▨

If $p > 0.10$: The probability associated with the random-variation explanation for the observed difference between the two sample variances is so high that this explanation cannot be excluded. Hence, there is insufficient evidence to conclude that the two population variances are different.

■ **EXAMPLE** "The variance of standardized test scores was not significantly different between urban schools with diverse student bodies and suburban schools with more homogeneous student populations (p = 0.45)." ▨

10.2
Using the
F-Table

The F-table provided in this book is an abbreviated version of standard F-tables provided in higher-level statistics texts. Typically, the larger versions of the F-table provide critical F-values for a range of α-values. The F-table in Appendix Table 5 provides critical F-values only for the Type I error rate $\alpha = 0.05$. You will generally use a statistics computer program to perform F-tests, and the print-out will provide an exact p-value. In those rare instances when you do this test by hand, Appendix Table 5 will allow you to determine whether or not the difference between sample variances is significant at the standard $p \leq 0.05$ level. My purpose in this section is to show you how to use an F-table. Should you require a more precise p-value than Appendix Table 5 provides, this procedure will enable you to use larger standard F-tables available in higher-level texts.

How to Use an F-Table

- Calculate the degrees of freedom for each sample variance, S^2_1 and S^2_2. The number of degrees of freedom is equal to the value in the denominator of the equation used to compute the variance. In the standard formula for a sample variance, this value is $n - 1$. The F_{test} statistic has *two* degrees of freedom values, one for the variance in the numerator and one for the variance in the denominator of the equation $F_{test} = S^2_1/S^2_2$.

- Find the appropriate column in the F-table. The column titles across the top of the F-table correspond to the *numerator* degrees of freedom. Find the column in the table that corresponds to the degrees of freedom for the larger of your two sample variances (which will always be in the numerator).

Note: If the exact number corresponding to the degrees of freedom for your data is not listed in the *F*-table, always *round down* the degrees of freedom to the tabled value.

- Find the appropriate row in the *F*-table. The row titles down the left-hand margin of the *F*-table correspond to the *denominator* degrees of freedom. Go down the column in the table that corresponds to your numerator degrees of freedom until you reach the row labeled with the appropriate denominator degrees of freedom.

- The *F*-value in the table at this position is the critical *F*-value for α = 0.05. If your computed F_{test} value is greater than or equal to this critical *F*-value, the *p*-value for your test of significance is $p \le 0.05$. If your computed F_{test} value is less than the tabled critical *F*-value, the *p*-value for your test is $p > 0.05$.

Example 10.1 illustrates how this procedure is done. Example 10.2 presents results from a biological experiment for which the two-sample *F*-test is shown.

▦ EXAMPLE 10.1

Using the *F*-table. Use the *F*-table to determine if the computed F_{test} values listed (with associated degrees of freedom) provide strong evidence ($p < 0.05$) that the population variances are not equal. DFN and DFD denote the degrees of freedom for the numerator and denominator, respectively. $F_{\alpha=0.05}$ is the critical *F*-value from Appendix Table 5

S^2_1	S^2_2	n_1	n_2	F_{test}	DFN	DFD	$F_{\alpha=0.05}$	*p*-Value
121	64	20	20	1.89	19	19	2.17	$p > 0.05$
365	64	12	84	5.70	11	83	1.91	$p < 0.05$
213	78	4	44	2.73	3	43	2.84	$p > 0.05$
5357	2139	80	120	2.50	79	119	1.42	$p < 0.05$

▦

▦ EXAMPLE 10.2

Application of the two-sample *F*-test. Based on the theory of evolution through natural selection, biologists predict that species found in a wide variety of environments will exhibit greater genetic variation than species that occur only in a single type of environment. Natural selection pressures will differ in different environments, selecting for individuals with different sets of genetically determined traits.

Plant ecologists measured mature plant size for several species found in a wide range of environments in California, from moderate coastal environments to extreme mountaintop environments. Seeds from plants growing in different environments were planted in a single research garden to determine the genetically controlled mean plant size (as distinct from environmental influences on plant growth). One widely distributed species was found in virtually all environments. Other species were found in only one environment. The investigators used the variance of mature plant size as a measure of genetic variability within each species. Suppose that the results for three plant species were as below:

Species	# Plants	Mean Height (cm)	Variance	Environment
A	45	100.2	194.7	(Lowland only)
B	37	15.4	23.2	(Alpine only)
C	81	73.4	305.5	(All elevations)

Do these results provide evidence that species found in a wide range of environments have greater genetic variability than species found only in specific environments? Two-sample F-tests for all pair-wise comparisons among the three species variances are given in the table below.

Species	S^2_1	S^2_2	DFN	DFD	F_{test}	$F_{\alpha=0.05}$	p-Value
A vs. B	194.7	23.2	44	36	8.39	1.74	$p < 0.05$
A vs. C	194.7	305.5	80	44	1.57	1.61	$p > 0.05$
B vs. C	23.2	305.5	80	36	13.2	1.65	$p < 0.05$

Conclusion: Widely distributed Species C had a significantly higher variance in mature plant height than Species B, which is found only at high elevation, in cold alpine meadows. However, Species C did not have significantly greater variance than Species A, which is found only in warm low elevation environments. The only generalization from these results seems to be that species with larger mean mature plant size have greater variance than species with small mean mature size. ▨

10.3
The Levene Test

1. **Application:** There is no entirely satisfactory test of significance for comparing measures of spread between two populations that have non-Normal distributions. The **Levene test** is an approximate test that is less sensitive to non-Normal population distributions than the F-test. That is, p-values from the Levene test provide better estimates of the true Type I error rate than are obtained from the F-test. Here I will describe the application of this test to compare the spread (variability) for two samples. In Chapter 11 I will describe the extension of this test to compare more than two samples. *Assumptions:* Data values represent independent observations from the population of interest, obtained by a randomized, unbiased study design.

2. **Statement of Hypotheses**

   - H_0: $\mu_{AD1} = \mu_{AD2}$
   - H_a: $\mu_{AD1} \neq \mu_{AD2}$

   where AD is the absolute difference between the individual data values and their mean. μ_{AD} is the true mean of the absolute differences between all individual values and their population mean. This parameter is estimated by $\bar{x}_{AD}$, the **mean absolute difference** between individual data values and their sample mean, computed as:

   $$\bar{x}_{AD} = \Sigma|x_i - \bar{x}|/n.$$

   Here we use the absolute value function to eliminate negative differences (instead of squaring the differences, as is done to compute variance). The absolute value function makes mean absolute deviation less sensitive to skewed distributions and outliers than is the variance. This makes the Levene test less sensitive to violation of the Normality assumption than is the F-test.

3. **Test Statistic**

   The two-sample version of the Levene test is simply the application of a two-sample t-test to the mean absolute deviations computed for two samples. The sample mean and standard deviation of the AD values are computed for each sample ($\bar{x}_{AD1}$, S_{AD1}, $\bar{x}_{AD2}$, S_{AD2}). The two-sample t-test is used to determine if the mean absolute deviation differs between the two groups. The t_{test} statistic is computed as follows:

$$t = \frac{\bar{x}_{AD1} - \bar{x}_{AD2}}{\sqrt{\dfrac{S_{AD1}}{n_1} + \dfrac{S_{AD2}}{n_2}}}$$

NOTE The t-test formula for the separate variance t-test is given here, but a pooled-variance t-test could be used if the variances of the absolute deviation values are sufficiently similar. ■

4. **Sampling Distribution**

This test statistic has a sampling distribution that is approximately a t-distribution with degrees of freedom computed in the same manner as described in Chapter 9 for the two-sample t-test. Absolute deviations are *not* Normally distributed. However, the mean of a sufficiently large number of absolute deviation values will have a sampling distribution that is approximately Normal, as described by the Central Limit Theorem.

5. **p-Value**

If $p \le 0.05$: The probability associated with the random-variation explanation for the observed difference in mean absolute deviation between the two samples ($\bar{x}_{AD1} - \bar{x}_{AD2}$) is sufficiently low to reject this explanation. If an appropriate study design has been correctly implemented, the only remaining likely explanation is that the two populations differ with regard to the amount of variability among individuals.

If $0.05 < p \le 0.10$: The probability associated with the random-variation explanation for the observed difference in mean absolute deviation between the two samples is sufficiently high that this explanation cannot be confidently excluded. However, this probability is sufficiently low that we may suspect that there is a difference between the populations. Some scientists would conclude that the data "suggest" that the populations differ with regard to the amount of variability among individuals, but more data are needed to verify this conclusion.

$p > 0.10$: The probability associated with the random-variation explanation for the observed difference in mean absolute deviation among the two samples is so high that this explanation cannot be excluded. Hence, there is insufficient evidence to conclude that the two populations differ with regard to the amount of variability among individuals.

Example 10.3 shows the Levene test used with two non-Normal distributions.

■ **EXAMPLE 10.3**

Application of the Levene test (two-sample case). The data values below were obtained by randomly sampling from two non-Normal distributions, one of which was positively skewed and the other flat. The boxplots and Normal quantile plots display the data distributions for two samples of $n = 40$ individuals.

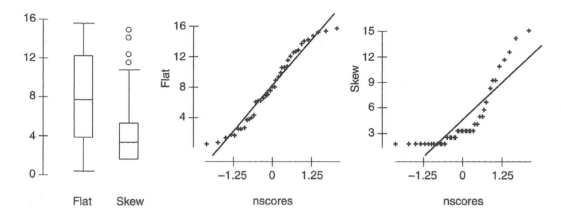

The values listed below are a subset of data values from each of the two samples of $n = 40$ individuals. This table is intended to give you an idea of the calculations used for the Levene test. The means and standard deviations listed below the data values were computed based on all $n = 40$ individuals in each sample.

| | **Original Data** | | **Absolute Deviations** | |
	Flat	Skewed	Flat	Skewed
1	1.3	5.8	$6.8 = \lvert 1.3 - 8.1 \rvert$	$1.2 = \lvert 5.8 - 4.6 \rvert$
2	3.7	9.2	$4.4 = \lvert 3.7 - 8.1 \rvert$	$4.6 = \lvert 9.2 - 4.6 \rvert$
3	6.3	1.7	$1.8 = \lvert 6.3 - 8.1 \rvert$	$2.9 = \lvert 1.7 - 4.6 \rvert$
4	12.0	3.3	$3.9 = \lvert 12 - 8.1 \rvert$	$1.3 = \lvert 3.3 - 4.6 \rvert$
.	.	.	.	.
.	.	.	.	.
40	6.3	2.5	$1.8 = \lvert 6.3 - 8.1 \rvert$	$2.1 = \lvert 2.5 - 4.6 \rvert$
n	40	40	40	40
$\bar{x}$	8.1	4.6	4.0	2.9
S	4.7	3.8	2.4	2.4

The t-test print-out below is a test of significance for the difference in mean absolute deviation between the two samples $(\bar{x}_{AD1} - \bar{x}_{AD2}) = (4.0 - 2.9) = 1.1$. The p-value from this two-sample t-test represents the probability of obtaining a difference in mean absolute deviation of 1.1 or more due solely to random sampling variation. Fl = flat distribution and Sk = skewed distribution.

Pooled t-Test of $\mu 1 - \mu 2$

- H_0: $\mu 1 - \mu 2 = 0$

- H_a: $\mu 1 - \mu 2 \neq 0$

AD-Fl − AD-Sk:

Test H_0: $\mu(AD\text{-}Fl) - \mu(AD\text{-}Sk) = 0$ vs. H_a: $\mu(AD\text{-}F1) - \mu(AD\text{-}Sk) \neq 0$

Difference between Means = 1.1177797 t-statistic = 2.091 with 78 df

$p = 0.0398$

Conclusion: Mean absolute deviation is greater in the flat population than in the skewed population ($p = 0.0398$). That is, there is greater variation among individuals in the flat population than among individuals in the skewed population.

Applying the *F*-test to determine if these data provided sufficient evidence to conclude that the population variances differ, we obtain the following results.

$$F_{test} = S^2_{larger}/S^2_{smaller} = 4.7^2/3.8^2 = 22.09/14.44 = 1.53$$

The *p*-value for $F_{test} = 1.53$, with DFN = 39 and DFD = 39, is p = 0.0943.

Conclusion: There is insufficient evidence to conclude the variances of these two populations differ ($p = 0.0943$).

> **NOTE** The Levene test provided sufficient evidence to conclude that the populations had different variance, while the *F*-test did not. However, the *F*-test was invalid (the *p*-value inaccurate) because the Normality assumption was violated. Hence, the results of the Levene test could be considered a more valid basis for drawing conclusions about the populations. As I have said before, it is essential that you select statistical analyses appropriate to your specific question and data. Otherwise, the results of these analysis do not provide a valid basis for drawing conclusions about the nature of the real world. ∎

CHAPTER SUMMARY

Two tests for comparing sample variances were covered in this chapter: the *F*-test and the Levene test.

1. The **F-test** is used to evaluate the significance of the ratio of two sample variances as evidence to test the Null hypothesis that two population variances are equal ($\sigma^2_1 = \sigma^2_2$). The F_{test} statistic is computed as the ratio of the larger of the two sample variances divided by the smaller variance. This test requires that both population distributions are Normal, and can be very sensitive to violation of this assumption.

2. The **Levene test** is used to evaluate sample data as evidence to test the Null hypothesis that two population variances are equal ($\sigma^2_1 = \sigma^2_2$). This test is based on the mean of absolute deviations between individual data values and their corresponding sample mean. This test also assumes that the population distributions are Normal, but it is less sensitive to violation of this assumption than is the *F*-test.

11 Tests for Comparing More Than Two Sample Means

INTRODUCTION The tests of significance for sample means described in Chapter 9 are appropriate only for the simplest of study designs, involving comparisons for only one or two sample means. Much real-world scientific research is more complex, and requires comparisons among more than two sample means. For example, many drug trials do not simply have a treatment and control group. Rather, they might include multiple groups that receive a range of doses of the drug. This study design would allow the investigators to determine both the effectiveness of the drug and the optimum dose of the drug for attaining the desired health effects. In a wildlife biology research context, an investigator may want to identify the characteristic(s) that distinguish the habitats of two animal species. The investigator would measure many different environmental variables in habitats where each of the two species are found. The investigator would then need to perform a two-sample t-test or a Mann-Whitney U-test for each of the variables measured to determine if the mean for that variable differs between the habitats for the two species. Whether the study design involves a single variable measured on more than two groups, or multiple variables measured on two or more groups, this increased complexity requires more complex statistical analyses to provide appropriate tests of significance. This chapter presents some of the data analysis options for these more complex study designs. ▪

GOALS AND OBJECTIVES

In this chapter you will learn how and when to use tests of significance for evaluating the differences among three or more sample means as evidence to test hypotheses about whether or not the corresponding population means are equal.

By the end of the chapter, given the study design and the nature of the study question and data, you will be able to:

1. Identify situations in which multiple tests of significance would test a single overall Null hypothesis and so would be affected by cumulative Type I error rate.
2. Identify the appropriate test of significance from among those listed below, implement the test, and draw appropriate conclusions with regard to the scientific question of interest:

    Bonferroni procedure
    Analysis of variance (ANOVA)
    Multiple comparisons test for means
    (Tukey's HSD test and Tukey-Kramer test)
    Planned contrasts for Means
    Kruskal-Wallis test
    Multiple comparisons test for medians

11.1
Multiple Tests of Significance and Cumulative Error Rate

It is common that multiple tests of significance will be done within the context of a single study. Each of these tests has a Type I error rate α, which is the probability of concluding a difference exists when, in fact, there is no difference. The more tests of significance one does in a study, the more likely that at least one test will incorrectly reject a Null hypothesis due only to random variation. This is particularly problematic if the multiple tests of significance all pertain to the same hypothesis.

Suppose a study is conducted to test for a difference in academic skills between male and female students. The study compares male vs. female test scores from 20 different types of academic achievement tests. A separate test of significance is done to compare mean scores from each type of test. Suppose the investigator decides that he will conclude a difference exists between males and females if any one of these 20 multiple tests of significance has a p-value ≤ 0.05. Remember, this is the probability that we would observe a difference sufficiently large to reject the Null hypothesis due merely to random sampling variation. Hence, for each separate test of significance there is an $\alpha = 0.05$ probability that the investigator would make this Type I error. Given this description, what is the probability the investigator would erroneously conclude that males differ from females when, in fact, there is no difference?

It should be obvious that the overall probability of making a one or more Type I errors out of 20 tests is much greater than $\alpha = 0.05$. In this example, the probability $P[\geq 1$ Type I error out of 20 tests$] = 0.6415$. That is, there is greater than a 64% probability that at least one of the 20 tests would cause us to conclude that males and females differ in academic ability due simply to random sampling variation when, in fact, they are *not* different. The phenomenon of increasing Type I error rate as the number of related tests of significance increases is called **cumulative Type I error rate**. Figure 11.1 displays the increase in cumulative Type I error rate with increasing number of tests. After all the discussion in earlier chapters about the need to control the probability of incorrectly claiming there is a difference when there really isn't, you should understand that this is an unacceptable situation.

Scientists may commonly perform many tests of significance during the course of any one research project, but the problem of cumulative error rate is especially important for multiple "related" tests of significance. **Related tests of significance**

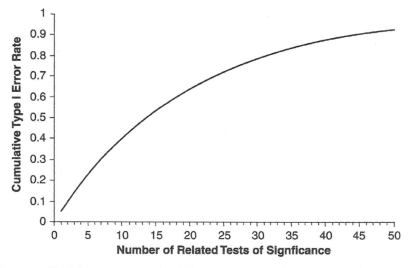

FIGURE 11.1   Multiple related tests of significance and cumulative error rate.   The probability of making one or more ($X \geq 1$) Type I errors is shown as a function of the number of related tests of significance. Tests are assumed independent, each with a Type I error rate of $\alpha = 0.05$ (i.e., if the p-value for a single test is ≤ 0.05, the Null hypothesis will be rejected). The y-axis displays the cumulative Type I error rate, computed as follows: $1 - (1 - \alpha)^{Ntests}$, with Ntests = the number of related tests of significance and α = the allowable Type I error rate for each test (usually 0.05).

all test the same overall Null hypothesis, and if any one of these tests produce a p-value less than the specified acceptable Type I error rate α, you would reject this Null hypothesis. Related tests have two general formats:

1. The related tests might compare the mean for a single characteristic or response variable among more than two groups. If any two means are significantly different, the investigator would conclude that the groups are different and that the treatment presumed to cause this difference had an effect.

    ■ **EXAMPLE** Compare the weight gain of four groups of white rats that were given four different doses of a drug. If mean weight for any two groups differs, conclude that the drug is effective.   ■

2. The related tests might compare the mean for two or more characteristics or response variables between the two or more treatment groups or populations. If a significant difference is found for any of these variables, the investigator would conclude that the populations are different.

    ■ **EXAMPLE** Compare mean concentrations of Ca, Mg, K, Na, N, P for soil samples taken at two sites. If the two site means for any one of these variables differ, conclude that the soils at these sites are different. The example of using 20 academic achievement tests to compare male and female students also falls into this category.   ■

Comparison-wise vs. Experiment-wise Type I Error Rate

When a study requires that multiple, related tests of significance be done, it is necessary to distinguish between "comparison-wise" versus "experiment-wise" Type I error rate. The **comparison-wise error rate** is the probability that a single test of significance will reject the Null hypothesis of "no difference" when the observed difference between the sample means is due merely to random sampling variability. Each test of significance represents a "comparison." When multiple related tests of significance are performed, this is the Type I error rate for each test. The **experiment-wise error rate** is the cumulative probability that at least one Type I error occurs as a result of two or more related tests of significance, leading to an erroneous conclusion to reject the overall Null hypothesis. This is the *overall* probability of a Type I error for all related tests of significance combined. For a fixed comparison-wise Type I error rate (α), as the number of related tests of significance (Ntests) increases, the experiment-wise error rate increases. To determine the probability that one or more of the Ntests tests of significance will incorrectly reject H_0 when it is true, with a comparison-wise Type I error rate set at α (usually 0.05) for each test, use the following formula:

$$\text{Experiment-wise Error Rate} = 1 - (1 - \alpha)^{\text{Ntests}}$$

■ **EXAMPLE** For the example of using Ntests-20 achievement tests to compare male and female academic ability, with a comparison-wise error rate $\alpha = 0.05$, the probability that one or more tests would incorrectly conclude there is a real difference when the observed difference is due only to random sampling variation is computed as $1 - (1 - 0.05)^{20} = 0.6415$. ▨

Controlling Experiment-wise Error Rate

There are two general approaches for controlling experiment-wise Type I error rate when more than two sample means must be compared to test a single, overall Null hypothesis:

1. Reduce the acceptable comparison-wise Type I error rate for each of the related tests of significance so that the cumulative, experiment-wise error rate adds up to $\alpha = 0.05$. This approach is called the *Bonferroni procedure*.

2. Perform a single, *simultaneous test* of the overall Null hypothesis that all population means are equal (i.e., H_0: $\mu_1 = \mu_2 = \mu_3 = \ldots = \mu_k$). Only after the overall Null hypothesis is rejected at $\alpha \leq 0.05$ by this single test of significance do you perform multiple tests to determine exactly which of the many population means differ. Simultaneous tests are more powerful than the Bonferroni procedure because only a single test of significance is required and the critical α for rejecting the overall Null hypothesis is 0.05. The critical α values of the Bonferroni procedure are generally much smaller than 0.05, making it more difficult to reject the Null hypothesis. However, if the overall Null hypothesis is rejected by a simultaneous test, we can only conclude that some of the population means are different; we do not know *which* means differ. Additional tests are needed to determine which means are different.

Bottom line: Simultaneous tests have greater statistical power, but are more complex than the Bonferroni procedure.

11.2
The Sequential Bonferroni Procedure

The **sequential Bonferroni procedure** can be used whenever multiple tests of significance are used to test an overall Null hypothesis. The two general scenarios that require multiple related tests of significance are:

1. Comparisons of a sample statistic for *one variable* among *more than two groups*, requiring multiple two-sample tests to compare all possible pairs of groups.

2. Comparisons of sample statistics for *more than one variable* among *two or more groups*, with a separate test of significance for each variable.

In the first case, you could use either the Bonferroni procedure or a simultaneous test like Analysis of Variance (described in Section 11.3). In the second situation, the Bonferroni procedure is the only appropriate analysis.

Performing the Sequential Bonferroni Procedure

These four steps will enable you to use the Bonferroni procedure for your analysis:

1. Perform the multiple tests of significance (e.g., two-sample *t*- or *Z*-tests, Mann-Whitney *U*-test) to make all the comparisons needed to test the overall Null hypothesis.

- If you are comparing means or proportions for a single variable across more than two groups, perform all the pair-wise two-sample tests of significance required to compare each group to all other groups.

- If you are comparing the means or proportions for two or more variables across two or more groups, perform all the tests of significance required to compare each variable among the different groups.

- *It is not necessary that all tests of significance be of the same type.* If the data for some comparisons meet the Normality assumption, but the data for others do not, you may use parametric and nonparametric tests, as necessary. If some variables are categorical and others quantitative, you many use two-sample Z-tests for proportions and *t*-tests for means.

2. List the results from the multiple tests of significance in a Bonferroni procedure table, *with the list of test results ordered from smallest to largest p-values.* (See Example 11.1 below.)

3. For each test of significance, compute the Bonferroni critical *p*-value required to reject the Null hypothesis for that comparison, using the formula below:

   Critical p-value $= 1 - (1 - \alpha)^{\frac{1}{Ntests-d}}$

   where α is the desired overall experiment-wise Type I error rate (usually 0.05), Ntests is the total number of tests of significance performed to test the overall Null hypothesis, and d is the number of tests of significance that precede a specific test in the Bonferroni procedure table. For the first (smallest) p-value at the top of the ordered list, $d = 0$; for the second p-value, $d = 1$; for the third p-value, $d = 2$; and for the last (largest) p-value, $d = $ Ntests $- 1$.

   NOTE You can increase the power of a sequential Bonferroni analysis by limiting the total number of tests of significance Ntests to only those comparisons that are essential to the main question of your study. Just because you collected data for 20 variables does not mean you must do tests of significance for all these variables. If you look at the formula above, you will see that the smaller that Ntests is, the larger will be the critical p-value for rejecting the Null hypothesis for any one comparison in the table. Reducing Ntests will make it easier to reject the comparison-wise Null hypotheses, increasing the power of the test. ■

4. Starting at the top of the list of ordered p-values, with the smallest p-value, if that p-value is less than or equal to the Bonferroni critical p-value, then the difference between the two sample means, medians, or proportions is significant at the level $\alpha = 0.05$. If the smallest p-value from the pair-wise tests of significance is larger than the Bonferroni critical p-value, then none of the comparisons in the table are significant at the overall experiment-wise error rate of α. If the first comparison in the table is significant, proceed to the second comparison. If the second smallest p-value is greater than the corresponding Bonferroni critical p-value, then the associated difference and all subsequent differences with larger p-values are not significant. If the difference with the second smallest p-value is significant, proceed to the third comparison in the list. If the p-value for the third comparison is larger than the Bonferroni critical p-value, the third and all subsequent differences in the ordered list are not significant. Continue this process until you reach the first p-value in the ordered

list that is greater than the Bonferroni critical p-value. This difference and all subsequent differences with larger p-values are not significant.

Examples 11.1 and 11.2 show two applications of the sequential Bonferroni procedure.

■ **EXAMPLE 11.1**

Application of the sequential Bonferroni procedure (comparing the mean for a single variable among more than two groups). In an experiment to determine the efficacy of a new drug to reduce blood cholesterol, study subjects were randomly assigned to four treatment groups, The control group received a placebo pill that did not contain the drug. Group 1 received the drug at a low dose, Group 2 received the drug at a moderate dose, and Group 3 received a high dose of the drug. The investigator wants to know if the drug reduces cholesterol, and if so, what is the lowest dose that produces the desired response. The mean blood cholesterol for the four experimental groups are:

Control	Group 1 (Low)	Group 2 (Moderate)	Group 3 (High)
534	473	335	307

Results of Two-Sample t-Tests for All Pair-wise Comparisons between Groups

Difference between Group Means	t-Test p-Value	Ntest	d	Critical p-Value	Conclusion
C − G3 = 227	0.0008	6	0	0.0085	Significant
C − G2 = 199	0.0069	6	1	0.0102	Significant
G1 − G3 = 66	0.0102	6	2	0.0128	Significant
G1 − G2 = 138	0.025	6	3	0.0170	Nonsignificant (N.s.)
C − G1 = 61	0.087	6	4		N.s.
G2 − G3 = 28	0.381	6	5		N.s.

Conclusion: The new drug is effective at reducing blood cholesterol ($p < 0.05$). This is the conclusion for the *overall* Null hypothesis. The moderate dose was just as effective for lowering cholesterol as the high dose; the difference between these two group means was not significant ($p = 0.381$). The low dose was not effective, as the mean for this group was not significantly different from the mean for the control group ($p = 0.087$).

> **NOTE** p-Values of this size have previously been interpreted to "suggest" there is a difference. However, with multiple tests and cumulative Type I error rate, most scientists would need stronger evidence before they would decide that the data suggests that a difference exists. ■

■ **EXAMPLE 11.2**

Application of the sequential Bonferroni procedure (comparing means for more than one variable between two groups). National Forest managers in the eastern United States have begun to use controlled fires to destroy shrubs and saplings of undesirable tree species so as to increase the abundance of seedlings and saplings of desirable, fire-tolerant tree species. However, an undesirable consequence of these fires is that they may degrade nesting habitat for certain bird species that are already experiencing population declines. An investigator performs a field experiment wherein she implements a controlled fire in one part of the forest and compares the abundance of these bird species between the burned area and an adjacent, similar but unburned area. If the abundance of any one of these species is lower in the burned area than in the unburned control area, she will conclude that the controlled fires have a negative impact on ground- and shrub-nesting bird species. Because the bird abundance data were not Normally distributed, she used a separate Mann-Whitney U-test to compare the abundance of each bird species between burned and unburned areas. The results of these tests, and the sequential Bonferroni analysis are presented below:

Bird Species	U-Test p-Value	Ntest	d	Critical p-Value	Conclusion
Ovenbird	0.005	5	0	0.0102	Significant
Hooded warbler	0.010	5	1	0.0128	Significant
Kentucky warbler	0.038	5	2	0.0170	Nonsignificant (N.s.)
Worm-eating warbler	0.074	5	3		N.s.
Black-and-white warbler	0.204	5	4		N.s.

Conclusion: The investigator concludes there is strong evidence (experiment-wise error rate $p < 0.05$) that the abundance of at least some of these bird species was lower in the burned areas than in the unburned control area. Based on results of the Bonferroni procedure, there is strong evidence that ovenbird and hooded warbler abundance is reduced in the burned area. Although the p-value for the Mann-Whitney U-test for the difference in Kentucky warbler abundance is less than 0.05, it is not less than the Bonferroni critical p-value that controls experiment-wise error rate to less than 0.05. Therefore, the results of this test would *not* be considered strong evidence of a difference in abundance of Kentucky warbler between burned and unburned areas. Some scientists would infer that the data "suggest" Kentucky warbler is negatively affected by the fires, but more data are needed to verify this. The abundance of worm-eating warbler and black-and-white warbler was not significantly different between burned and unburned areas.

11.3

Analysis of Variance (ANOVA)

1. **Application: Analysis of variance** (**ANOVA**) is a simultaneous test used to determine with a single test of significance whether any of *three or more population means for a single variable* differ from each other. If the Null hypothesis of "no difference" is rejected, a second test must be done to determine which of the population means are different. In many experimental studies, randomly formed groups (also called *cells*) of individuals will be exposed to different levels of a single treatment or "factor," and a response will be measured on each individual (e.g., see Figure 1.2). ANOVA is used to compare the mean response among three or more cells as evidence for the effect of the treatment. *Assumptions:* The data are obtained in a randomized, unbiased manner, so the expected value for each sample mean is the corresponding population mean. The population distributions for individuals exposed to the different treatment levels have equal variances and are Normal.

Evaluating Whether the Assumptions for ANOVA Are Satisfied

a. *Study design:* In experimental studies, the subjects should have been randomly assigned to treatment groups, and extraneous factors should have been controlled. In studies that compare means computed from field samples, the sampling should have been randomized. If the comparison of samples from a natural experiment is intended to provide evidence of the effect of a particular variable on the populations of interest, differences in other, extraneous factors should be minimized.

b. *Normality assumption:* Use separate boxplots and Normal quantile plots for the data values in each of the groups to be compared. The data distribution for each group should be approximately Normal, with no outliers. Even if the boxplots of group data values are symmetric, this assumption is violated if the Normal quantile plots indicate that the data distribution deviates from Normal by being more "peaky" or "flat." The Central Limit Theorem does *not* apply to ANOVA.

c. *Equal variances assumption:* Compute summary statistics for each group, including either the variance or the standard deviation. Perform the F_{max} test to determine if the sample variances provide sufficient evidence to conclude that the population variances are *not* equal.

F_{max} Test for Equality of Variances

- Compute the ratio $F_{max} = S^2_{largest}/S^2_{smallest}$.

NOTE F_{max} is the ratio of the largest over the smallest group *variance*, the square of the standard deviation. ■

- Look up the computed F_{max} value in Appendix Table 7, the probability distribution of F_{max} values, with k = the number of group means that will be compared by the analysis of variance and df = $n - 1$, where n is the smaller sample size of the two groups whose variances are used to compute F_{max}.

- If the computed F_{max} is greater than or equal to the critical F_{max} from Table 7, the sample variances are sufficiently different that you can conclude the population variances are unequal, and the assumption is violated.

Important: If the Normality assumption for ANOVA is violated, the Normality assumption for the F_{max} test for equal variances is also violated. An extension of the Levene test (described in Chapter 10) to more than two samples is less sensitive to violation of the Normality assumption than the F_{max} test. This Levene test will be described later in this chapter. Some statistics software will automatically provide a Levene test as part of the ANOVA print-out.

Sensitivity of ANOVA to violations of assumptions: ANOVA is reasonably "robust" against modest violations of the Normality and equal variances assumptions *if the samples sizes for the groups are equal.* That is, the p-value from ANOVA will not be much affected by moderate violations of these assumptions. However, ANOVA is "sensitive" to violation of the equal variances assumption if sample sizes are unequal, particularly if the larger variance is associated with a group that has a smaller sample size. In this case, violation of the assumption results in a p-value that is smaller than the true Type I error rate α. For example, the p-value from ANOVA is reported as $p = 0.02$, but the actual probability that the observed difference is due to random variation is $\alpha = 0.075$. This situation can lead to incorrect conclusions and should be avoided. In such circumstances, an alternate test of significance should be used (e.g., the Kruskal-Wallis test described later in this chapter).

2. **Statement of Hypotheses**

- H_0: $\mu_1 = \mu_2 = \mu_3 = \ldots = \mu_k$

  where: k = the number of groups to be compared

- H_a: Two or more population means differ from each other

3. **The ANOVA F_{test} Statistic**

We first show the F_{test} as a quotient of the between-group variance and the within-group variance; we then discuss each of these in detail:

The ANOVA F_{test} Statistic

$$F_{test} = \frac{\text{Between-Group Variance}}{\text{Within-Group Variance}}$$

Between-group variance: If the Null hypothesis is true, the true population means for all groups are equal. Hence, all group means estimate the same value, and any differences among the sample means are due only to random sampling variation. The data for all groups are combined to compute an estimate of this "overall mean $\bar{x}_{overall}$." The calculation of **between-group variance** begins with the calculation of the variance among group means, as shown below:

$$S_{\bar{x}}^2 = \frac{\sum_{i-1}^{k}(\bar{x}_i - \bar{x}_{overall})^2}{k-1}$$

where $\bar{x}_i$ is the mean for each group and k = the number of group means

This is similar to empirical estimates of the standard deviation of the mean that we used in Chapter 6. (The standard deviation of means from repeated samples taken from the same population provides an empirical estimate of $S_{\bar{x}}$.) If the Null hypothesis is true, the data for groups being compared using ANOVA would be analogous to "repeated samples from the same population."

Between-group variance in ANOVA is an estimate of the population variance σ^2 that is based on this variance among the group means. The relationship between the standard deviation of the mean $\sigma_{\bar{x}}$ and the population standard deviation σ and sample size n is:

$$\sigma_{\bar{x}} = \sigma/\sqrt{n}$$

This can be expressed in terms of variances as follows:

$$\sigma_{\bar{x}}^2 = \sigma^2/n$$

The relationship can be reorganized to compute the population variance σ^2 from the standard deviation of the mean $\sigma_{\bar{x}}^2$ and sample size n:

$$n\sigma_{\bar{x}}^2 = \sigma^2$$

The variance among the group means $S_{\bar{x}}^2$, computed as shown above, can be used to compute an estimate of the population variance as follows:

Between-Group Variance

$$S_{between}^2 = \frac{n\sum_{i=1}^{k}\left(\bar{x}_i - \bar{x}_{overall}\right)^2}{k-1} = nS_{\bar{x}}^2$$

This is the "between-group variance" in the numerator of F_{test} above. In this equation, n = the sample size *in each group* (*not* the combined sample sizes of all groups). I present this formula for computing between-group variance only to help you understand the conceptual basis for ANOVA. You will typically use computer statistics software to perform ANOVA and would rarely implement this formula by hand.

Within-group variance: If the equal-variances assumption is valid, the samples for groups being compared using ANOVA were taken from populations that have the same variance σ^2. Therefore, any differences in the sample variances S_1^2, S_2^2, ..., S_k^2, are due only to random sampling variation. As we did for the pooled-variance two-sample t-test, we can obtain a better estimate of the true variance σ^2 if we pool the data from the k groups to compute a single pooled estimate of the population variance S_p^2. This is done by simply extending the formula presented for the pooled-variance t-test to include more than two sample variances:

Within-Group Variance

$$S_p^2 = \frac{(n_1 - 1)S_1^2 + (n_2 - 1)S_2^2 + (n_3 - 1)S_3^2 + \cdots + (n_k - 1)S_k^2}{n_1 + n_2 + n_3 + \cdots + n_k - k}$$

Remember, each of the group variances represents the variation among individuals who were all exposed to the same treatment and other conditions. Hence, the sole source of variation is random variation among individuals within each group. Therefore, the value for S_p^2 is not only the best estimate of the true population variance σ^2, it is also the measure of **within-group variance**. This is the denominator in for F_{test}. Hence:

$$F_{test} = \frac{\text{Between-Group Variance}}{\text{Within-Group Variance}} = \frac{nS_{\bar{x}}^2}{S_p^2}$$

Interpreting the F_{test} Statistic

a. *If the Null hypothesis is true ($\mu_1 = \mu_2 = \mu_3 = \ldots = \mu_k$):* The group means $\bar{x}_1, \bar{x}_2, \bar{x}_3, \ldots, \bar{x}_k$ are all independent estimates of the same overall mean μ and therefore represent means from repeated samples from the *same population*. Consequently, the between-group variance value is a valid estimate of σ^2, the true population variance. Within-group variance is always a valid estimate of the population variance, so long as the equal-variances assumption is valid. Hence, both the numerator and the denominator terms are valid estimates of the true population variance σ^2, and the value of F_{test} should be close to 1.0.

b. *If the Null hypothesis is false:* The within-group variance is still a valid estimate of the true population variance. However, the between-group variance no longer just reflects random sampling variation. If the Null hypothesis is false, some of the populations means are different from others, presumably due to the "treatment effect" under study. Hence, we expect the samples means to also differ. This variation due to the fact that the population means are not equal (the treatment effect) increases the size of between-group variation above what would be expected due only to random sampling variation. The resulting F_{test} values will be greater than 1. The larger the deviation from the expected value of 1, the stronger the evidence against the overall Null hypothesis.

NOTE: All of the differences among the k group means are combined into a single number, between-group variance. The larger the differences among the group means, the larger this variance will be. The F_{test} statistic compares this single between-group variance to a value that represents the expected amount of variation due only to random sampling variation. With a single test statistic, the ANOVA *simultaneous test* is used to determine if there are any significant differences among the group means. ■

4. **Sampling Distribution for F_{test}**

The sampling distribution for the ANOVA F_{test} statistic is an **F-distribution,** the probability distribution for the ratio of two sample variances computed from two independent samples from the same population (described in Chapter 10). If the two sample variances are both estimates of the same population variance, the expected value of the F_{test} statistic = 1.0. That is, under a Null hypothesis the center of the probability distribution for F_{test} is 1.0. However, random sampling variation will generally cause the F_{test} value to deviate somewhat from 1.0. As is the case for all other tests of significance described in this book, we must determine the probability of obtaining the observed deviation from the Null hypothesis expected value due only the random variation (the p-value).

Like the t-distribution, there are different F-distributions for each combination of degrees of freedom for the between- and within-group variances. Unlike the t-test, the F-test has two degrees of freedom values, one for between-group variance and one for within-group variance. The $df_{between}$ and df_{within} are determined as shown below:

$df_{between} = k - 1$ where k = the number of group means. This is the denominator of the variance formula for computing $S_{\bar{x}}^2$, as described above.

$df_{within} = N - k$ where N is the total number of data values in the study. df_{within} is equal to the denominator of the pooled variance formula for within-group variance $(n_1 + n_2 + n_3 + \ldots + n_k - k)$. *Note:* A degree of freedom is lost for each group mean computed.

5. **p-Value**

a. The p-value for the computed F_{test} statistic is obtained from the F-distribution, with DFN $= k - 1$ and DFD $= N - k$. I described how to obtain critical F-values for $\alpha = 0.05$ from Appendix Table 5 in Chapter 10. However, you will usually do ANOVA with a computer program that will provide exact p-values.

b. Because the computation of between-group variance does not distinguish between positive and negative differences among group means, *the ANOVA F-test is always a two-tailed test* for differences among the sample means. The Alternative hypothesis is always that two or more population means are "not equal."

c. *Stating conclusions:*

If $p \leq 0.05$: The probability associated with the random-variation explanation for the observed differences among the sample or treatment group means is sufficiently low to reject this explanation. If an appropriate study design has been correctly implemented, the only remaining likely explanation is that some of the population means are actually different. However, because ANOVA is a simultaneous test, at this point you have no basis for knowing *which* of the population means are different from each other. It is entirely possible that some population means differ, while others do not. If the ANOVA p-value is less than or equal to 0.05, you must implement a "multiple comparisons test" (see Section 11.4) to determine which sample means are significantly different. Typically, the nature of your conclusion statement will depend on the outcome of the multiple comparisons test.

If $0.05 < p \leq 0.10$: The probability associated with the random-variation explanation for the observed differences among the sample means is sufficiently high that we cannot confidently exclude this explanation. However, the probability is sufficiently low that we might suspect there is a difference among some population means. Some scientists would conclude that the results "suggest" that a difference may exist among some of the population means, but more data are needed to verify this. However, in this circumstance a multiple comparisons test would *not* be done.

■ **EXAMPLE**   "The differences in water pH among similar lakes that receive different levels of acid rain suggest that acid rain may affect lake pH ($p = 0.079$), but more data are needed to verify this."

If $p > 0.10$: The probability associated with the random-variation explanation for the observed differences among the sample means is so high that this explanation cannot be excluded. Hence, there is insufficient evidence to conclude that there is a difference among the population means.

■ **EXAMPLE**   "There were no significant differences in mean water pH among similar lakes that receive different levels of acid rain ($p = 0.22$). Therefore, acid rain does not appear to affect lake pH."

Example 11.3 presents a researcher's use of ANOVA to compare exercise regimes.

■ **EXAMPLE 11.3**

Application of ANOVA. A researcher in the field of physical fitness training and health education does an experiment to determine if several short bouts of exercise provide the same health benefits as one longer bout of exercise. He obtains 48 volunteers, all of whom are overweight female university students. All subjects are given the same dietary counseling and instructed to maintain a similar diet regime. The subjects are randomly assigned to four exercise treatment groups: (1) Control = no exercise other than what they were already doing, (2) 1×30 = subjects exercise once per day for 30 minutes on each of 5 days per week, (3) 2×15 = subjects exercise twice per day for 15 minutes on each of 5 days per week, and (4) 3×10 = subjects exercise three times per day for 10 minutes on each of 5 days per week. All subjects were supervised while they exercised to ensure compliance with the standardized exercise regime. The response measure for each subject was change in weight (ΔWt) between the beginning of the study and after 12 weeks of the exercise regime. The researcher wishes to compare mean ΔWt among the four treatment groups to identify the best regime for stimulating weight loss.

Exploratory Data Analysis for ΔWt

Group	Sample Size	Mean	Median	StdDev	Min	Max
Control	12	0.29	−0.33	2.29	−2.99	4.17
1 × 30	12	−2.33	−3.07	3.31	−7.38	3.37
2 × 15	12	−3.71	−3.22	3.36	−10.5	2.43
3 × 10	12	−4.66	−5.73	3.47	−9.43	0.07

Normal Quantile Plots for ΔWt Data

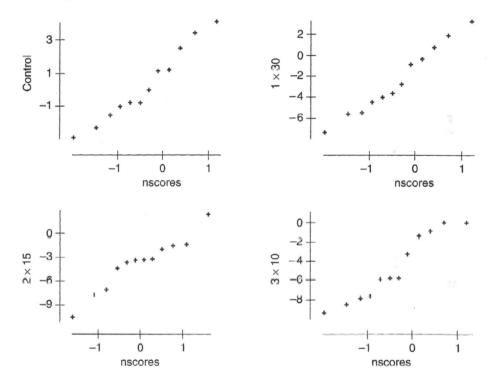

Boxplots for ΔWt

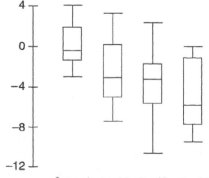

Assessing ANOVA Assumptions

a. *Randomized, unbiased study design:* Subjects were randomly assigned to treatment groups. All subjects received same diet counseling and their exercise was supervised to ensure compliance with the treatments. This assumption appears to be valid.

b. *Population distributions are Normal:* The Normal quantile plots and boxplots indicate modest deviations from Normality. However, with equal sample sizes in each group, ANOVA should be robust to such minor violations of the Normality assumption.

c. *Population variances are equal:* The F_{max} test is used here to assess the validity of this assumption, even though the Normality assumption for this F-test is violated. Again, with equal sample sizes the F_{max} test should be robust to the minor deviations from Normality apparent in the data distributions for the four treatment groups. (At the end of this example I show the results for an extension of the Levene test to more than two groups. The Levene test is less affected by violation of the Normality assumption and should be used instead of F_{max} in such cases.)

$$F_{max} = \frac{S^2_{largest}}{S^2_{smallest}} = \frac{(3.47)^2}{(2.27)^2} = 2.34$$

$F_{critical}$ with $k = 4$, $df = 12 - 1 = 11$, rounded down to $10 = 5.67$

Conclusion: The sample variances are sufficiently similar that the researcher can assume that the population variances are equal.   ▓

The Analysis of Variance Table

The ANOVA print-out below for the comparison of mean ΔWt among the four treatment groups is typical of output from statistics computer programs. Below the following computer print-out of the ANOVA table, I provide a description of the different columns of the table.

Analysis of Variance

Source	df	Sums of Squares	Mean Squares	F-Value	p-Value
Group	3	166.819	55.6064	5.6251	0.0024
Error	44	434.956	9.88537		
Total	47	601.775			

Source: This column identifies the sources of variation among the data values (between-group variance and within-group variance, which sum to the total variance). The between-group variance will usually be on the first line in the table (labeled with the treatment or group variable name in this example, also called *Model* in the print-out from other statistics software). Within-group variation will be on the second line (called *Error*, but sometimes labeled "Within"). The word "error" is used in statistics contexts to mean "random variation." The bottom line of the ANOVA table will always contain the total variance (sometimes called "Corrected Total"), which is the variation of all individuals for all groups combined about the overall mean.

df: This column contains degrees of freedom for each source of variation. If you are uncertain about which lines in the ANOVA table correspond to the between- and within-group variance, look at the degrees of freedom. The between-group variance line will have $df = k - 1$ (k = number of treatment groups) and the within-group variance line will have $df = N - k$ (N = total number of data values for all groups combined). The df_{total} is equal to the sum of df_{group} and df_{error}.

Sums of Squares (SS): This column contains the values for "sums of squared deviations," which are the numerators in the formulae for computing within-group and between-group variances [e.g., $\Sigma(x_i - \bar{x})^2$]. Total sums of squares is equal to the sum $SS_{group} + SS_{error}$.

Mean Squares (MS): This column contains the "mean squared deviations." This is another name for variance that is used in ANOVA contexts. While I regret the additional jargon, this terminology is universally used.

$$MS_{group} = SS_{group}/df_{group} = \text{Between-group variance}$$
$$MS_{error} = SS_{error}/df_{error} = \text{Within-group variance}$$

F-value: This column contains the F_{test} statistic $= MS_{group}/MS_{error}$. In this example, the F-statistics indicates that between-group variance MS_{group} is more than 5.6 times larger than within-group variance MS_{error}. Under the Null hypothesis, the expected value for F_{test} is 1.0.

p-value: This is the probability associated with the random-variation explanation for the observed differences among the sample means. The probability of getting the observed $F_{test} = 5.6$ if the differences among the sample means were due only to random sampling variation is $p = 0.0024$. Because the random-variation explanation is unlikely, the researcher rejects the Null hypothesis and concludes that at least two of the exercise regimes differ with regard to the mean weight loss of subjects on those regimes.

The Levene Test for Equality of Variances for More Than Two Groups

Now that you can interpret an ANOVA print-out, you can implement the Levene test for equality of variances for more than two groups. In Chapter 10, I described the two-sample Levene test for comparing variances when the Normality assumption of the F-test is violated. A two-sample t-test was used to compare mean absolute deviation between two samples. When performing ANOVA, you need to extend this test to deal with more than two groups. You compute the absolute deviation between each data value and its group mean. Instead of using a two-sample t-test to compare two mean absolute deviation values, you use ANOVA to compare mean absolute deviation among three or more groups. The print-out for the Levene test applied to the data in Example 11.3 is given on the next page:

Analysis of Variance for AbDiff

Source	df	Sums of Squares	Mean Square	F-Ratio	p-Value
Group	3	0.000165323	0.0000551078	0.0000055747	1.0000
Error	44	434.956	9.88537		
Total	47	434.956			

Conclusion: Mean absolute deviation does not differ among the four groups. The equal variances assumption for ANOVA is valid for these data. *Note:* This is the same conclusion we made based on the F_{max} test. When sample size is equal for all groups, the F-test is not much affected by violation of the Normality assumption.

11.4
Multiple Comparisons Tests for Means

When the ANOVA F-test produces a p-value ≤ 0.05, you have strong evidence that at least two of the population means are different. However, you do not know *which* group means are significantly different from each other. There is no single best procedure for determining which means are different after the ANOVA F-tests rejects the overall Null hypothesis. There is a wide variety of such tests, and the one that is best for your particular study depends on a number of factors:

If you were *not* able to state specific Alternative hypotheses at the beginning of your study, you must use an **unplanned multiple comparisons procedure (UMCP)**. The UMCP that is most appropriate for a specific study will depend on several factors:

- Are group sample sizes equal or not?
- Do you want to compare a number of treatment groups against a single control group, or do you want to do all possible pair-wise comparisons of group means?
- Are group variances equal or unequal?
- Do you want to control the comparison-wise error rate or the experiment-wise error rate? (We usually want to control the latter.)
- What multiple comparison tests are available in your statistics computer program?

One approach to comparing multiple group means is to compute "simultaneous" confidence intervals such that two means can be determined to be different at a specific significance level α if their confidence intervals do not overlap. This is particularly useful if you plan to use a graph to display differences among your group means. Some multiple comparisons tests allow for the computation of such a simultaneous confidence interval, others do not. For a detailed discussion of planned and unplanned multiple comparisons tests, see a review by R. W. Day and G. P. Quinn.[1] Based on this review, I present in this book Tukey's honestly significant difference (HSD) test and the Tukey-Kramer test for unequal sample sizes. These tests are relatively simple, have good statistical power, and are widely available in computer statistics programs.

If you were able to state, *before you collect your data,* specific Alternative hypotheses about how the population means might differ, the most powerful procedure

---

[1]Day, R. W. and Quinn, G. P. (1989). Comparisons of treatments after an analysis of variance in ecology. *Ecological Monographs* 59: 433–463.

for testing differences among group means is **planned contrasts** (see the later discussion in this section). Each "contrast" tests a separate hypothesis about specific predicted differences between group means. In this case, you can do each planned contrast at the comparison-wise error rate $\alpha = 0.05$. Planned contrasts have greater statistical power than any other method for testing the significance of differences among more than two sample means.

Tukey's Honestly Significant Difference Test

Tukey's honestly significant difference (HSD) test is used to determine which of three or more sample means are significantly different *after* an ANOVA F-test has indicated there is sufficiently strong evidence to reject the overall Null hypothesis that all population means are equal. This particular multiple comparisons test is appropriate only if the sample sizes for the groups are equal and the assumptions of ANOVA were fulfilled.

The following steps outline the Tukey's HSD procedure.

1. Look up the critical value $Q_{\alpha,[k,df]}$ from the Studentized range table (Appendix Table 6), where α is the allowable experiment-wise Type I error rate (usually set at 0.05), k = number of group means to be compared, and df = the degrees of freedom for the mean square error (MSE) term in the ANOVA table ($= N - k$). Use the number of group means k to determine the appropriate table column, and the df for MSE to determine the appropriate row.

2. Compute the honestly significant difference (HSD):
   $$\text{HSD} = Q_{\alpha,[k,df]} \sqrt{\text{MSE}/n_i}$$
   where n_i = the sample size *per group*. (In Example 11.3, $n_i = 12$.)

3. If the difference between any pair of means is greater than or equal to this HSD value, the difference is significant at the $p \le \alpha$ level. Tukey's HSD test is always two-tailed.

NOTE In Table 6, as the number of means to be compared (k) increases, the critical Q-value also increases. Increasing the Q-value increases the critical HSD value. Hence, as the number of comparisons increases, a larger difference between cell means is required to be deemed significant at the $p = 0.05$ level. This is how Tukey's HSD test controls experiment-wise Type I error rate. However, the greater the number of means compared, the lower the power of the test to detect any difference. You can maximize the power of your study by performing only comparisons critical to the question of interest. ■

Presenting results of the Turkey's HSD test. Typically, the statistics computer program that you utilize to obtain the ANOVA table will allow you to ask for a multiple comparisons test. The format of the print-out of results from such a test varies from one program to another. In most cases, the group means are listed in order from smallest to largest, and some notation is used to display which means are significantly different. I describe some of the more common formats below:

a. *Underline Format:* Cell means that are not significantly different are connected by an underline, while means that are different are *not* connected by an underline.

■ **EXAMPLE**   Suppose you did a Tukey's HSD test for the following four means and HSD = 1.5.

Group 2	Group 1	Group 3	Group 4
2.5	3.1	4.5	5.1 Means listed in order

The differences between means for Groups 1 and 2, 1 and 3, and 3 and 4 are not significant at the $p \leq 0.05$ level. The mean for Group 2 is different from the means for Groups 3 and 4. The mean for Group 1 is different from the mean for Group 4 only. The mean for Group 3 is different from the mean for Group 2 only. The mean for Group 4 is different from the means for Groups 1 and 2.

b. *Letter Format:* Letters are placed after each group mean. If any two means have at least one letter in common, they are not significantly different. Using the example above:

Group 2	Group 1	Group 3	Group 4
2.5 a	3.1 ab	4.5 bc	5.1 c

c. *Matrix Format:* The means are listed as the rows and columns of a matrix. If the mean for a row group significantly differs from the mean for a column group, an * is placed in that position. Using the same example as above:

Group 2	Group 1	Group 3	Group 4
2.5 Group 2			
3.1 Group 1			
4.5 Group 3	*		
5.1 Group 4	*	*	

d. *Confidence Interval Format:* Compute "simultaneous confidence intervals" for the k group means $\bar{x}_i$ as:

Lower confidence limit = $\bar{x}_i - (1/2)$ HSD
Upper confidence limit = $\bar{x}_i + (1/2)$ HSD

Plot the group means in a graph and bracket the plot symbols for the means with the upper and lower bounds of their simultaneous confidence interval. Any two group means are significantly different if their confidence intervals do not overlap. In the graph below, X indicates the value of the cell mean and the dashed lines display the simultaneous confidence intervals. Using the same example as above (HSD = 1.5):

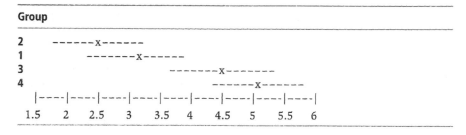

Example 11.4 shows the Tukey's HSD for the ANOVA in Example 11.3.

■ **EXAMPLE 11.4**

Application of Tukey's honestly significant difference test. Look at the ANOVA table presented in Example 11.3. There were four treatment groups ($k = 4$), with $n_i = 12$ experimental subjects in each group, and the MSE = 9.89, with $df_{error} = 44$. Tukey's HSD test for this ANOVA is computed as follows:

1. Critical Q-value from the Studentized range distribution (Appendix Table 6), with experiment-wise error rate $\alpha = 0.05$, $k = 4$ groups, and $df_{error} = 44$ is $Q = 3.791$.

   NOTE The degrees of freedom were rounded down to df = 40. ■

2. Compute Tukey's honestly significant difference:

   $$\text{HSD} = Q_{0.05,[3,63]} \sqrt{\text{MSE}/n_i} = 3.791 \sqrt{9.89/12}$$
   $$= 3.971 \ (0.908)$$
   $$= 3.61$$

3. List group means in order from smallest to largest. If the difference between two group means is greater than or equal to the HSD value, indicate that those group means are significantly different at the $p \leq 0.05$ level.

   a. *Underline Format:*

	Control	**1 × 30**	**2 × 15**	**3 × 10**
Group Means	0.291	−2.33	−3.71	−4.66

   b. *Letter Format:*

	Control	**1 × 30**	**2 × 15**	**3 × 10**
Group Means	0.291a	−2.33 ab	−3.71b	−4.66b

   c. *Matrix Format:*

		Control	**1 × 30**	**2 × 15**	**3 × 10**
0.291	**Control**				
−2.33	**1 × 30**				
−3.71	**2 × 15**	*			
−4.66	**3 × 10**	*			

   d. *Confidence Interval Format:*

Group	
C	---------x---------
1 × 30	---------x---------
2 × 15	----------x---------
3 × 10	----------x---------
	--|----|----|----|----|----|----|----|----|----|
	2 1 0 −1 −2 −3 −4 −5 −6 −7

Interpretation: Mean weight loss for subjects in the 2 × 15 and 3 × 10 exercise groups was greater than for subjects in the control group. There were no differences in mean weight loss among the three exercise regimes; one long bout of exercise was equally effective as multiple shorter bouts of exercise for reducing weight. ■

Tukey-Kramer Test

The **Tukey-Kramer test** is used under the same circumstances as the Tukey's HSD test, but it is used when the samples sizes for the groups are *not* equal. Follow this procedure:

1. Create a table that lists all pair-wise differences among the group means, often arranged with the largest differences at the top and smallest at the bottom.

2. For each comparison of a pair of sample means, compute a minimum significant difference (MSD) value based on the average of the two sample sizes, $\bar{n}$ computed as $(n_i + n_j)/2$. MSD is calculated as shown below:

$$\text{MSD}_{i,j} = Q_{\alpha,[k,\text{df}]} \sqrt{\text{MSE}/\bar{n}}$$

   where i and j indicate that this MSD applies only for the two cell means $\bar{x}_i$ and $\bar{x}_j$. The Q-value is obtained from Appendix Table 6 in the same way as described for the Tukey's HSD test.

3. If the observed difference between the group means $(\bar{x}_i - \bar{x}_j)$ is greater than this $\text{MSD}_{i,j}$ value, the means are significantly different at the $p \leq \alpha$ level for experiment-wise Type I error rate. The Tukey-Kramer test is always two-tailed.

4. Because the sample sizes are not equal, this calculation must be repeated for each pair-wise comparison among the group means, with different values for $\bar{n}$ in each comparison.

Example 11.5 applies the Tukey-Kramer test to the weight-loss experiment described in Example 11.3.

■ **EXAMPLE 11.5**

Application of the Tukey-Kramer test. In Example 11.3 I described an experiment to evaluate the efficacy of three exercise regimes for weight loss. Suppose that some of the study subjects had dropped out before the end of the study so that at the end of the experiment there were unequal sample sizes in the four treatment groups. The resulting summary statistics and ANOVA print-out are given below. Let's assume that the data fulfill the assumptions of Normality and equal variances.

Summary Statistics

Group	Count	Mean	StdDev
Control	12	0.291	2.28896
1 × 30	12	−2.33	3.31438
2 × 15	10	−3.71	3.70513
3 × 10	9	−4.66	3.38695

Analysis of Variance

Source	df	Sums of Squares	Mean Squares	F-Ratio	p-Value
Group	3	150.128	50.0426	4.9561	0.0052
Error	39	393.792	10.0972		
Total	42	543.92			

Conclusion: Mean weight change differed among subjects in the different exercise regime groups ($p = 0.0052$). The Tukey-Kramer test must be used to determine which groups had different weight change.

Tukey-Kramer test.

Pair-wise Comparisons	$\bar{n}$	MSD = $Q_{\alpha=0.05,[4,39]}\sqrt{MSE/\bar{n}}$
3×10 vs. Control $-4.66 - 0.291 = -4.951$	$(9 + 12)/2 = 10.5$	$3.845\sqrt{10.1/10.5} = 3.77$
2×15 vs. Control $-3.71 - 0.291 = -4.001$	$(10 + 12)/2 = 11$	$3.845\sqrt{10.1/11} = 3.68$
1×30 vs. Control $-2.33 - 0.291 = -2.621$	$(12 + 12)/2 = 12$	$3.845\sqrt{10.1/12} = 3.53$
3×10 vs. 1×30 $-4.66 - (-2.33) = -2.33$	$(9 + 12)/2 = 10.5$	$3.845\sqrt{10.1/10.5} = 3.77$
2×15 vs. 1×30 $-3.71 - (-2.33) = -1.38$	$(10 + 12)/2 = 11$	$3.845\sqrt{10.1/11} = 3.68$
3×10 vs. 2×15 $-4.66 - (-3.71) = -0.95$	$(9 + 10)/2 = 9.5$	$3.845\sqrt{10.1/9.5} = 3.96$

Conclusion: The differences in mean weight loss between the control group and the 3×10 and 2×15 exercise regime groups were larger than the computed minimum significant difference (MSD). Therefore, these differences are significant at the $p \leq 0.05$ level. The difference in mean weight loss between the control group and the 1×30 exercise regime group was smaller than the computed MSD, so these means are not significantly different. Mean weight loss was not significantly different among any of the three exercise regime groups.

Planned Contrasts between Group Means

Planned contrasts are performed when the investigator can make specific predictions about expected differences among means for different treatment levels under the Alternative hypotheses *before the data are collected*. Each predicted difference among specific group means is called a *planned contrast*. For example,

- H_a: $\mu_C < (\mu_{T1} = \mu_{T2} = \mu_{T3} = \mu_{T4})$ Treatment means greater than control mean
- H_a: $(\mu_1 = \mu_2) < (\mu_3 = \mu_4 = \mu_5)$ First two population means less than other three means

Of course, the Null hypothesis is still H_0: $\mu_1 = \mu_2 = \mu_3 = \mu_4 = \mu_5$. Each planned contrast is tested by a separate test of significance, with a comparison-wise Type I error rate usually set at $\alpha = 0.05$. This would seem to be a situation where multiple tests lead to a cumulative Type I error rate that is greater than α. However, this is allowed because (1) the H_a's were stated *beforehand* and (2) each H_a tests a separate, independent hypothesis. Given these circumstances, planned contrasts have higher power to detect differences than UMCPs. Unlike UMCPs, it is not required that you perform ANOVA and reject the overall Null hypothesis before doing a planned contrast. However, most researchers perform the ANOVA as a simple means to obtain the MS_{error} value used in the calculations of the planned contrast. The planned contrasts procedure is discussed next, followed by an application (Example 11.6).

Planned contrasts procedure.   Using the following rules, assign *coefficients* (a_i) to each of the group means. These coefficients represent the nature of the comparison or contrast that you want to test.

1. Means proposed to be different in H_a are assigned different coefficients. Means proposed to be equal in H_a have the same coefficient. If a group mean is not part of a specific comparison (contrast), it is given a coefficient of zero.

2. The sum of all the coefficients = 0 (i.e., some must be positive and some negative).

3. When a subset of the group means is proposed to be equal, its coefficients should be the equivalent of taking the average of the subset of means.

4. The means that are hypothesized to be smaller are given negative coefficients and those hypothesized to be larger are given positive coefficients.

   ■ **EXAMPLE** To perform the contrast $(\mu_1 = \mu_2) < (\mu_3 = \mu_4 = \mu_5)$:

   a. The coefficient for groups 1 and 2 would be the same value and the coefficients for groups 3, 4, and 5 would be the same value.

   b. The coefficients for groups 1 and 2 would be negative and sum to -1.0, and the coefficients for groups 3, 4, and 5 would be positive and sum to $+1.0$.

   c. For the coefficients of groups 1 and 2 to be identical and sum to -1.0, they must both be -0.5. Multiplying the two group means by 0.5 and adding the product is the equivalent of computing the pooled estimate of the population mean under the hypothesis that $(\mu_1 = \mu_2)$. That is, $(\bar{x}_1 + \bar{x}_2)/2 = 0.5 \bar{x}_1 + 0.5 \bar{x}_2$.

      For the coefficient for groups 3, 4, and 5 to be identical and sum to $+1.0$, they must be $+0.33$. Multiplying the three group means by 0.33 and adding the product is the equivalent of computing the a pooled estimate of the population mean under the hypothesis that $(\mu_3 = \mu_4 = \mu_5)$. That is, $(\bar{x}_3 + \bar{x}_4 + \bar{x}_5)/3 = 0.33\bar{x}_3 + 0.33\bar{x}_4 + 0.33\bar{x}_5$.

A *contrast* (designated by the Greek symbol "psi," ψ) is a "linear combination" of population means equal to the sum of the products of the means times their coefficients:

$$\psi = \Sigma^k_{i=1} (a_i\mu_i) \text{ (for } k \text{ group means)}$$

which is estimated using the sample group means.

Sample Contrast (C)

$$C = \Sigma^k_{i=1} a_i\bar{x}_i$$

For the alternative hypothesis H_a: $(\mu_1 = \mu_2) < (\mu_3 = \mu_4 = \mu_5)$, the contrast C would be computed as:

$$C = -0.5\bar{x}_1 - 0.5\bar{x}_2 + 0.33\bar{x}_3 + 0.33\bar{x}_4 + 0.33\bar{x}_5$$

For the alternative H_a: $\mu_C < (\mu_{T1} = \mu_{T2} = \mu_{T3} = \mu_{T4})$, the contrast would be between the control group mean and the mean of the four treatment group means.

$$(\mu_C)/1 < (\mu_{T1} + \mu_{T2} + \mu_{T3} + \mu_{T4})/4$$
$$C = -1.0\bar{x}_C + 0.25\bar{x}_{T1} + 0.25\bar{x}_{T2} + 0.25\bar{x}_{T3} + 0.25\bar{x}_{T4}$$

Important: When H_0 is true (all population means are equal), the expected value of the linear combination (contrast) will be $E(C) = \psi = 0$. When the group means are *not* equal, the contrast C will be much larger or smaller than 0. To determine the significance of a contrast C, we must compare the size of the difference $(C - 0)$ to the standard error of the contrast S_C, using a one-sample t-test. ■

Compute the **standard error of the contrast** of group means (S_C): If we can assume that the group variances are equal, we can use the pooled estimate of the variance = MS_{error} from the ANOVA table to compute the variance of the sample contrast S^2_C:

$$S^2_C = MSE [(a_1)^2/n_1 + (a_2)^2/n_2 + (a_3)^2/n_3 + (a_4)^2/n_4 + (a_5)^2/n_5]$$

For the contrast $(\mu_1 = \mu_2) < (\mu_3 = \mu_4 = \mu_5)$ described in the example above:

$$S^2_C = MSE [(-0.5)^2/n_1 + (-0.5)^2/n_2 + (0.33)^2/n_3 + (0.33)^2/n_4 + (0.33)^2/n_5]$$

The standard error of the contrast is the square root of the variance.

$$S_C = \sqrt{S_C^2}$$

Compute the t_{test} **statistic for the contrast** and determine the p-value using the degrees of freedom associated with MS_{error}. The test statistic is computed as:

$$t_{test} = C/S_C$$

This t_{test} statistic represents how far the sample contrast value C is away from the expected value under the Null hypothesis (zero) in "standard error units." If the value of C is quite different from zero, this would constitute strong evidence that the true population means are different in the manner specified by the contrast. The p-value is obtained from the t-table with df = df_{error} from the ANOVA Table = $(N - k)$.

Suppose that the researcher who performed the exercise experiment described in Example 11.3 had been able to make specific predictions prior to the experiment. In Example 11.6 I present a plausible prediction and its associated planned contrast.

■ **EXAMPLE 11.6**

Application of planned contrasts. *Prediction:* Individuals in the three exercise regime groups will lose more weight over the 12-week study than individuals in the control group. The appropriate hypotheses and contrast would be:

- H_0: $\mu_C = \mu_{1 \times 30} = \mu_{2 \times 15} = \mu_{3 \times 10}$
- H_a: $\mu_C > \mu_{1 \times 30} = \mu_{2 \times 15} = \mu_{3 \times 10}$

> **NOTE** Since weight loss is indicated by negative mean ΔWt, the mean for the control group is expected to be greater than the means for the exercise groups. ■

Sample Contrast
$$C = 1.0\bar{x}_C - 0.33\bar{x}_{1 \times 30} - 0.33\bar{x}_{2 \times 15} - 0.33\bar{x}_{3 \times 10}$$
$$= 1.0(0.29) - 0.33(-2.33) - 0.33(-3.71) - 0.33(-4.66)$$
$$= 3.82$$

Standard Error of the Contrast
$$S_C = \sqrt{MSE \, \Sigma a_i^2/n_i}$$
$$= \sqrt{9.89[(1^2/12) + (0.33^2/12) + (0.33^2/12) + (0.33^2/12)]}$$
$$= \sqrt{9.89(0.1106)}$$
$$= 1.05$$

t-Test Statistic for the Contrast
$t_{test} = C/S_C = 3.82/1.05 = 3.64$
$P[t_{df=44} \geq 3.64] = p < 0.0005$

NOTE The direction of the inequality symbol in $P[t_{df=44} \geq 3.655]$ is the same as specified in H_a. ■

Conclusion: Mean weight loss for individuals in the three exercise regime groups was greater than for individuals in the control group ($p < 0.0005$). ▨

11.5
Test for Comparing More Than Two Medians When the Normality Assumption Is Violated

If the sample sizes for the groups being compared are equal, ANOVA is relatively robust against violations of the Normality and equal variances assumptions. This means you can proceed with ANOVA even if the data values exhibit modest skewness or the F_{max} test determines that there is evidence the population variances are not all equal. However, if the sample sizes are *not* equal, or the data values exhibit extreme skewness or outliers, the assumptions for ANOVA are not fulfilled, and conclusions based on the ANOVA *p*-value may be invalid. In such situations, you have two options for comparing the groups.

1. *Transform the data:* Nonlinear transformation of the data will often correct violations of both Normality and equal variances assumptions. You must apply the same transformation to the data values in all groups, and then perform ANOVA, multiple comparisons tests, or planned contrasts using the transformed data values. Conclusions from tests applied to the transformed data values are equally applicable to the original data. A more detailed description of the use of nonlinear transformations for tests of significance is given in Chapter 9. When the test of significance is done on transformed data, you may still want to report the effect sizes using nontransformed means, as this scale of measurement is usually more meaningful. However, this may be problematic if the means and medians are quite different due to outliers or skewness. In such cases, you should report the effect size based on differences among the group medians of the original data values.

2. *Perform a nonparametric test:* Nonparametric tests do not require the population distributions to be Normal, and they are less sensitive to unequal variances than the parametric *F*-test used for ANOVA.

Kruskal-Wallis (K-W) Test

1. **Application:** The **Kruskal-Wallis (K-W) test** is the nonparametric equivalent of a one-way analysis of variance (ANOVA). This test is used to compare the centers of the distributions for three or more groups with regard to one treatment variable. The K-W test is applied when the Normality assumption for ANOVA is violated, no single transformation corrects this violation for all groups, and the sample sizes are unequal (so ANOVA is sensitive to violations of this assumption).

   If the shape of the distributions of data values for the groups are similar, the K-W test can be used to assess whether or not there is evidence that the population *medians* are equal. If the distributions of the data values for the groups are *not* similar, the K-W test can still be used to assess the similarity of the data values for different groups.

NOTE The K-W test is *not* a "cure-all" for violations of the equal variances assumption of ANOVA. The K-W test will misrepresent the Type I error rate if sample sizes and group variances are highly unequal, but not as much as ANOVA. ■

In the worst-case scenario (unequal sample sizes, unequal variances, and non-Normal data distributions), it may be necessary to utilize more complex analytical procedures, which are beyond the scope of this text.

The K-W test is an extension of the two-sample Mann-Whitney U-test to handle comparisons among three or more groups. Although the K-W test can be applied to data that meet the assumptions for parametric ANOVA, it is not as powerful as ANOVA *when the assumptions are fulfilled*. Hence, if the data meet the Normality and equal variances assumptions of parametric ANOVA, the parametric test is most appropriate as it will minimize both Type I and Type II error rates. Alternatively, when the ANOVA assumptions are violated, the K-W test may be more powerful (i.e., be more likely to detect a real difference than an invalid ANOVA).

Assumptions: (1) The data were obtained by an unbiased, randomized study design, and (2) the population distributions being compared have similar shape and spread.

NOTE The K-W test is less affected by violations of assumption (2) than is ANOVA. ■

2. *Statement of Hypotheses*

   If the data distributions indicate that it is valid to assume that the population distributions have similar shape and spread, the Kruskal-Wallis procedure tests the hypothesis that the population *medians* are equal:

   - H_0: $M_1 = M_2 = M_3 = \ldots M_k$
   - H_a: One or more populations have a different *median* than other populations.

   If the data distributions do not have similar shape and spread, this procedure tests the hypothesis that randomly chosen data values from one population are equally likely to be greater or less than randomly chosen values from the other populations.

3. **The Kruskal-Wallis Test Statistic**

   There are different ways to perform the Kruskal-Wallis test, one based on an extension of the Mann-Whitney U_{test} calculations and another based on ANOVA calculations applied to the ranks of the data values. Statistics computer programs typically use the extension of Mann-Whitney U-test calculations to compute an H_{test} statistic, as follows:

   $$H_{test} = \frac{12}{N(N+1)} \sum_{i=1}^{k} \frac{\Sigma R_i^2}{n_i} - 3(N+1)$$

   where N is the total number of observations for all groups combined, n_i is the number of observations in group i, and ΣR_i is the sum of the rank values for group i.

   If your statistics computer program does not provide the K-W test, but does provide analysis of variance, applying ANOVA to the *ranks* of the data values is an alternate way to implement this test.

a. Combine all the data values for the various groups and assign each data value a rank based on its size relative to all other data values. Assign rank 1 for the smallest value, 2 for the second smallest and rank N (= total number of observations for all groups combined) for the largest value.

NOTE Most computer statistics programs and spreadsheets have a feature that will automatically create a variable containing the ranks of the data values. However, you must organize the data so that values for all groups are in a single column (variable), with a second column (variable) that identifies the group affiliation for each data value. ■

b. Perform an analysis of variance on the rank values to obtain the F_{test} statistic and associated p-value.

4. **Sampling Distribution of H_{test} and F_{test}**

   The sampling distribution for the Kruskal-Wallis H_{test} statistic is approximately a χ^2 distribution (pronounced "Ki-squared," Appendix Table 8), with df $= k - 1$, where k = the number of groups being compared. The χ^2 distribution is similar to the F-distribution in that the range of possible values includes only positive numbers, and the shape of the distribution is positively skewed (see Chapter 12 for additional description of this probability distribution). To obtain a p-value from Appendix Table 8, use the same procedure as described for the t-table.

   The sampling distribution for the Kruskal-Wallis F_{test} statistic is approximately an F-distribution with $df_{between} = k - 1$ and $df_{within} = N - k - 1$. As described in Section 11.3, the F-distribution has an expected value of 1.0 if the Null hypothesis is true. The larger the computed F_{test} value, the less likely this outcome is due to random sampling variation alone, the stronger the evidence that at least two of the populations are different.

5. **p-Value**

   a. The K-W test does not distinguish between positive and negative differences. Hence, the K-W test is always a two-tailed test.

   b. *Stating conclusions:*

   If $p \leq 0.05$: The probability associated with the random-variation explanation for the observed differences among the sample or treatment group *medians* is sufficiently low to reject this explanation. If an appropriate study design has been correctly implemented, the only remaining likely explanation is that some of the population medians are actually different. Because this is a simultaneous test, this is the only conclusion you can make at this point. You have no basis for knowing *which* of the population medians are different from each other. If the K-W test p-value is ≤ 0.05, you will need to implement a multiple comparisons test for medians (see the next section) to determine which sample medians are significantly different. The nature of your conclusion will depend on the outcome of the multiple comparisons test.

   If $0.05 < p \leq 0.10$: The probability associated with the random-variation explanation for the observed differences among the sample medians is sufficiently high that we cannot confidently exclude this explanation. However, the probability is sufficiently low that we might suspect there is a difference among the population medians. Some scientists would conclude that the results "suggest" that a difference may exist among some of the population medians, but more data are needed to verify this. However, a multiple comparisons test would typically *not* be done.

■ **EXAMPLE**   "The differences in median weight loss among the exercise regimes suggest that multiple short bouts of exercise are more beneficial than one longer bout (p = 0.079), but more data are needed to verify this." ■

If p > 0.10: The probability associated with the random-variation explanation for the observed differences among the sample medians is so high that this explanation cannot be excluded. Hence, there is insufficient evidence to conclude that there is a difference among the population medians.

■ **EXAMPLE**   "There were no significant differences in median weight loss among the three exercise regimes (p = 0.22)." ■

Multiple Comparisons Tests for Medians

If the K-W test provides sufficient evidence to reject the Null hypothesis and conclude that the medians for at least two of the populations are different, you must next determine *which* medians differ. There are a variety of multiple comparisons tests for medians. To be consistent, I present here the test for medians that is analogous to the Tukey's HSD test for means. This Tukey-type test is based on a comparison of the mean ranks for the different groups.

Procedure for equal group sample sizes.

1. Compute the **mean rank** ($\bar{R}_i$) for each of the k groups ($= \Sigma R_i/n_i$).

    NOTE If you use a statistics computer program that performs the K-W test, mean ranks for each group will usually be part of the print-out. If you use the "ANOVA on ranks" procedure, also request summary statistics for the rank variable by group to obtain mean ranks. To facilitate comparisons in Step 4 below, mean ranks for the groups should be listed in order from smallest to largest. ■

2. Compute the **standard error** (**SE**) of the difference between mean ranks:

$$SE = \sqrt{\frac{k(N + 1)}{12}}$$

    where k = the number of groups and N = the total sample size (all groups combined).

3. Compute the minimum significant difference between two medians:

    MSD $= Q_{\alpha = 0.05, k, df = \infty}$ (SE)

    where: The Q-value is obtained from the Studentized range table (Appendix Table 6), with k = the number of groups being compared and df = ∞.

4. If the difference between any two mean ranks ($\bar{R}_1 - \bar{R}_2$) is greater than or equal to the minimum significant difference, this is strong evidence that the corresponding populations have different medians.

Procedure for unequal group sample sizes.

1. Compute all the pair-wise differences between group mean ranks ($\bar{R}_1 - \bar{R}_2$).
2. For each pair-wise difference, compute the standard error of the difference.

$$\text{SE} = \sqrt{\frac{N(N+1)}{12}\left(\frac{1}{n_1} + \frac{1}{n_2}\right)}$$

where N is the total sample size, and n_1 and n_2 are the sample sizes for the two groups being compared.

3. Select the **critical *q*-value** to control cumulative Type I error rate to $\alpha = 0.05$ from the table below (k = number of groups to be compared).

Critical *q*-values $\alpha = 0.05$ (k = **number of groups being compared**)

$k = 2$	3	4	5	6	7	8	9	10	11	12	13	14
1.96	2.39	2.64	2.81	2.94	3.04	3.12	3.20	3.26	3.32	3.37	3.41	3.46

4. Compute the minimum significant difference (MSD).

   $$\text{MSD} = q_{k,\alpha}(\text{SE})$$

5. If the difference between any two group mean ranks is greater than or equal to the critical MSD value computed for those two groups, this provides sufficient evidence to conclude that the corresponding two population medians differ.

The following two examples (Examples 11.7 and 11.8) apply the Kruskal-Wallis test and the data of the exercise study.

▪ **EXAMPLE 11.7**

Application of the Kruskal-Wallis test for non-Normal data (test based on F_{test} statistic with equal sample sizes). Suppose that the data from the exercise study described in Example 11.3 were as shown in the exploratory data analyses below.

Summary Statistics

Group	Count	Mean	Median	StdDev	Min	Max
Control	12	0.875	−0.33	3.61	−2.99	10.17
1 × 30	12	−2.87	−4.07	4.41	−8.38	7.37
2 × 15	12	−4.80	−6.19	4.19	−9.50	4.43
3 × 10	12	−4.45	−5.73	4.78	−12.4	5.57

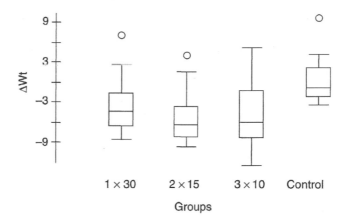

Assessing ANOVA Assumptions

a. *Randomized, unbiased study design:* Valid, as discussed in Example 11.3.

b. *Normality assumption:* All four data distributions exhibit some degree of positive skewness and three have outliers. This assumption is violated.

c. *Equal variances assumption:* Given that the Normality assumption is violated, the F_{max} test is not appropriate for assessing the equal variances assumption using these data. The results of the Levene test listed below indicate that the population variances can be assumed to be equal.

Analysis of Variance for Absolute Differences (Levene Test)

Source	df	Sums of Squares	Mean Square	F-ratio	p-Value
Group	3	8.52007	2.84002	0.4085	0.7477
Error	44	305.899	6.95224		
Total	47	314.419			

Let's suppose that no one transformation is able to make all four data distributions even approximately Normal. The only remaining option is the K-W test. Let us further suppose that your statistics computer program does not provide the K-W test. You would convert the original data values into ranks and then perform an ANOVA on the ranks. The results are provided below.

Note: Requesting descriptive statistics for the rank values in each group is an easy way to obtain the mean rank values needed for a multiple comparisons test of the four group medians. ■

One-Way ANOVA: Rank ΔWt versus Group

Source	df	Sums of Squares	Mean Square	F-ratio	p-Value
Group	3	2592.33	864.111	5.7436	0.0021
Error	44	6619.67	150.447		
Total	47	9212			

Conclusion: With $p = 0.0021$, the probability associated with the random-variation explanation for the observed differences among the groups is sufficiently unlikely for us to reject this explanation. The researcher concludes that median weight loss differs among women in the four exercise regime groups. A table of the descriptive statistics for *ranks* of ΔWt is shown below:

Group	Count	Mean	Median	Min	Max
Control	12	36.6	37.0	26	48
1 × 30	12	23.9	21.5	5	47
2 × 15	12	17.4	13.5	2	44
3 × 10	12	20.1	16.5	1	46

Multiple Comparisons Test for Medians

$$SE = \sqrt{k(N + 1)/12} = \sqrt{4(48 + 1)/12} = 4.04$$

$$Q_{k=4, \alpha = 0.05} = 3.633$$

$$MSD = Q_{k=4, \alpha = 0.05} (SE) = 3.633 (4.04) = 14.68$$

	2 × 15	3 × 10	1 × 30	Control
Mean Ranks	17.4	20.1	23.9	36.6

Conclusion: Median weight loss for the 2 × 15 and 3 × 10 exercise regime groups was greater than for the control group. Weight loss did not differ between the 1 × 30 and control groups. There were no significant differences in weight loss among the exercise regimes. ▦

■ **EXAMPLE 11.7**

Kruskal-Wallis test (based on H_{test}) and multiple comparisons for medians with unequal sample sizes. Suppose that some of the participants in the exercise experiment described in previous examples had dropped out of the study, resulting in unequal sample sizes in the four treatment groups. Further suppose that the data distributions were similar to those described in Example 11.7, so that the K-W test is the appropriate analysis. The following is a print-out of a K-W test based on the H_{test} statistic obtained from a statistics computer program.

Kruskal-Wallis Test on ΔWt

Group	N	Median	Mean Rank	Z
1 × 30	11	−4.5110	20.5	−0.33
2 × 15	10	−6.1858	15.4	−1.80
3 × 10	9	−5.7455	17.1	−1.21
Control	12	−0.3314	30.8	3.12
Overall	42		21.5	

$H_{test} = 10.65$, df $= 3$, $p = 0.014$

Conclusion: Median ΔWt differs among at least two of the treatment groups ($p = 0.014$). The remaining question in this analysis is *which* groups differ with regard to median ΔWt. The multiple comparisons test below is necessary because of the unequal sample sizes.

Multiple Comparisons Test for Medians with Unequal n_i($N = 42$)

Comparison	$(\bar{R}_1 - \bar{R}_2)$	n_1	n_2	SE	MSD = $q_{k=4}$ SE	Conclusion
Control − 2 × 15	30.8 − 15.4 = **15.4**	12	10	5.25[a]	2.64 (5.25) = **13.9**	Significant difference
Control − 3 × 10	30.8 − 17.1 = 13.7	12	9	5.41	2.64 (5.41) = 14.3	Not significant (N.s.)
Control − 1 × 30	30.8 − 20.5 = 10.3	12	11	5.12	2.64 (5.12) = 13.5	N.s.
1 × 30 − 2 × 15	20.5 − 15.4 = 5.1	11	10	5.36	2.64 (5.36) = 14.1	N.s.
1 × 30 − 3 × 10	20.5 − 17.1 = 3.4	11	9	5.51	2.64 (5.51) = 14.5	N.s.
3 × 10 − 2 × 15	17.1 − 15.4 = 1.7	9	10	5.64	2.64 (5.64) = 14.9	N.s.

[a]Example calculation of SE: $= \sqrt{[N(N + 1)/12][1/n_1 + 1/n_2]}$

$$= \sqrt{[42(42 + 1)/12][1/12 + 1/10]} = 5.25$$

Conclusion: Median weight loss for the 2 × 15 exercise regime group was greater than for the control group. Weight loss was not significantly different among any of the other groups.

CHAPTER SUMMARY

1. When multiple tests of significance are required to come to a conclusion about a single Null hypothesis, the overall probability that a Type I error will occur increases as the number of tests increases (*cumulative error rate*). This requires that we distinguish **comparison-wise error rate** (probability that a Type I error will occur for one test of significance) from **experiment-wise error rate** (probability that a Type I error will occur if the overall Null hypothesis is rejected due to one or more Type I errors among the multiple tests of significance).

2. When multiple tests of significance are required because means for *multiple variables* must be *compared among two or more groups*, the **Bonferroni procedure** can be used to control experiment-wise error rate. This procedure reduces the acceptable comparison-wise Type I error rate for each test of significance so that the overall experiment-wise error rate does not exceed a specified level (typically $\alpha = 0.05$). However, reducing the allowable Type I error rate for each test of significance makes it more difficult to detect a difference, reducing power.

3. When multiple tests of significance are required because means for *one variable* must be compared among more than two groups, **analysis of variance (ANOVA)** provides a single, simultaneous test of the Null hypothesis that all group means are equal. With a single test, experiment-wise and comparison-wise error rate are the same. This simultaneous test is accomplished by summing all differences among group means into a single value for **between-group variance**. An *F*-test is used to compare this variance to **within-group variance**, which reflects random variation among individuals within the groups being compared. If between-group variance is sufficiently larger than within-group variance, this provides evidence that some of the group means differ.

4. If the results of analysis of variance indicate that some group means differ from others, **multiple comparisons tests** or **planned contrasts** are used to determine *which* group means are different.

5. The **Kruskal-Wallis test** is a nonparametric test that is used to evaluate sample data to test the Null

hypothesis that *medians* for three or more groups are equal. This test is used when the Normality assumption of analysis of variance is violated and nonlinear transformations are not effective for making the data distributions approximately Normal.

6. If the results from a Kruskal-Wallis test indicate that two or more group medians differ, a **multiple comparisons test for medians** should be performed to determine *which* medians differ.

12 Tests for Comparing Two or More Sample Proportions

INTRODUCTION In Chapter 8, the discussion of tests of significance for proportions was limited to situations in which the categorical variable had only two classes (e.g., male/female, live/dead, heads/tails). However, in many situations the categorical variable of interest has more than two classes. For example, blood type has four classes (A, B, AB, O), flower color could have three classes (red, pink, white), and taxonomic group (e.g., species, genus) is a variable that could have tens or hundreds of possible classes. In Chapter 8, the two-sample Z-test was used to compare the proportion of individuals in a single class between two samples. However, many studies require comparisons of the proportions of individuals in two or more classes of a categorical variable among more than two samples. In this chapter I will describe tests of significance for these more complex study designs.

To make comparisons involving more than two classes of a categorical variable, or comparisons among more than two samples using one-sample or two-sample Z-tests would require multiple tests of significance. In Chapter 11, I introduced the concept of cumulative Type I error rate associated with multiple related tests of significance. When more than one test of significance is performed to evaluate a single "overall" Null hypothesis, the probability increases that an apparently significant difference will be found due only to random sampling variation when there really is no difference. Although the probability of Type I error is usually held to $p \le 0.05$ for individual tests of significance, when multiple tests are used to evaluate the same hypothesis, the overall Type I error rate accumulates to unacceptably high levels. In Chapter 11, I described two approaches for preventing this accumulation of Type I error rate: the Bonferroni procedure and simultaneous tests of significance.

1. *Bonferroni procedure:* Control cumulative Type I error rate by using a smaller critical p-value for each of the multiple tests of significance. This procedure can be

used in all situations for which multiple tests of significance are required, but it often has unacceptably low power. The Bonferroni procedure is based solely on p-values, the number of tests of significance done (Ntests), and the desired overall Type I error rate (α). The nature of the multiple tests of significance does not enter into the Bonferroni calculations. The procedure is the same for multiple tests involving sample proportions as for tests involving sample means or medians. Hence, I will not repeat the description of the procedure here, and you should refer to Chapter 11 for a description of this analysis.

2. *Simultaneous test of significance:* In a single test of significance with $\alpha = 0.05$, test the overall Null hypothesis that the sample proportions for all classes are equal to the values stipulated by the overall Null hypothesis. If the overall Null hypothesis is rejected, this provides evidence that the true proportion for at least one class or population differs from some Null hypothesis value. However, additional analyses are required to determine which group proportions are significantly different. χ^2 tests (pronounced "Ki-squared" and sometimes written "chi-squared") are the simultaneous tests for comparing two or more sample proportions to test an overall Null hypothesis.

Even though the format of χ^2 tests of significance appears quite different from other tests you have already learned, the underlying principles will be the same. These tests are based on a test statistic that measures the degree to which the sample data deviate from expected values stipulated by the overall Null hypothesis. The larger the difference between sample test statistic and the Null hypothesis expected value, the less likely this difference is due simply to random sampling variation, and the stronger the evidence against the Null hypothesis. ■

GOALS AND OBJECTIVES

In this chapter you will learn when and how to use χ^2 tests of significance for evaluating: (1) differences between sample proportions for two or more classes of a categorical variable, computed from a single sample, and hypothetical values for the unknown population proportions, and (2) differences among sample proportions for two or more classes of a categorical variable, computed from two or more samples, as evidence of differences among their corresponding population proportions.

Given the study design and characteristics of the data, by the end of the chapter you will be able to select the ap-

propriate χ^2 test of significance, correctly implement the test, and draw appropriate conclusions with regard to the original scientific question. You are responsible for the following procedures:

- χ^2 goodness-of-fit test
- χ^2 test of homogeneity (of population proportions)
- Subset χ^2 test (to identify which proportions differ)
- Multiple comparisons tests for proportions used when the χ^2 test rejects the overall Null hypothesis.

12.1
χ^2 Goodness-of-Fit Test

1. **Application:** The χ^2 **goodness-of-fit test** is used when a *categorical variable has k ≥ 2 classes* and you wish to compare the two or more sample proportions computed from a *single sample* to hypothetical population proportion values as specified by a Null hypothesis. *Assumptions:* (1) The individuals that constitute the sample from the population were chosen by a randomized, unbiased procedure, so that the expected values of the sample proportions for each of the k classes $E(\hat{p}_1, \hat{p}_2, \hat{p}_3, ..., \hat{p}_k)$ equal the true population proportions $P_1, P_2, P_3, ..., P_k$, respectively, (2) the population is sufficiently large relative to the sample size ($N > 100n$) that the proportion of individuals in each class of the categorical variable (P_i) can be assumed constant, and (3) the sample size is large enough that $n/k \geq 2$ and $n^2/k \geq 10$, where n is the number of individuals in the sample and k is the number of classes for the categorical variable.

2. **Statement of Hypotheses**

   - H_0: $P_1 = P_{H0,1}$ and $P_2 = P_{H0,2}$ and $P_3 = P_{H0,3}$ and . . . $P_k = P_{H0,k}$

     The proportion of individuals in each of k classes of the categorical variable is equal to the value stipulated in a Null hypothesis.

   - H_a: One or more of the population proportions for the k classes of the categorical variable is *not* equal to the value stipulated in the Null hypothesis.

3. **Test Statistic**

   The χ^2_{test} statistic is based on the difference between the "observed" *count* of individuals that fall into a class of the categorical variable (O_i) and the "expected" count of individuals that should be in that class, given the sample size n and that the Null hypothesis value $P_{H0,i}$ ($E_i = nP_{H0,i}$). Deviations between the observed and expected counts are summed across all k classes of the categorical variable to obtain the χ^2_{test} statistic, as shown below:

   $$\chi^2_{test} = \sum_{i=1}^{k} \frac{(O_i - E_i)^2}{E_i}$$

   NOTE The difference ($O_i - E_i$) is squared, so χ^2_{test} can take only positive values. If you divide the squared difference by the expected value E_i, classes with small E_i values are given equal weight with classes that have large E_i values. The greater the differences between observed and expected counts of individuals in the k classes, the larger will be the value of

χ^2_{test} and the less likely these differences are due merely to random sampling variation. This test is a "simultaneous test" because the values for $(O_i - E_i)^2/E_i$ computed for the k classes are summed to obtain a single test statistic for an overall Null hypothesis of "no differences." ■

4. Sampling Distribution for the χ^2_{test} Statistic

The sampling distribution for the χ^2_{test} statistic under the Null hypothesis is approximated by a continuous, asymmetric probability distribution called the χ^2 distribution. Like the t-distribution and F-distribution, there are multiple χ^2 distributions, one for each value of degrees of freedom. Given the way that χ^2 is computed, the range of possible values for this test statistic extends from zero to positive infinity (no negative values). Like the t-distribution and F-distribution, the shape of the χ^2 distribution depends on the number of degrees of freedom (df) in the sample data. The probability density curve for the χ^2 distribution is very positively skewed when df is a small number. As df increases, the center of this distribution shifts to larger values and the shape becomes progressively more symmetric. For the χ^2 goodness-of-fit test, the number of degrees of freedom is computed as df $= k - 1$ (where $k =$ the number of classes for the categorical variable). Looking at Figure 12.1, you can see that as df increases, larger χ^2_{test} values are required to attain a p-value ≤ 0.05. The reason is that χ^2_{test} is a *sum* across the k classes of the categorical variable. The more classes the categorical variable has,

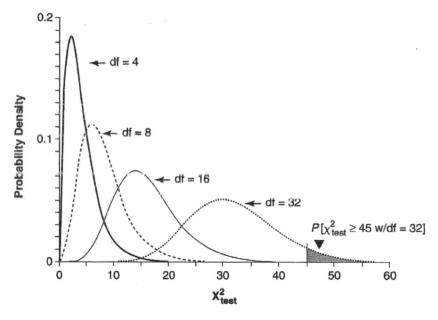

FIGURE 12.1 Probability distribution of χ^2_{test} for a range of degrees of freedom.   The greater the number of classes for a categorical variable, or cells in a two-way table, the larger will be the sum of $(O - E)^2/E$ values, even if the differences are due only to random sampling variation. This expectation is shown above; notice how the center of the sampling distribution of χ^2_{test} under the Null hypothesis shifts to larger values as the number of degrees of freedom increases. Hence, as the value of the degrees of freedom increases, the critical χ^2 value required to reject the Null hypothesis with $\alpha = 0.05$ also increases. Because χ^2_{test} can only have positive values, the p-value is represented by the area in the right tail of the sampling distribution, even though χ^2 tests are always two-tailed.

the greater the number of $(O_i - E_i)^2/E_i$ values that are summed, the larger χ^2_{test} will be, even if the differences $(O_i - E_i)$ are due only to random sampling variation.

Approximate p-values for the χ^2_{test} statistic can be obtained from Appendix Table 8. Find the row with the df value appropriate for your analysis, then read across that row to the two table values that bracket your computed χ^2_{test} value. The approximate p-value is between the two probabilities listed at the top of these two columns.

Exact p-values for the χ^2_{test} statistic can be obtained from many computer spreadsheet programs using the function **CHIDIST**(χ^2_{test},**df**).

■ **EXAMPLE**   In MS Excel, **= CHIDIST(4.8,2)** returns the exact p-value 0.0907. In Correl QuattroPro, **@CHIDIST(4.8,2)** returns the same p-value.   ■

5. **Test of Significance**

   a. Because the computation of χ^2_{test} does not distinguish between positive and negative differences $(O_i - E_i)$, the p-value for this test is always two-tailed.

   b. *Stating conclusions:*

   If $p \leq 0.05$: The probability associated with the random-variation explanation for the observed differences between sample proportions $\hat{p}_i$ for k classes of a categorical variable and the corresponding Null hypothesis values for the population proportions P_i, is sufficiently low to reject this explanation. If an appropriate study design has been correctly implemented, the only remaining likely explanation is that one or more of the true population proportion values differ from those stipulated by the Null hypothesis. However, because the χ^2 test is a simultaneous test, at this point you have no basis for knowing *which* proportions are different. It is possible that the proportions for some classes differ from the Null hypothesis values, while others do not. There is no "multiple comparisons test" for the goodness-of-fit test. The following section describes two procedures for exploring which proportions differ from the Null hypothesis values. The nature of your conclusion statement will depend on the outcome of these procedures.

   If $0.05 < p \leq 0.10$: The probability associated with the random-variation explanation for the observed differences between sample proportions and the Null hypothesis values is sufficiently high that we cannot confidently exclude this explanation. However, the probability is sufficiently low that we might suspect that there really are differences. Some scientists would conclude that the results "suggest" a deviation from the Null hypothesis proportions, but more data are needed to verify this conclusion. In such cases, no addition analysis is done to determine which proportions differ from the Null hypothesis values.

   ■ **EXAMPLE**   "The observed proportions of individuals who exhibited various combinations of two traits suggest that the traits are not independent, but rather, are controlled by genes linked on the same chromosome ($p = 0.091$). However, more data are needed to verify this."   ■

   If $p > 0.10$: The probability associated with the random-variation explanation for the observed differences between sample proportions and the Null hypothesis values is so high that this explanation cannot be excluded. Hence, there is insufficient evidence to conclude that the true population proportion values for the k classes differ from the values stipulated by the Null hypothesis.

■ **EXAMPLE** "The observed proportion of individuals who exhibited various combinations of two traits did not differ from expected values if the two traits are controlled by genes on separate chromosomes ($p = 0.87$)."   ■

6. **Determining Which Proportions Differ from Null Hypothesis Values**

If the overall Null hypothesis of the goodness-of-fit test is rejected, the next obvious step is to determine which of the k class proportions differ from their respective Null hypothesis values. There is no multiple comparisons test for this situation. Two options for investigating which class proportions differ from their Null hypothesis value include:

a. *Bonferroni approach:* Perform multiple one-sample Z-tests that evaluate each difference $(\hat{p}_i - P_{H0,i})$.

b. *Subset χ^2 analysis:* Rerun the χ^2 test multiple times, deleting a different class of the categorical variable from the analysis each time. If the χ^2 test with class i missing is not significant, but the test with class i included is significant, this is evidence that $\hat{p}_i$ is significantly different from $P_{H0,i}$. This procedure is most efficiently done by first dropping classes with the largest values for $(O_i - E_i)^2/E_i$.

NOTE These two procedures can help identify the classes whose proportions are most different from the Null hypothesis values. However, the strength of this evidence is not as great as if you had hypothesized specific differences *before* looking at the data. You should obtain new data to perform an independent test to verify the specific differences identified by these procedures. ■

Example 12.1 shows how the χ^2 goodness-of-fit test is applied to a genetics experiment.

■ **EXAMPLE 12.1**

Application of the χ^2 goodness-of-fit test.

1. **Application:** Dihybrid crosses are used by geneticists to determine if two genes are located on the same chromosome. Controlled matings are made between males and females whose genetic characteristics are known. If the two genes are located on separate chromosomes, the expected numbers of offspring who exhibit the four possible combinations of traits controlled by these two genes will be in a ratio 9:3:3:1. This ratio statement can be re-expressed in terms of expected proportion values under the Null hypothesis as follows: $P_1 = 9/16 = 0.5625$, P_2 and $P_3 = 3/16 = 0.1875$, and $P_4 = 1/16 = 0.0625$.

A student performs an experiment to determine if the gene that codes for wing length of fruit flies is on a different chromosome than the gene that codes for body color. The two classes for wing length are "long" (L) and "short" (S). The two classes for body color are "gray" (G) and "black" (B). The matings produce 480 offspring with observed frequencies for the four possible combinations of traits shown below. If the genes are on separate chromosomes and passed to offspring independently, the expected numbers of individuals in the combination classes LG, LB, SG, and SB should be in the ratio 9:3:3:1, respectively. If the observed numbers deviate substantially from this expected ratio, this is evidence that the genes for wing length and body color are on the same chromosome, and *not* inherited independently.

Assessing assumptions:

a. The study design provided for random mating of fruit flies with known genetic characteristics. We will assume the student performed the procedures correctly.

b. In this kind of study, the population is all fruit flies and can be assumed to be sufficiently large.

c. The sample size is large enough for the sampling distribution of the χ^2_{test} statistic to approximate a χ^2 distribution with $k - 1$ degrees of freedom. With $n = 480$ and $k = 4$, $n/k = 480/4 = 120 \geq 2$, and $n^2/k = (480)^2/4 = 57600 \geq 10$.

2. **Statement of Hypotheses:**

- H_0: $P_{LG} = 0.5625$ $P_{LB} = 0.1875$ $P_{SG} = 0.1875$ $P_{SB} = 0.0625$
- H_a: For one or more classes, $P \neq P_{H_0}$

3. χ^2 **Test:**

Class	LG	LB	SG	SB	Total
Observed	260	95	91	34	480
P_{H_0}	0.5625	0.1875	0.1875	0.0625	
Expected	270	90	90	30	
	480(.5625)	480(.1875)	480(.1875)	480(.0625)	
$(O - E)^2/E$	0.37	0.28	0.01	0.53	
χ^2_{test}	1.19				
df = (k − 1)	3				
p-**Value**	$p > 0.25$ (From Appendix Table 8)				

Conclusion: The observed counts (and proportions) of offspring displaying the four classes of traits are *not* significantly different from the expected counts ($p > 0.25$). Therefore, the two genes are not linked (i.e., are located on separate chromosomes). ▥

12.2
χ^2 Test of Homogeneity

1. **Application:** The χ^2 **of homogeneity test** is used when you wish to *compare sample proportions for $k \geq 2$ classes* of a categorical variable among $m \geq 2$ samples. Requirements for this test are: (1) The individuals that constitute the samples from the m populations were chosen by a randomized, unbiased procedure, so that you can assume the expected values for the sample proportions for class i in population j $E(\hat{p}_{i,j})$ equal the true population proportions $P_{i,j}$, (2) the populations are sufficiently large that the values for $P_{i,j}$ can be assume fixed ($N > 100n$), and (3) the sample sizes are sufficiently large that the average expected frequency across all classes and populations $[N/(k \times m)]$ is ≥ 6. For example, if a categorical variable had $k = 5$ classes, and you wished to compare $m = 3$ samples, the minimum overall sample size for all samples combined (N)

required to perform this test of significance could be computed as: $N/(5 \times 3)$ = 6, or $N = 6(5 \times 3) = 90$.

2. **Statement of Hypotheses** (for k classes and m samples)

- H_0: $P_{1,1} = P_{1,2} = P_{1,3} \ldots = P_{1,m}$ and
  $P_{2,1} = P_{2,2} = P_{2,3} \ldots = P_{2,m}$ and
  $\ldots$ and
  $P_{k,1} = P_{k,2} = P_{k,3} \ldots = P_{k,m}$

  For each class (1 to k) of the categorical variable, the proportion of individuals that fall into that class is equal across all populations 1 to m.

- H_a: For one or more of the k classes of the categorical variable, the proportion of individuals in that class differs among two or more of the m populations.

3. **Test Statistic**

This test also uses the χ^2_{test} statistic, but differs from the goodness-of-fit test with regard to how the expected values are computed. In the goodness-of-fit test, the E_i values were computed based on values for the population proportions, P_i, derived from the Null hypothesis. In the test of homogeneity, the $E_{i,j}$ values are computed based on P_i values estimated from the data under the assumption that the P_i value for each class is equal across the m populations being compared. Calculations of the E_i values are facilitated if the observed counts of individuals in each class and sample are arranged in a matrix (called a **two-way table**), with a different row for each class of the categorical variable, and a different column for each sample.

The two-way table below displays the general format for such a table to compare the proportion of individuals in k classes of a categorical variable across m samples. The count in the upper-left corner, $n_{A,1}$, is the number of individuals in Sample 1 who fall into class A of the categorical variable. The Row Total n_A is the total number of individuals in class A for all m samples combined. The Column Total n_1 is the total number of individuals (all classes combined) in Sample 1. The value N in the lower-right corner is the total number of individuals with all samples and classes combined. The values $\hat{p}_{pool}$ for rows 1 to k are the proportions of individuals that fall into each class, based on data from all samples combined (e.g., $\hat{p}_{pool,A} = n_A/N$).

Class	Sample 1	Sample 2	Sample m	Row Total Row $\hat{p}_{pool}$
A	$n_{A,1}$	$n_{A,2}$	$n_{A,m}$	n_A $\hat{p}_{pool,A}$
B	$n_{B,1}$	$n_{B,2}$	$n_{B,m}$	n_B $\hat{p}_{pool,B}$
C	$n_{C,1}$	$n_{C,2}$	$n_{C,m}$	n_C $\hat{p}_{pool,C}$
.	.	.	.	.
k	$n_{k,1}$	$n_{k,2}$	$n_{k,m}$	n_k $\hat{p}_{pool,k}$
Column Total	n_1	n_2	n_m	N

Compute the Expected Values $E_{i,j}$

- Sum the total number of individuals in each class across all the samples (Row Total), and sum the total number of individuals in each sample across all classes of the categorical variable (Column Total). The overall sample size across all samples (N) is obtained by summing either the row totals or the column totals.

- Compute the proportion of all N individuals in the sample that fall into each class 1 to k of the categorical variable ($\hat{p}_{pool}$ values in the preceding matrix). These values represent pooled estimates of the proportion of individuals that are expected to be in each class of the categorical variable for all m samples *if the Null hypothesis is true* (e.g., $P_{A,1} = P_{A,2} = P_{A,3} = \ldots = P_{A,m} = P_A$).

- For each sample, compute the expected count of individuals $E_{i,j}$ for each class of the categorical variable by multiplying the $\hat{p}_{pool}$ values for each class times the sample size for each sample n_i. For example, the expected number of individuals in class A for sample 1 is computed as: $E_{A,1} = n_1 \hat{p}_{pool,A}$.

Compute the χ^2_{test} Statistic

- For each class of the categorical variable (row in the matrix) and each sample (column in the matrix) you should have an observed count of individuals that fall in that "cell" and an expected value computed as described above.

- Compute the value $(O_{i,j} - E_{i,j})^2/E_{i,j}$ for each cell in the matrix. To facilitate identification of specific differences should the overall χ^2 test reject the Null hypothesis, you should list these values along with the observed and expected count values in each cell of the matrix.

- Sum the $(O_{i,j} - E_{i,j})^2/E_{i,j}$ values across all cells in the matrix. This is the χ^2_{test} statistic.

This one χ^2_{test} statistic represents the sum of all deviations between observed counts of individuals in each sample/class combination and the expected counts if the Null hypothesis of "no difference" is true. Because of random sampling variation, the observed counts will rarely be exactly equal to the expected values. However, if the Null hypothesis is true, we expect these deviations, and the value of χ^2_{test}, to be small. The larger the value of χ^2_{test}, the less likely this "sum of deviations from expected" is due only to random variation and the stronger the evidence against the Null hypothesis.

4. **Sampling Distribution of the Test Statistic**

   The sampling distribution for the χ^2_{test} statistic under the assumption that the Null hypothesis is true is a χ^2 distribution with degrees of freedom computed as:

   df $= (k - 1)(m - 1)$

   For example, if the categorical variable had $k = 5$ classes, and you wanted to compare the proportion of individuals in each class among $m = 3$ samples, df $= (5 - 1)(3 - 1) = 8$.

5. *p*-Value

   a. If you are doing the test by hand, use Appendix Table 8 to obtain an approximate *p*-value for the χ^2_{test} statistic. Statistics or spreadsheet computer programs can be used to obtain an exact *p*-value (see Chapter 12). Because the χ^2 test does not distinguish positive or negative differences ($O_{i,j} - E_{i,j}$), the *p*-value is always two-tailed.

b. *Stating conclusions:*

If $p \le 0.05$: The probability associated with the random-variation explanation for the observed differences between observed counts of individuals and expected counts is sufficiently low to reject this explanation. If an appropriate study design has been correctly implemented, the only remaining likely explanation is that the true proportions of individuals that fall into one or more classes of the categorical variable really differ among the populations sampled. Because the χ^2 test is a simultaneous test, at this point you have no basis for knowing *which* populations differ. You also do not know for which specific class(es) of the categorical variable the proportions differ among the populations; the proportions for some classes may differ among two or more populations, while others do not. If the p-value from the χ^2 test is ≤ 0.05, you will need to implement a multiple comparisons test to determine which populations and classes have different proportions. The nature of your conclusion statement will depend on the outcome of this second analysis.

If $0.05 < p \le 0.10$: The probability associated with the random-variation explanation for the observed differences between observed and expected counts is sufficiently high that we cannot confidently exclude this explanation. However, the probability is sufficiently low that we might suspect that there really are differences. Some scientists would conclude the results "suggest" that two or more populations differ with regard to the proportion of individuals in one or more classes of the categorical variable. However, more data would be needed to verify this conclusion. In such cases, a multiple comparisons test would usually *not* be done.

■**EXAMPLE**   "The results suggest that the proportions of individuals who have blood types A, B, AB, and O differ among Native American tribes from eastern and western regions ($p = 0.075$), but more data are needed to verify this conclusion."   ■

If $p > 0.10$: The probability associated with the random-variation explanation for the observed differences between observed and expected counts is so high that this explanation cannot be excluded. Hence, there is insufficient evidence to conclude that the proportions of individuals in the classes of the categorical variable differ among the populations being compared.

■**EXAMPLE**   "The proportions of individuals who have blood types A, B, AB, and O were not significantly different among western tribes of Native Americans ($p > 0.25$)."   ■

6. **Determining Which Class Proportions Differ Among Which Populations**

When the overall Null hypothesis for a χ^2 test for homogeneity is rejected, the investigator must often consider two questions: (1) For which class of the categorical variable are the population proportions different? and (2) Which populations are different? If you are comparing only two populations, or if the categorical variable has only two classes, the answer(s) to these questions are obvious. However, if the number of classes and the number of populations are both greater than two, these questions can be addressed only by additional analyses, as described below.

a. *Subset χ^2 analysis* is used to identify which class proportion(s) differ among the populations. Rerun the χ^2 test multiple times, deleting a different class of the categorical variable from the analysis each time. If the χ^2 test with class i removed is not significant, but the test with class i included is significant, this could be taken as evidence that there is a difference among the populations in the proportion of individuals in this class. This procedure is most efficiently done by first deleting those classes with the largest values for $(O_i - E_i)^2/E_i$.

b. *Multiple comparisons test* is used to identify which populations differ with regard to the proportion of individuals in class i. Once you identify the class(es) for which the proportions differ among the populations sampled, perform this test to identify which populations are different (assuming there are more than two).

Procedure for Equal Sample Sizes

- For the class i identified by the subset χ^2 analysis, compute the proportion of individuals in each of the m samples that fell into that class ($\hat{p}_{i,j} = O_{i,j}/n_j$), and list these proportions in order from smallest to largest.

- Apply the following nonlinear transformation to the $\hat{p}_{i,j}$ values to compute $p'_{i,j} = \arcsin \sqrt{\hat{p}_{i,j}}$. This transformation makes the sampling distribution of the $p'_{i,j}$ values approximately Normal.

  NOTE On some calculators, the arcsin function is listed as SIN^{-1}. This function returns the degrees in an angle that has a sine equal to $\sqrt{\hat{p}_{i,j}}$. This transformation does not work well for $\hat{p}$ values that are very close to 0 or 1. Different versions of this transformation should be used for such $\hat{p}$ values.[1] ■

- Compute the standard error (SE) of the difference ($p'_{i,1} - p'_{i,2}$).

$$SE = \sqrt{\frac{820.7}{n + 0.5}}$$

  where n = equal sample size for each of the samples.

- Compute the minimum significant difference (MSD):

  $MSD = Q_{0.05,k} \, (SE)$

  where the critical value $Q_{0.05,k}$ is obtained from the Studentized range distribution (Appendix Table 7), with df = ∞ (infinity) and k = number of samples being compared.

- Any difference between two *transformed* sample proportions ($p'_{i,1} - p'_{i,2}$) that is greater than or equal to the MSD value is significant at the $p \leq 0.05$ level.

---

[1]See J. H. Zar. 1996. Biostatistical Analysis. 3rd Edition. Prentice Hall, Upper Saddle River, NJ.

Procedure for Unequal Sample Sizes

- Compute the proportions of individuals in Class i for each of the m samples and apply the arcsin transformation to these sample proportions.
- Compute all pair-wise differences among the transformed sample proportions ($p'_{i,j}$) and list these in a table in order from the largest difference to the smallest.
- Compute the standard error for each of the differences ($p'_{i,1} - p'_{i,2}$)

$$SE = \sqrt{\frac{410.35}{n_1 + 0.5} + \frac{410.35}{n_2 + 0.5}}$$

where n_1 and n_2 are the sample sizes from populations 1 and 2.

- For each pair-wise comparison, compute the minimum significant difference:

MSD = $Q_{0.05,k}$ (SE), again with df = ∞

- If the observed difference ($p'_{i,1} - p'_{i,2}$) is greater than or equal to the MSD, the difference is significant at the $p \leq 0.05$ level.

NOTE The subset χ^2 analysis, with its repeated tests, does not really provide control over cumulative Type I error rate. Therefore, you should obtain independent data from a new study that can be used to test for the specific differences you identified by these procedures. ▪

Example 12.2 applies the χ^2 test of homogeneity to a water quality study.

▤ **EXAMPLE 12.2**

Application of the χ^2 test of homogeneity. A student does an ecology lab project on the impact of urban pollution on water quality in a river that passes through town. She studies the abundance of various kinds of stream bottom invertebrates at three locations in the river, upstream of the city (presumably good water quality), just downstream of the city (suspected poor water quality), and several miles further downstream of the city (intermediate water quality). Upstream of the city the river flows through rural agricultural areas with no major pollution sources. The student anticipates that whatever pollution comes from the city will be most concentrated just downstream of the city limits. The further downstream from the city, the more diluted the pollutants become, and she expects the water quality to improve somewhat. Based on knowledge of the biology of stream bottom invertebrate taxonomic groups, the student knows that tubificid worms and midgefly larvae are tolerant of pollution and common in polluted waters. Mayfly and caddisfly larvae are sensitive to pollution, and abundant only in unpolluted waters. Other taxonomic groups are less useful as indicators of water quality. The student collects invertebrates at the three locations. She is careful that all samples are taken from places with similar water depth, flow velocity, and stream bottom structure (as much as possible, the only difference between the three sites is the difference in water quality). The two-way table that follows contains the matrix of counts, expected values, sample proportions, and $(O - E)^2/E$ values for this study. Sample calculations for the first cell in the first row of the two-way table are shown just below the table.

Two-way table for Example 12.2

Species	Good Site	Intermediate Site	Poor Site	Row Total $\hat{p}_{pool}$
Mayfly				
(Obs)	149	213	9	371
(Exp)	98.4	180.8	91.7	0.335
$(\hat{p}_{i,j})$	0.507	0.394	0.033	
$(O - E)^2/E$	26.02	5.73	74.58	
Caddisfly				
(Obs)	78	65	13	156
(Exp)	41.4	76.0	38.6	0.141
$(\hat{p}_{i,j})$	0.265	0.120	0.047	
$(O - E)^2/E$	32.89	1.59	16.98	
Clams				
(Obs)	6	34	34	74
(Exp)	19.6	36.1	18.3	0.067
$(\hat{p}_{i,j})$	0.02	0.063	0.124	
$(O - E)^2/E$	9.45	0.12	13.47	
Snails				
(Obs)	14	7	52	73
(Exp)	19.4	35.6	18.1	0.066
$(\hat{p}_{i,j})$	0.048	0.013	0.190	
$(O - E)^2/E$	1.50	22.98	63.49	
Midgefly				
(Obs)	43	156	134	333
(Exp)	88.4	162.3	82.3	0.301
$(\hat{p}_{i,j})$	0.146	0.289	0.489	
$(O - E)^2/E$	23.32	0.24	32.48	
Worms				
(Obs)	4	65	32	101
(Exp)	26.8	49.2	25.0	0.091
$(\hat{p}_{i,j})$	0.014	0.120	0.117	
$(O - E)^2/E$	19.40	5.07	1.96	
Column Total	**294**	**540**	**274**	**1108**

Sample Calculations (for mayflies at the "good" site)

Observed count $(X_{mayfly,good})$ = 149

Expected value $(E_{mayfly,good})$ = $n_{good} \times \hat{p}_{pool,mayflies}$ = 294 × 0.335 = 98.4

Observed proportion $(\hat{p}_{mayfly,good})$ = $X_{mayfly, Good}/n_{good}$ = 149/294 = 0.507

$(O - E)^2/E$ = $(149 - 98.4)^2/98.4$ = 26.02

Conclusion: The results of the χ^2 test provided strong evidence (χ^2_{test} = 351.27, $p <$ 0.0001) that the proportions of invertebrates in one or more taxonomic classes differed among the three sites sampled. However, to infer that these differences were due to water pollution rather than any of a number of other possible reasons, the student must determine that species tolerant of pollution had higher abundance at the "poor" site and species that require clean water had higher abundance at the "good" site. To further explore the specific differences among the three areas sampled, the student must determine *which* species differed among the samples and *which* samples differed.

Subset χ^2 analysis to determine which species differ: The student progressively deleted invertebrate taxa beginning with those that had the highest sum of $(O - E)^2/E$ values, proceeding in order of decreasing values. A list of the χ^2_{test} statistics and p-values from these subset χ^2 tests is given below.

Taxon Deleted	χ^2_{test}	df	p – value
None	351.0	10	< 0.001
Mayfly	190.8	8	< 0.001
Mayfly, Caddisfly	56.1	6	< 0.001
Mayfly, Caddisfly, Snail	13.8	4	0.008
Mayfly, Caddisfly, Snail, Worms	1.7	2	0.438

Conclusion: Based on this analysis, the student concluded that the significant differences among the three sites were for the taxonomic classes Mayfly, Caddisfly, Snails, and Worms. Differences for Clams and Midgefly were not significant.

Unplanned multiple comparison: Having determined that the proportions of several taxonomic classes differ among the three sites, the student now must determine *which* of the three sites differ for each class. Separate Tukey-type multiple comparisons tests can be used for each of the taxonomic classes identified by the subset χ^2 analysis. An example of this test for the snail class is shown below:

	Poor Site (P)	Good Site (G)	Intermediate Site (I)
Sample sizes (n_j):	274	294	540
Ordered $\hat{p}$ values:	0.19	0.048	0.013
$p' = \text{Sin}^{-1}\sqrt{\hat{p}}$:	25.84	12.66	6.55

Pair-wise Comparisons	$(p'_1 - p'_2)$	SE[1]	$Q_{0.05, k=3}$	MSD	Conclusion
Good − Poor	−13.18	1.70	3.314	5.63	$P_G \neq P_P$
Good − Intermediate	+6.11	1.47	3.314	4.87	$P_G \neq P_I$
Poor − Intermediate	+19.29	1.50	3.314	4.97	$P_P \neq P_I$

[1]SE $= \sqrt{[410.35/(n_1 + 0.5)] + [410.35/(n_2 + 0.5)]}$, with $n_G = 294$, $n_I = 540$, and $n_P = 274$

Conclusion: Based on this multiple comparisons test, the student concludes that the proportion of snails in the stream bottom invertebrate communities differs among all three sites.

NOTE Like all other Tukey-type multiple comparisons tests, this is a two-tailed test. ■

CHAPTER SUMMARY

1. When multiple tests of significance are required for you to come to a conclusion about a single Null hypothesis regarding population proportions, overall experiment-wise error rate must be limited to less than the maximum allowable probability of a Type I error. This can be done in one of two ways:

   a. *Use the Bonferroni procedure.* Reduce the allowable comparison-wise error rate of each test so that the cumulative, experiment-wise error rate is held equal to α or less. The Bonferroni procedure is applied to tests for proportions in the same manner as described for means in Chapter 11.

   b. *Perform a single simultaneous test* to determine if the overall Null hypothesis should be rejected.

2. The χ^2 **test for goodness-of-fit** is used when a categorical variable has more than two classes and you wish to compare multiple sample proportions to their corresponding Null hypothesis values to test an overall Null hypothesis. With a single simultaneous test, the experiment-wise and comparison-wise error rates are the same.

3. If the overall Null hypothesis is rejected based on the χ^2 test for goodness-of-fit, multiple one-sample Z-tests for proportions might be done to identify which sample proportions are significantly different from their corresponding Null hypothesis value. The Bonferroni procedure should be used to reduce the comparison-wise error rates so that the overall experiment-wise error rate is less than or equal to α (usually 0.05).

4. The χ^2 **test of homogeneity** is used when you wish to compare sample proportions for two or more classes of a categorical variable among two or more samples to test the Null hypothesis that the proportion of individuals in each class of the categorical variable is equal across all of the populations.

5. If the overall Null hypothesis is rejected based on the χ^2 test of homogeneity, a two-step process may be used to identify which populations differ with regard to the proportion of individuals in specific classes of the categorical variable.

   a. *Subset analysis* involves systematically deleting classes of the categorical variable from the two-way table to determine which classes have different proportions of individuals in the multiple populations.

   b. *Multiple comparisons tests* determine which populations differ with regard to the proportion of individuals in a specific class, as identified by the subset analysis.

13 Tests of Association Between Two Quantitative Variables

INTRODUCTION   There is an *association* between two variables if the values of one variable tend to change in a systematic manner when the values of the other variable change. For example, taller people tend to weigh more than shorter people. This does not mean that all taller people are heavier than all shorter people. Rather, the association between height and weight means that *on average* the weight of taller people is greater than the weight of shorter people. This is the systematic *pattern* of the association. The *noise* of this association relates to both random variation among individuals and factors other than height that also influence weight (e.g., body fat, muscle mass, bone density). The result is that taller *individuals* might weigh less than shorter individuals, in spite of the general association between height and weight.

In Chapter 2, I described graphical methods for displaying the nature and strength of the association between two variables. The most common graphic used for this purpose is the scatterplot. In Figure 2.12, I showed a variety of scatterplots and described how the patterns of points in the plot are used to describe the nature and strength of the association (e.g., positive vs. negative, strong vs. weak, linear vs. nonlinear associations). *Strong associations* are indicated when the points of a scatterplot are tightly clustered around a line with either a positive or negative slope. *Weak associations* are indicated when the points are scattered randomly in a "cloud," or when the points fall close to a line that has zero slope (value of the *Y*-variable is the same for all values of the *X*-variable). Even though scatterplots provide detailed information about the nature of the association, assessment of scatterplots is somewhat subjective. We must be able to quantitatively measure the strength of associations if we are to test hypotheses about associations.

In this chapter, I present methods for describing and quantifying the association between two quantitative variables measured on the same study subjects. I will present summary statistics and tests of significance that are used to evaluate sample data for evidence of true associations

in the real world. Although the nature of the sample statistics for measuring associations may be different from means and proportions, the format of the tests of significance for associations is the same. We will describe the sampling distributions for association statistics under the Null hypothesis that there is *no* association. We will use these sampling distributions to determine the probability of obtaining the observed value of the association statistic. If this probability is very small, we will conclude there is strong evidence that an association exists in the real world. The ultimate objective of analyses of associations is often to identify cause-effect relationships between the two variables. However, the procedures described here are only the first step in a more complex process that is required to fully support such conclusions about true cause and effect relationships.

Statistical analyses of associations among quantitative variables fall into two classes, depending on the purpose of a study. **Correlation analyses** compute a sample statistic that represents the strength and nature (positive/negative) of linear associations. Correlation analysis is relatively simple and is efficient for exploring associations among many variables. **Linear regression analysis** also evaluates the strength and nature of linear association. However, the primary purpose of regression analysis is to provide a means for using the statistical association between two variables to *predict* values for one variable based on values for the associated variable. Regression analysis is more complex than correlation analysis and so less efficient for exploring associations among many variables. ■

GOALS AND OBJECTIVES

In this chapter you will learn to quantify the nature and strength of associations among quantitative variables and to use sample data as evidence to assess whether or not there is a real association. You will understand that a statistically significant association between two variables does not necessarily mean that the variables have a cause-effect relationship.

By the end of the chapter, you will be able to:

1. Explain the concept of association and describe how correlation coefficients quantitate this concept. You will be able to interpret Pearson's and Spearman's correlation coefficients, determine which of these two coefficients is appropriate in specific circumstances, and perform tests of significance for these two correlation coefficients to assess evidence against the Null hypothesis that there is no association.

2. Distinguish between statistical association and true cause-effect relationship, and be able to describe the rules of evidence for inferring the existence of a relationship from evidence of a statistical association.

3. Interpret standard computer print-outs of linear regression analysis with regard to the nature and strength of the association, as indicated by the regression coefficients, the ANOVA table, the standard-error-of-the-regression model, and the coefficient of determination (R^2).

4. Compute and interpret regression-based confidence intervals for the mean of Y, given X, and prediction intervals for an individual value of Y given X.

13.1 Correlation Analysis

Pearson's Correlation Coefficient (r)

Our discussion of correlation analysis will start with Pearson's linear correlation coefficient and proceed to analyses that can be used when the Normality and/or linearity assumptions of Pearson's correlation are violated.

1. **Application:** Pearson's correlation coefficient r is used to evaluate sample data as evidence that a *linear* association exists between two quantitative variables. This statistic is used to test the Null hypothesis that there is *no* association between the variables X and Y.

   Assumptions: (1) Data for the two variables X and Y were obtained by a randomized, unbiased study design, so the expected value of r is the true correlation for the population. (2) Both variables have an approximately Normal distribution, as determined using boxplots, Normal quantile plots, or statistical tests of Normality. (3) The "true" association among the variables X and Y is linear. A scatterplot of the X and Y data values is used to evaluate validity of the linearity assumption. If the cloud of points in the scatterplot appears to have a linear pattern, the assumption is fulfilled. If the cloud of points has a nonlinear pattern and/or the X or Y variable is not Normally distributed, you will need to assess the association using alternate methods (see the next section).

2. **Sample Statistic**

   Pearson's correlation coefficient (r) (also called *Pearson's product moment*) is a summary statistic that represents the strength and nature (positive or negative) of linear association between two variables. Pearson's r is dimensionless (i.e., not expressed in any unit of measurement) and ranges between -1 and $+1$. A value of $+1$ corresponds to a scatterplot wherein all the points fall on a straight line with a positive slope (perfect positive association). A value of -1 corresponds to a scatterplot wherein all the points fall on a straight line with a negative slope (perfect negative association). A correlation value of zero indicates there is no association between the two variables. (See Figure 13.1.)

   Pearson's r provides an accurate measure of the strength of *linear* association only. If the cloud of points in a scatterplot has a curvilinear pattern, Pearson's

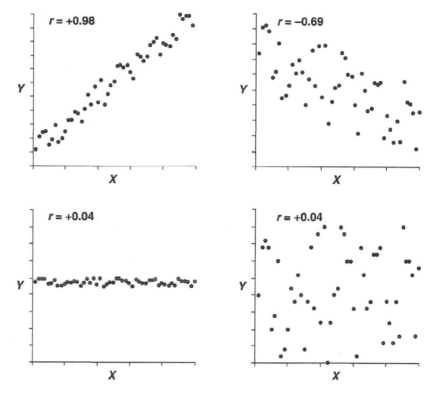

FIGURE 13.1 Scatterplots and correlation coefficients. The four panels in this figure display scatterplots that exhibit different amounts of pattern and noise, along with correlation coefficients that quantitatively represent the strength and nature of the association.

r will tend to underestimate the strength of the association. The correlation co-efficient r can be computed using the following formula:

Pearson's r

$$r = \frac{\sum_{i=1}^{n} (x_i - \bar{x})(y_i - \bar{y})}{(n-1)\, S_x S_y}$$

However, no one actually uses this formula to compute Pearson's r. Correlation analysis is typically done using a calculator or computer program. I present this "conceptual formula" here to help you understand how this correlation co-efficient quantifies the nature and strength of the association.

Interpretation of the conceptual formula.   In this formula, $\bar{x}$ and $\bar{y}$ are the sample means for the X- and Y-variables, S_x and S_y are the sample standard devi-ations for the two variables, and n is the sample size (number of paired obser-vations of X and Y). The numerator is the sum of *cross products*. The point in a scatterplot defined by the means for the two variables X and Y ($\bar{x}, \bar{y}$) falls in the center of the cloud of points. This point defines four quadrants within the scat-terplot. (See Figure 13.2, part A.) If a point in the scatterplot (x_i, y_i) falls in the upper-left quadrant, x_i is less than $\bar{x}$, and the value of the quantity $(x_i - \bar{x})$ will be negative. The value of y_i is greater than $\bar{y}$, and $(y_i - \bar{y})$ will be positive. The cross product $(x_i - \bar{x})\,(y_i - \bar{y})$ is negative. If a point falls in the upper-right quadrant, both the values $(x_i - \bar{x})$ and $(y_i - \bar{y})$ will be positive, as will their cross product. (See Figure 13.2, part A.) The numerator term in the formula for Pearson's r sums these positive and negative cross products. If the majority of the cross products are positive (points in the scatterplot tend to fall in the lower-left and upper-right quadrants), Pearson's r will have a large positive value (Figure 13.2, part B). If the majority of the cross products are negative (points in the scatterplot tend to fall in the upper-left and lower-right quadrants), Pearson's r will have a large negative value (Figure 13.2, part C). If there are equal numbers of points in all quadrants, the positive and negative cross products will "cancel out" when they are summed, and Pearson's r will have a value close to zero (Figure 13.2, part D). The denominator term in the equation for Pearson's r includes the standard deviations for variables X and Y to make the value of r a dimensionless

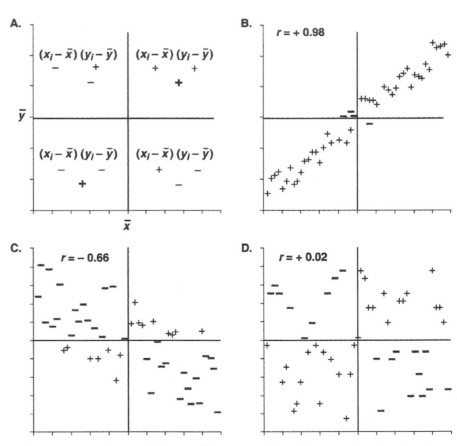

FIGURE 13.2   How do correlation coefficients quantitate the strength and nature of an association?   Small $+/-$ symbols under each subtraction indicate the sign of the resulting difference. Large $+/-$ symbols in each quadrant indicate the sign of the cross product of the differences.

index of association that is not influenced by the different scales of measurement for the two variables.

3. **Statement of Hypotheses**

- One-tailed test:
  - H_0: $\rho = 0$
  - H_a: $\rho < 0$ *or* $\rho > 0$
- Two-tailed test
  - H_0: $\rho = 0$
  - H_a: $\rho \neq 0$

where ρ is the lowercase Greek letter "rho," used here to indicate the "true" correlation between variables X and Y that would be obtained if all individuals in the population were measured and these data were used to compute the correlation.

4. **Sampling Distribution of Pearson's r**, assuming H_0 is true

Center: $\rho_0 = 0$

Spread: $S_r = \sqrt{\dfrac{1 - r^2}{n - 2}}$

Shape: t-distribution (df $= n - 2$)

See Figure 13.3 for examples of empirical sampling distributions that display random sampling variation in the value of the correlation coefficient r.

5. **Test Statistic**

$$t_{\text{test}} = \frac{r - \rho_0}{\sqrt{\dfrac{1 - r^2}{n - 2}}}$$

$$= \frac{r}{S_r}$$

NOTE The test of significance for the sample correlation coefficient r is a one-sample t-test for the difference between the observed sample r and the Null hypothesis value for $\rho =$ zero. The t_{test} statistic represents the size of the difference between the r-value and zero in "standard error units" that quantify the expected amount of random sampling variation in the value of r. Random sampling variation is determined by sample size n and the value of r itself. Because r can only take values between -1 and $+1$, r-values close to either end of this range tend to be less variable than r-values in the middle of this range.   ■

6. **p-Value**

   a. *One-tailed test:* The probability $P(t \geq +t_{\text{test}})$ or $P(t \leq -t_{\text{test}})$, from the t-distribution, is the p-value for the one-tailed test of significance. A one-tailed test is justified only if there was a strong, well-documented basis for predicting that the correlation would be positive (or negative) *before* the data were collected. This p-value is the probability of obtaining values of the sample correlation coefficient that are greater than or equal to (or $\leq$) the observed r-value, if there really was no association between the X- and Y-variables.

b. *Two-tailed test:* The probability $2 \times P(t \geq |t_{test}|)$ is the p-value for the two-tailed test of significance. The term $|t_{test}|$ is the absolute value of t_{test} (any $-$ sign is deleted). In a two-tailed test, the sign of t_{test} is irrelevant, as there was no prior expectation that the correlation would be positive or negative. The p-value is the probability of getting a sample correlation coefficient greater than or equal to $|r|$ if the true correlation was zero.

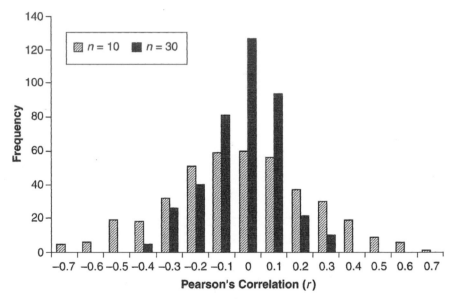

FIGURE 13.3 Empirical sampling distribution for Pearson's correlation coefficient. This graph is based on repeatedly creating two variables that are random numbers taken from a standard Normal distribution and computing the correlation between the two sets of random numbers. The expected correlation between two sets of random numbers is $\rho = 0$. This graph shows the random sampling variability of the statistic r. This simulation was performed 500 times with $n = 10$ individuals per sample (gray bars) and another 500 times with $n = 30$ individuals (black bars). Several observations can be made from these simulations:

a. The shape of the sampling distribution of r is approximately bell-shaped.

b. Sampling variability of the correlation coefficient r decreases when n increases.

c. Even when H_0 is true ($\rho = 0$), as was the case here, it is still possible for r to have fairly large values simply due to random sampling variability. With $n = 10$, the smallest value of r that would be statistically significant ($p \leq 0.05$) is $|0.632|$. That is, values of $r \leq -0.632$ or $r \geq +0.632$ would occur by random chance in 5% or less of repeated samples when the true value of $\rho = 0$. With $n = 30$, the smallest value of r that is significant ($p \leq 0.05$) is $|0.361|$. You can see in the graph that among the 500 repeated samples of random numbers, statistically significant r-values occur several times, even though the two variables are simply random numbers.

c. *Stating conclusions:*

If $p \leq 0.05$: The probability associated with the random-variation explanation for the observed correlation coefficient r-value is sufficiently low to reject this explanation. If an appropriate study design has been correctly implemented, the only remaining likely explanation supported by the data is that there really is an association between the two variables. Regardless of whether the test was one-tailed or two-tailed, your conclusion should state whether the observed association was positive or negative.

■ **EXAMPLE**   "There was a positive correlation between blood concentration of low-density-lipids and blood sugar ($r = 0.61, p = 0.011$)."   ■

Important: Beware of describing a correlation as "strong" or "weak." You should always report the actual value of the correlation coefficient along with the p-value. In analyses of associations, the value of the correlation coefficient is the measure of effect size (i.e., the strength of the association). By presenting r, you allow the reader to determine for themselves the strength of the association. Never confuse "statistical significance" with the strength of the association. If the sample size is sufficiently large, even a small r value is unlikely to be due to random variation alone (small p-value). For example, with $n = 1000$, a correlation of $r = 0.10$ would have a p-value < 0.01. The correct interpretation of such a result is that there is strong evidence ($p < 0.01$) for a weak association ($r = 0.1$).

If $0.05 < p \leq 0.10$: The probability associated with the random-variation explanation for the observed correlation is sufficiently high that we cannot confidently exclude this explanation. However, the probability is sufficiently low that we might suspect that there really is an association. Some scientist would conclude that the data "suggest" there is an association between the two variables, but more data are needed to verify this conclusion.

■ **EXAMPLE**   "The results suggest there is a positive correlation between blood concentration of low-density lipids and blood sugar ($r = 0.25, p = 0.089$), but more data are needed to verify this."   ■

If $p > 0.10$: The probability associated with the random-variation explanation for the observed value of the correlation coefficient is so high that this explanation cannot be excluded. Hence, there is insufficient evidence to conclude that there is an association between the two variables.

■ **EXAMPLE**   "There was no correlation between concentration of low-density lipids and blood sugar ($r = -0.04, p = 0.82$)."   ■

In Example 13.1, Pearson's correlation analysis is used for a study of the association between the percentage of body fat and age.

Test of significance for Pearson's correlation coefficient. An investigator wants to study the association between an adult's age and the percentage of body weight accounted for by fat. A study of 50 randomly selected subjects produced the following results:

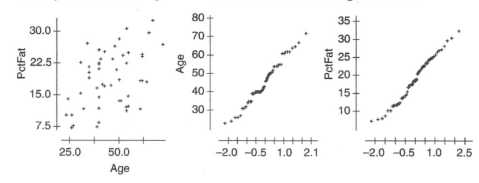

Scatterplot of PctFat vs. Age   **Normal Quantile Plots of Age and PctFat**

The scatterplot indicates a linear association, and the Normal quantile plots indicate that both variables are approximately Normally distributed. Hence, Pearson's correlation coefficient is the appropriate measure of the strength of the association. The value of the correlation coefficient for these data is $r = 0.436$. Is there an association between age and body fat?

Hypotheses

- H_0: $\rho = 0$
- H_a: $\rho \neq 0$

Sampling distribution of r, assuming H_0 is true

Center: $E(r) = 0$

Spread: $S_r = \sqrt{\dfrac{1 - r^2}{n - 2}}$

$= \sqrt{\dfrac{1 - 0.436^2}{50 - 2}}$

$= 0.13$

Shape: t-distribution (df = 48)

$r = +0.436$

−0.52 −0.39 −0.26 −0.13 0 0.13 0.26 0.39 0.52

This sampling distribution displays the likely values for Pearson's r if there really was *no* association, and r deviated from zero due only to random sampling variation. The preceding sampling distribution indicates that the observed $r = 0.436$ is very unlikely if there were no association.

Test of Significance

$$t_{\text{test}} = \frac{r}{S_r} = \frac{0.436}{0.13} = \mathbf{3.354}$$

Because the study question did not specify whether the association was expected to be positive or negative, the test of significance is two-tailed. The p-value from Appendix Table 4 is:

$$P[r \geq +0.436 \text{ if } \rho = 0] = P[t_{\text{df}=48} \geq 3.354] = p < 0.001 \times 2 = \mathbf{p < 0.002}$$

This p-value indicates that the random-variation explanation for the observed correlation between age and percentage body fat is extremely unlikely.

Conclusion: There is a positive correlation between adult age and the percentage of body fat ($r = +0.436$, $p < 0.002$).

Correlation Analysis When the Normality and/or Linearity Assumptions Are Violated

When exploratory data analyses indicate a violation of the Normality and/or linearity assumptions for Pearson's r, you have two options for analysis of the association:

1. Transform the data so that linearity and Normality assumptions are fulfilled; then use Pearson's r to assess the association between the transformed variables.

2. Use a nonparametric procedure that does not have Normality and linearity assumptions with regard to the original data values.

Using nonlinear transformations, you can transform one or both variables; you do not need to apply the same transformation to both X- and Y-variables. If data for one of the two variables is Normally distributed, but not for the other, you need only transform the non-Normal variable. See Figure 13.4 for examples of the use of transformations. When you assess nonlinear associations based on applying Pearson's r to transformed data, you should always explicitly describe what you did, along with any conclusion about the association.

Nonparametric procedures for assessing associations are based on the correlation between the *ranks* of the data values. I describe Spearman's rank correlation coefficient in the following section.

Spearman's Correlation Coefficient r_s

1. **Application:** The test of significance for **Spearman's correlation coefficient r_s** is used to evaluate data as evidence of an association between two quantitative variables when (1) there is reason to believe that the distributions for the two variables X and Y are *not* Normal, or (2) the scatterplot indicates a nonlinear, but monotonic association. (By "monotonic" we mean that the trend is only increasing or only decreasing.) This procedure tests the Null hypothesis that the true correlation between the *ranks* of the data values is zero (i.e., that there is *no* association). *Assumptions:* (1) The data were obtained by a randomized

A. Distribution of *Y* + Skewed, Distribution of *X* Normal: Log-transform *Y*

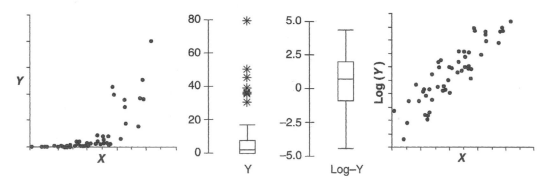

B. Distribution of *X* + Skewed, Distribution of *Y* Normal: Transform *X* $\sqrt[5]{X}$

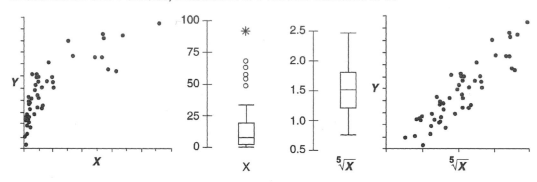

C. Distribution of Both *X* and *Y* + Skewed, Log-transform Both *X* and *Y*

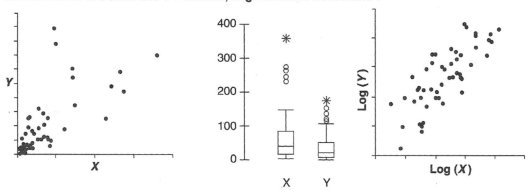

FIGURE 13.4   Use of transformations to assess nonlinear associations.   In parts A, B, and C of the figure the graphs on the left of the boxplots display nonlinear associations. The graphs on the right of the boxplots display the same association after a nonlinear transformation has been applied to the data.

unbiased study design, and (2) the *true* association among the *ranks* of variables *X* and *Y* is linear. You should evaluate validity of this assumption using a scatterplot of data value *ranks*. Most spreadsheet and statistics programs have a function in the form RANK (*variable name*) to automatically create new variables that contain the ranks for quantitative variables.

NOTE Although you could apply Spearman's rank correlation to data that meet the assumptions for Pearson's correlation analysis, the analysis based on correlations among ranks is not as powerful as correlations among the original data values. That is, Spearman's rank correlation is less likely to detect a real association than Pearson's correlation. If the data meet the assumptions, you should use Pearson's correlation analysis to minimize both Type I error rate (probability of claiming there is an association when there is not) and Type II error rates (probability of failing to detect an association that really exists). ■

For example, the value of Spearman's rank correlation coefficient for the data described in Example 13.1 is $r_s = 0.391$. Pearson's correlation coefficient for these same data was $r = 0.436$.

2. **The Sample Statistic**

   The Spearman's rank correlation coefficient (r_s) is computed using the same formula as given above for Pearson's r, but the formula is applied to the *ranks* of the data values rather than to the original data values. Ranks are assigned independently to the values for each variable *X* and *Y* (smallest data value for each variable is given rank 1, second smallest rank 2, and the largest rank n = sample size). See Figure 13.5 for examples of nonlinear associations and how they are affected by converting the original data values to ranks. Note that some nonlinear associations remain nonlinear even after the data are converted to ranks. As a general rule, ranks of data values for *X* and *Y* will have a linear pattern if the scatterplot of the original data values displays a monotonic nonlinear trend. If the scatterplot indicates a more complex pattern of increasing *and* decreasing *Y* as the value of *X* increases, the scatterplot of the ranks will also display this nonlinear pattern. In this case, Spearman's correlation is *not* an appropriate measure of the strength of association.

3. **Statement of Hypotheses**
   - One-tailed test:
     - H_0: $\rho_s = 0$
     - H_a: $\rho_s < 0$ *or* $\rho_s > 0$
   - Two-tailed test:
     - H_0: $\rho_s = 0$
     - H_a: $\rho_s \neq 0$

   where ρ_s is used here to indicate the *true* Spearman's correlation between the ranks of variables *X* and *Y* that would be obtained if all individuals in the population were measured and these data used to compute the correlation.

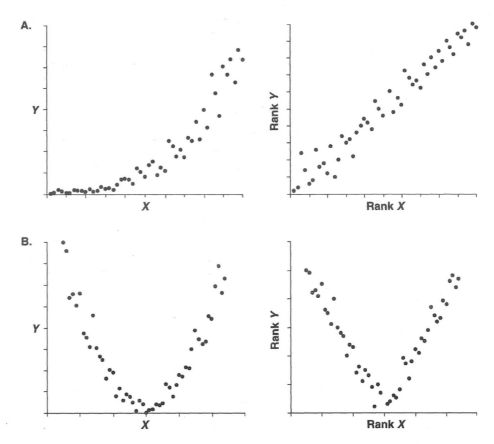

FIGURE 13.5 **Converting the original data values to ranks can make a nonlinear association into a linear association, but not always.** In parts A and B, the graphs on the left display scatterplots for original data values that have a nonlinear association. The graphs on the right display scatterplots of the *ranks* of these same data values. Part A shows that when the nonlinear trend is monotonic, converting data values to ranks can make a nonlinear association linear. Part B shows that when the nonlinear trend includes both increasing and decreasing segments, converting the data values to ranks does not make the association linear.

4. **Sampling Distribution and Test Statistic for Spearman's *r*,** assuming H_0 is true

As was the case with Pearson's *r*, the expected value of Spearman's r_s if there is no association between the two variables is $E(r_s) = \rho_s = 0$. However, rank values do not have a Normal distribution unless the sample size is larger than 30 to 50. (Different authors suggest different sample sizes for the Normal approximation.) If the sample size is smaller than this, the sampling distribution of r_s is such that special tables are required to determine the *p*-value of this statistic. However, if the sample size is sufficiently large, the sampling distribution of r_s is approximately the same as described for Pearson's *r*. You will likely use statistics computer programs that provide exact p-values whenever you perform the test of significance for Spearman's r_s. Hence, you will not need to worry about this distinction between the sampling distributions used to obtain the p-value.

5. ***p*-Value**

Interpretation of the p-value from Spearman's rank correlation analysis and statements of conclusions are similar to those described for Pearson's rank correlations. The only adjustment you should make is to be clear that the evidence for an association was based on *rank* correlation. In most scientific communication, you are required to describe which statistical analyses you used. Hence, you should clearly state that Spearman's rank correlation was used instead of Pearson's correlation.

13.2
Statistical Associations and Cause-Effect Relationships

Correlation analyses only provide evidence for or against the presence of an *association* between two variables. In this context, the word "association" refers to a systematic pattern of covariation in values of two variables that is sufficiently strong that it is unlikely to be due only to random variation. In contrast, the word "relationship" should be used more restrictively to refer to an actual cause-effect linkage between two variables. A common mistake made by inexperienced scientists is equating evidence of an association between two variables as "proof" of a cause-effect relationship. This is a serious mistake, and can have unfortunate consequences when the results of data analyses drive real-world decisions. *Evidence of a statistical association between two variables is never sufficient, in and of itself, to conclude that variable X causes change in variable Y, no matter how strong the association.*

There are at least two possible alternative explanations for a statistical association between two variables that do not involve a direct cause-effect relationship:

1. *Lurking variables:* Variables that are not measured or included in the study may nonetheless have a cause-effect relationship with one or both of the X and Y variables. A **lurking variable** Z may simultaneously influence the values of both X and Y, causing them to exhibit coincident variation. Although this coincident variation of X and Y results in a statistical association among these variables, this association is due to their common cause-effect linkage with Z and is *not* due to a direct linkage between X and Y. Lurking variables can either: (1) cause a correlation between otherwise unrelated X and Y variables by influencing both variables (see Figure 13.6) or (2) obscure a relationship between two variables by affecting one, but not the other, of X and Y.

2. *Chance coincidence in time:* Many variables in the real world exhibit change over time, some variables increasing and others decreasing. The coincidence that two variables change over the same time period can produce a statistical association, even though they are in no way related to each other. This type of "spurious correlation" is most common when two variables exhibit long-term trends over the same interval of time (see Figure 13.7). In contrast, repeated, coincident short-term variation in the values of two variables is much less likely to be due to spurious correlation, and so provides stronger

evidence for a relationship. (For example, see Figure 2.13, June rainfall vs. tree-ring width.)

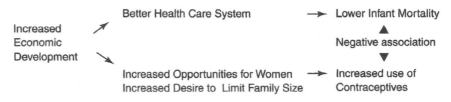

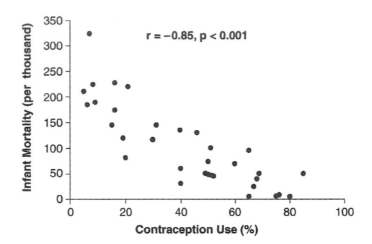

FIGURE 13.6   Contraception and infant mortality: Lurking variables?   This scatterplot displays data from 33 countries regarding use of contraception (percentage of married women who use some form of contraception) and infant mortality rates (number of children out of 1000 who die before their fifth birthday). The Pearson's correlation coefficient of $r = -0.85$ and associated p-value < 0.001 indicates there is strong evidence for a negative association. Countries where contraception use is high tend to have lower infant mortality rates than countries in which contraception is used by few women. *Interpretation:* One interpretation of this association is that women who use contraception are better able to space-out their children, to breast feed them longer, and so, delay the child's exposure to water and food-borne diseases. But this is not necessarily the only interpretation. These 33 countries spanned a wide range of economic development (United States to Bangladesh). High rates of contraception use by women is associated with higher levels of education for women, which is associated with higher economic development, including availability of quality of health care, which would also reduce infant mortality. Hence, a country's economic development status may be a lurking variable that is the true cause of the observed correlation between contraception use and reduced infant mortality rates.

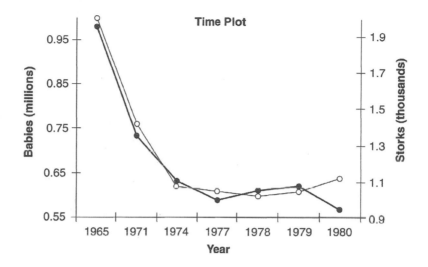

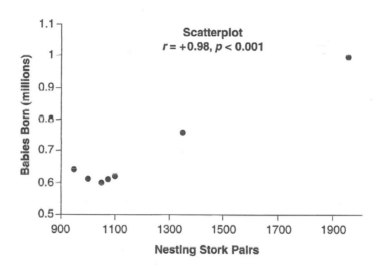

FIGURE 13.7 Coincident variation in time is not proof of a cause-effect relationship. These graphs display the association between two variables based on data obtained in Germany during the time period 1965 to 1980. One variable is the number of pairs of nesting storks. The other variable is the number of babies born in Germany in each of the years when the stork census was done. There is clearly coincident variation over time in these two variables; the stork population decreased, as did the number of babies born in Germany. The scatterplot and correlation of r = 0.98 provide strong evidence for an association between the numbers of storks and numbers of babies born. However, there obviously is no cause-effect relationship between numbers of storks and babies, unless you still believe the old children's stories about where babies come from.

Rules of evidence for claims of cause-effect relationships.   The following rules describe the evidence that is both necessary and sufficient to substantiate claims of cause-effect relationships between two variables. A significant statistical association is only part of a larger body of evidence required for such claims.

1. *A statistically significant association between the particular X and Y variables must be documented in multiple, independent studies,* preferably in multiple locations and by different investigators. This reduces the probability that the association is due merely to coincidence in time trends. For example, many studies done throughout the eastern United States have reported that white oak tree-ring widths are positively correlated with June precipitation and negatively correlated with June temperature.

2. *The possibility that the association is due to the action of a lurking variable must be explicitly eliminated,* usually through controlled experiments that eliminate the influence of all variables except the proposed "cause" variable. The action of lurking variables can also be reduced (but rarely eliminated entirely) through controlled sampling in a natural experiment. For example, the investigators who studied the association between contraception use and infant mortality (Figure 13.6) could have restricted their analysis to include only countries at a similar stage of economic development.

3. *There must be a plausible causal mechanism* that can explain *how* variation in one of the variables could cause a change in the other variable. It must be possible to document this mechanism through observation or experimentation. Such causal mechanisms are usually based on prior knowledge of how the study subjects "work." For example, there is plenty of plant physiological information that documents the effect of drought on the growth of trees. It has also been established in many underdeveloped countries that extended breast feeding increases infant survival, and that this is possible only if the time between births is extended. However, simply being able to devise a plausible cause-effect scenario is *not* sufficient in and of itself. The other two criteria must be fulfilled, and the proposed cause-effect mechanism must be amenable to a critical test through observation and/or experimentation.

13.3
Linear Regression Analysis

1. **Application: Regression** is used for two general purposes. Sometimes the *Y*-variable represents the *response* of the study subjects to the *X*-variable (the *cause*). **Regression analysis** describes the nature and strength of the association between the cause and response variables in a form that provides for predicting future responses. For example, we might perform an experiment to describe the association between atmospheric ozone concentration and the yield of a crop plant such as corn. Based on this association, we could make predictions about how corn yields might be influenced if more rigorous clean air laws were passed and air pollution was reduced.

Another common use of regression is to predict the value of a variable that is difficult to measure (*Y*) based on measurement of an associated variable that is easier to measure (*X*). In this second situation, we measure each study subject for both variables, and we use regression analysis to describe the nature of the association. If there is a strong association between the two variables, we subsequently need only make the easier measurements for variable *X* on other study subjects and then use the regression equation to estimate the values for variable Y. For example, health professionals are interested in the percentage of a person's body mass that is fat as one risk factor for cardiovascular disease. Directly measuring a person's percentage body fat can be a laborious procedure. However, other measurements of body dimensions are correlated with body fat, and are relatively easy to measure (e.g., waist circumference). A health science researcher might obtain a sample of people that encompasses a wide range of waist circumference and percentage body fat, measure both variables on this sample, and then use regression analysis to describe the association between these two variables. Once the results of this regression analysis are published, other health professionals need only measure waist circumference and then use the regression equation to estimate percentage body fat, saving much measurement effort.

Least-squares regression (LSR) produces an equation that describes the linear association between two quantitative variables X and Y. This process is done by fitting a straight line through the middle of points in a scatterplot in a manner that minimizes the distances between the data points and the line (i.e., produces a "best-fit" line). The equation of the LSR line includes sample estimates for the slope and Y-intercept of the line. Using this equation, you can predict values for the Y-variable from values of the X-variable.

Assumptions for least-squares regression analysis.

a. The data for X and Y were obtained by a randomized, unbiased study design.

b. *The true association between the X- and Y-variable is linear* A preliminary assessment of the linearity assumption can be done based on a scatterplot of X- and Y-variables. If the cloud of points forms an approximately linear pattern, the linearity assumption can be considered valid. (See Figure 13.8A and B.)

c. *The variance of the Y-variable is constant across the range of the X-variable.* A scatterplot can be used to assess the equal-variance assumption. If the spread of the points in the vertical direction is more or less constant across the range of the X-axis, the equal variance assumption can be considered valid. (See Figure 13.8A and C.)

d. *Both the X- and Y-variable are Normally distributed.* This assumption can be assessed using boxplots or Normal quantile plots for the X- and Y-variables.

2. **Sample statistics for least-squares regression.** Suppose that you were given a scatterplot that displays the association between a person's height and weight and you were asked to predict the weight of someone who was 70 inches tall. Figure 13.9 describes how you might go about using this scatterplot to obtain a *specific* predicted weight value. If you drew by hand the line that passes through the middle of the cloud of points in the scatterplot, you could probably get reasonable predictions of weight based on height. However, this graphical approach for obtaining predicted values for weight is laborious and subject to error. The calculation of a "best-fit" regression line that passes *exactly* through the middle of the cloud of data points provides a more accurate and efficient means for obtaining predicted values.

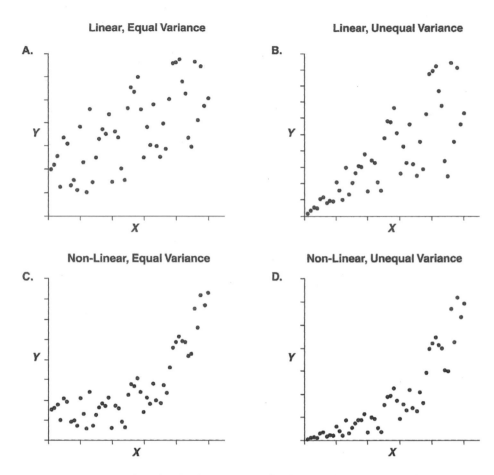

FIGURE 13.8 **Scatterplots that display examples of violations of the linearity and equal-variances assumptions for least-squares regression.**

Two summary statistics are necessary to define a specific best-fit regression line, the slope and Y-intercept. The **slope** represents the average amount of change in the value of the Y-variable associated with a one-unit change in the X-variable. For example, the slope describes the average change in an individual's weight associated with a one-inch change in height. If there is no association between the X- and Y-variable, the slope of the regression line is zero. The **Y-intercept** is the average value of the Y-variable when the value of the X-variable is zero. That is, it is the point where the best-fit line intersects the Y-axis.

Calculating the slope for the LSR line. Different authors present the equation of a straight line with various notations (e.g., $Y = mX + b$, with $m =$ slope and $b = Y$-intercept, or $Y = a + bX$ with $a = Y$-intercept and $b =$ slope). I use the notation $Y = b_0 + b_1X$, where b_0 is the Y-intercept and b_1 is the slope. In that notation, the slope of the least-squares regression line is computed as:

Slope

$$b_1 = \frac{\sum_{i=1}^{n} (x_i - \bar{x})(y_i - \bar{y})}{\sum (x_i - \bar{x})^2}$$

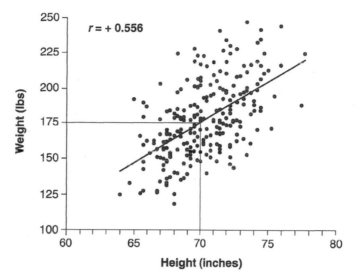

FIGURE 13.9   Graphic approach for predicting weight based on the association with height.
This scatterplot displays a positive association between a person's height and weight.
Suppose you wanted to use this association to predict the weight of a specific person who
was 70 inches tall. From the scatterplot, individuals in this sample who were 70 inches tall
had weights that ranged from approximately 150 to 225 lbs. A "best estimate" of the
weight of a person who was 70 inches tall might be somewhere in the middle of this range.
One way to facilitate using this graph to predict weight from height would be to draw a line
that runs through the middle of the cloud of points in the scatterplot. To predict a weight
value for any given height value X within the range of the data, you could simply draw a
vertical line up from the X-axis to the line; then draw a horizontal line to the Y-axis to
determine the predicted value of Y. From this method, a person who was 70 inches tall
would have a predicted weight of approximately 175 lbs.

Note that the numerator in this equation is the same "sum of cross products" I described for the calculation of Pearson's correlation coefficient (r). The description I
provided in Section 13.1 regarding how correlation coefficients measure associations is equally applicable to the slope of the regression line. The difference between
correlation and slope is that correlation coefficients are dimensionless indices of association, whereas the slope is expressed as units of change in Y per unit change in
X. Again, few people actually use this formula to compute the slope of the regression line. Rather, regression analysis is typically done on a statistical calculator or
using a statistics computer program.

Calculation of the Y-intercept for the LSR line is based on the stipulation that
this line *always* goes through the middle point in the cloud of data, defined by ($\bar{x}$,
$\bar{y}$). Given the point on the line ($\bar{x}$, $\bar{y}$) and the slope computed as described above,
then the Y-intercept is computed as follows:

Y-Intercept

$$b_0 = \bar{y} - b_1\bar{x}$$

For the least-squares regression (LSR) line, the distances between the scatterplot points and the line are minimized. Because regression lines are used to predict Y-values, the LSR line minimizes the *vertical* distances between the points and the regression line. The "least-squares" descriptor means that the slope and Y-intercept of the regression line are computed so as to minimize the sum of the squared differences between the scatterplot points and the regression line.

3. **Hypotheses for the LSR line.**

   a. *Hypothesis about the slope:* The primary hypothesis tested in LSR analysis is that the slope of the regression line is zero. This Null hypothesis means that the average of Y is the same for all values of X, so there is *no association* between these two variables (e.g., see bottom two panels of Figure 13.1). The statistical hypotheses are stated as:

   - $H_0: \beta_1 = 0$
   - $H_a: \beta_1 \neq 0$

   where β_1 (pronounced "Beta 1") is the unknown true slope of the LSR line that would be obtained if data for all individuals in the population were used to compute the line. The sample slope b_1 estimates this population parameter.

   b. *Hypothesis about the Y-intercept:* Standard regression analyses also test the hypothesis that the Y-intercept is equal to zero. The statistical hypotheses are stated as:

   - $H_0: \beta_0 = 0$
   - $H_a: \beta_0 \neq 0$

   where β_0 is the unknown Y-intercept of the true regression line. The sample statistic b_0 estimates this parameter. The test for this hypothesis has *nothing* to do with the primary question regarding whether or not there is an association between X and Y. Rather, this test simply addresses the question of whether or not the mean of the Y-variable is zero when the value of X is zero. In many real-world situations, it is reasonable to expect that the value of Y should be zero when X is zero. For example, a person whose age or height is zero should have a weight of zero as well. Sometimes, the estimated value of the Y-intercept makes no sense (e.g., a large negative number for a Y-variable like weight or height that cannot take negative values). This nonsense Y-intercept value provides for a "best-fit" line through the data points. However, a nonsense Y-intercept value indicates that the equation of the LSR line should not be used to predict values of Y for X-values outside the range of the data values. A significant negative Y-intercept commonly occurs when a positive association between X and Y is curvilinear beyond the range of X used to compute the regression line (see Figure 13.10).

4. **Sampling distributions for b_0 and b_1.** The Y-intercept (b_0) and the slope (b_1) of the LSR line are sample statistics that exhibit random sampling variation, just like the sample mean $\bar{x}$, proportion $\hat{p}$, and correlation r. If repeated samples of size n subjects were taken from the same population, each subject measured

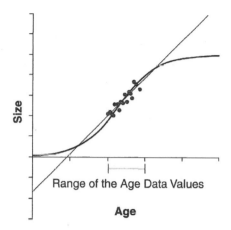

FIGURE 13.10   Sigmoid associations and negative Y-intercepts.   Nonsense negative Y-intercept values often occur when the nature of the association is linear within the restricted range of the X data values, but nonlinear over the entire range of the X-variable. In this graph, the points display the data values used to compute the LSR line (the straight line), while the smooth S-shaped curve displays the true nature of the association between age and body size across the full range of the age variable.

for variables X and Y, and a regression line computed for each independent sample, the values of b_0 and b_1 computed from these repeated samples would exhibit sampling variability (see Figure 13.11). Hence, we use the sample estimates b_0 and b_1 as "uncertain evidence" from which we draw conclusions about the population parameters for the true Y-intercept β_0 and slope β_1. As was the case with all other sample statistics, we use sampling distributions defined under the assumption that the Null hypothesis is true to determine if the sample data provide sufficient evidence to conclude that either β_0 or β_1 differ from zero. The characteristics of the sampling distributions under the Null hypotheses for these parameters are given below.

	Y-Intercept	Slope
Center:	$E(b_0) = 0$	$E(b_1) = 0$
Spread:	$S_{b0} = s\sqrt{\dfrac{1}{n} + \dfrac{\bar{x}^2}{(n-1)S_x^2}}$	$S_{b1} = s\sqrt{\dfrac{1}{(n-1)S_x^2}}$
Shape:	$t_{\text{df}=n-2}$ distribution	$t_{\text{df}=n-2}$ distribution

NOTE The calculation of the standard errors for the regression coefficients S_{b0} and S_{b1} also includes the value for the standard error of the regression (s). This value quantifies the deviations of the data points from the regression line, and represents the random noise variation in the association between the X- and Y-variables. This quantity is always provided in regression print-outs from statistics computer programs. I describe s in more detail below. The value S_x^2 is the sample variance for the X data values. ■

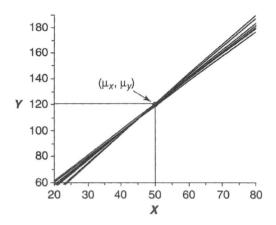

FIGURE 13.11 Random sampling variation for regression lines. Regression lines computed from data obtained by multiple, independent random samples from the same population. Note the greater sampling variation at the ends of the lines as compared to the middle of the lines, near the point (μ_X, μ_Y).

5. **Test statistics.** Tests of significance for both the Y-intercept b_0 and the slope b_1 are one-sample t-tests. The t_{test} statistics are computed as follows:

   Y-Intercept | Slope

   $t_{\text{test}} = b_0/S_{b0}$ $t_{\text{test}} = b_1/S_{b1}$

   NOTE The t_{test} statistics for the Y-intercept and slope both represent the differences between the observed values for b_0 and b_1 and the expected values under the Null hypothesis (zero). These differences are measured in "standard error units" that quantify the expected amount of random sampling variation in the values of these coefficients. ■

6. *p*-Value

   a. Statistics computer programs generally present the t-tests for both the Y-intercept and slope coefficients as *two-tailed tests*.

   b. *Stating conclusions for the slope:*

   If $p \leq 0.05$: The probability associated with the random-variation explanation for why b_1 deviates from zero is sufficiently low to reject this explanation. If an appropriate study design has been correctly implemented, the only remaining likely explanation is that the true slope $\beta_1 \neq 0$. This provides sufficient evidence to conclude that there is an association between the X- and Y-variables. Even though the test is two-tailed, your conclusion should indicate whether the association was positive or negative. You should also provide the equation for the LSR line.

■ **EXAMPLE**   "There was a negative association between exercise (number of hours per week) and blood pressure [BP = 140 − 2.13 (Hrs of exercise), $p = 0.021$].   ■

NOTE: Even when the data provide strong evidence for a statistical association between two variables, this is *not* sufficient to conclude there is a cause-effect relationship. Additional evidence is required to support such a claim, as described in Section 13.2.   ■

If $0.05 < p \leq 0.10$: The probability associated with the random-variation explanation for why b_1 deviates from zero is sufficiently high that we cannot confidently exclude this explanation. However, the probability is sufficiently low that we might suspect there really is an association. Some scientists would conclude that the data "suggest" there is an association, but more data are required to verify this conclusion. In this case, the regression equation would usually not be presented.

■ **EXAMPLE**   "The results suggest there is a negative association between exercise and blood pressure ($p = 0.075$), but more data are needed to verify this."   ■

If $p > 0.10$: The probability associated with the random-variation explanation for why b_1 deviates from zero is so high that we cannot exclude this explanation. Hence, there is insufficient evidence to conclude there is an association between the X- and Y-variables.

■ **EXAMPLE**   "There was no association between exercise and blood pressure ($p = 0.56$)."   ■

c. *Stating conclusions for the Y-intercept:* The general rules and forms for conclusions regarding the Y-intercept are the same as for the slope. However, the results from the test of significance for the Y-intercept can be used only to draw conclusions regarding whether or not the true intercept, β_0, is equal to zero. Whether or not the value of the Y intercept is zero tells us *nothing* about whether or not there is an association between the X and Y variables. In my experience, investigators rarely even mention the Y-intercept in their conclusions. Rather, if the Y-intercept is significant, this coefficient is presented when the equation for the LSR line is presented. If the Y-intercept is *not* significant, the investigator will often recompute the equation for the LSR line with the stipulation that the Y-intercept equals zero. This second equation will then be presented in the results of the study.

7. **Using the LSR line to predict values for Y, given values of X.**

For each possible value of the X-variable, the equation for the LSR line can be used to compute a **predicted value** of Y. To distinguish these predicted values of Y from the actual values for Y in the data, I use the symbol $\hat{y}$ (pronounced "y-hat") for predicted values, and the symbol y for actual data values. To compute predicted values for Y, you insert values for the slope, Y-intercept, and the specific value of the X-variable x_i into the LSR equation and solve for $\hat{y}_i$. The subscript i indicates that the X-value for a specific individual is entered into the equation to compute a predicted value $\hat{y}$ for that individual.

Predicted Value of y

$$\hat{y}_i = b_0 + b_1 x_i$$

The LSR line itself represents all possible predicted values of $\hat{y}$ for all X-values.

Examples 13.2 and 13.3 show the way the LSR line is used to make predictions.

Important: You cannot reverse the regression equation $Y = b_0 + b_1 X$ to predict values for X from values for Y. Even though it might seem reasonable to algebraically reverse the equation to predict X from values of Y [i.e., $X = (Y - b_0)/b_1$], this formula will *not* provide a least-squares estimate of X. A LSR line computed to predict X from Y would arrive at different values for b_0 and b_1 than the line computed for a LSR of Y on X.

8. **Residuals.**   Residuals are the differences between actual data values for the Y-variable (y_i) used to compute the LSR line and the predicted values ($\hat{y}_i$) computed for their associated x_i values. Residuals are computed as ($y_i - \hat{y}_i$) the actual Y–value minus the predicted Y–value. Residuals represent the vertical distances between the points of a scatterplot and the LSR line that passes through the middle of the points. The sum of the residuals for all data values x_i, y_i used to compute the LSR line is always zero. (The line passes through the middle of the data points, so half the residuals are positive and half negative.) The method used to compute the LSR line minimizes the sum of squared residuals $\sum_{i=1}^{n} (y_i - \hat{y}_i)^2$ for all data values used to compute the line. If the association between X and Y is strong, the sum of squared residuals will be small, indicating the points fall close to the LSR line. This is evidence that most variation of the Y-variable is associated with the pattern described by the LSR line, with relatively little noise or random variation. If the sum of squared residuals is large, indicating the points of the scatterplot vary widely about the LSR line, this is evidence that the pattern of the association is weak relative to the random noise.

9. **Standard error of the regression.**   The standard error of the regression (s) is the mean deviation between the points in a scatterplot of the data and the LSR line through those points, computed as follows:

Standard Error of the Regression

$$s = \sqrt{\frac{\Sigma(y_i - \hat{y}_i)^2}{n - 2}}$$

This statistic quantitates the noise component of the association and is used in the tests of significance for the sample slope and Y-intercept and for confidence intervals for predicted Y (described in a later section of this chapter).

■ **EXAMPLE 13.2**

Using a least-squares regression line to predict an individual's percentage of body fat from that individual's age. The scatterplot in Example 13.1 displays a positive association between age and percentage of body fat for 50 adults. The least-squares regression line that describes this association is:

$$\text{\% Body fat} = 8.27 + 0.2345 \text{ (Age)}.$$

> **NOTE** The value for the Y-intercept indicates that at birth (age zero), predicted percent body fat is 8.27%. The slope indicates that percentage of body fat increases by 0.2345% per year of age. ■

The table below displays predicted body fat values computed from age values using the equation for the LSR line, and residuals for several individuals from this sample.

Age (x_i)	Actual % Body Fat (y_i)	Predicted % Body Fat ($\hat{y}_i$)	Residual ($y_i - \hat{y}_i$)
26	10.4	14.4	−4.0
35	21.8	16.5	+5.3
50	13.8	20.0	−6.2
67	32.6	24.0	+8.6

■ **EXAMPLE 13.3**

Using a least-squares regression line to predict a chemical's concentration from a colorimetric test. Colorimetric tests are commonly used in chemistry to determine the concentration of a chemical in a solution. This procedure involves adding a substance to the solution that changes color in the presence of the chemical of interest. The higher the concentration of the chemical of interest, the more intense the color of the solution.

The intensity of the color is measured with a spectrophotometer. Light of a wavelength specific to the color is passed through the solution. The more intense the color of the solution, the more light is absorbed. However, it is still necessary to determine the association between these absorbance values and the actual concentration of the chemical of interest in the solution.

To do this a set of standard solutions is prepared, with a range of known concentrations of the chemical. The colorimetric analysis is performed on each standard solution and absorbance values are obtained for each of the known concentrations of the chemical. A regression line is then computed so that concentration of the chemical can be predicted from absorbance values. The following is an example of results that might be obtained from such an analysis.

The data:

Concentration (Y)	1	2	3	4	5	6	7	8	9	10
Absorbance (X)	.11	.21	.24	.36	.50	.55	.69	.70	.81	.95

Before proceeding with the regression analysis to compute the LSR line that will be used to estimate concentration from absorbance values, we must first verify that the assumptions for this analysis are valid.

Exploratory Data Analysis

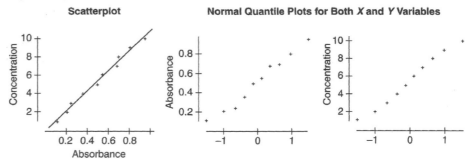

Summary Statistics for *X*- and *Y*-Variables

Summaries
No Selector

Variable	Count	Mean	StdDev
Concentration	10	5.5	3.02765
Absorbance	10	0.512651	0.278691

> **NOTE** Always compute the mean and standard deviation for the *X*- and *Y*-variables as part of the exploratory data analysis. You will need these summary statistics later in the analysis. ■

Assessment for Assumptions for LSR

a. The association is linear. The scatterplot displays a linear pattern.

b. The variance of *Y* (concentration) is equal across the range of *X* (absorbance). The variation among the points in the scatterplot in the vertical direction is the same across the *X*-axis.

c. Both variables are Normally distributed. The Normal quantile plots both display a linear pattern, indicating the distributions for both *X* (absorbance) and *Y* (concentration) variables are Normal.

Regression Print-out   (The numbered labels are explained after the print-out.)

Dependent variable is: Concentration
R-squared $= 98.9\%$ R-squared (adjusted) $= 98.8\%$
$7 \longrightarrow s = 0.3371$ with $10 - 2 = 8$ degrees of freedom

Source	Sums of Squares	df	Mean Square	F-ratio
Regression	81.5909	1	81.5909	718
Residual	0.909134	8	0.113642	

Variable	Coefficient	S.E. of Coeff	t-ratio	Probability
Constant	−0.038578	0.2326	−0.166	0.8724
Absorbance	10.8038	0.4032	26.8	≤0.0001

1 { (Constant, Absorbance)

↑ 2   ↑ 3   ↑ 4   ↑ 5   ↑ 6

Explanation of labels in the regression print-out:

1. This is the table of regression coefficients. This part of the print-out provides the estimates of the Y-intercept and slope coefficients and gives the tests of significance for these sample statistics.

2. The **Variable** column identifies the coefficients. "Constant" on the first line refers to the Y-intercept. "Absorbance" on the second line indicates that this line refers to the slope coefficient, which is multiplied by the variable absorbance.

3. The **Coefficient** column contains the actual estimates of b_0 (top line) and b_1 (bottom line). Based on these estimates, the equation for the LSR line that predicts concentration from absorbance is:

   Concentration $= -0.038578 + 10.8038$ (Absorbance).

4. The column labeled **S.E. of Coeff** lists the standard errors of the regression coefficients (S_{b0} and S_{b1}).

5. The t ratio column lists the t_{test} statistics for tests of significance for the Null hypotheses H_0: $\beta_0 = 0$ (top line) and H_0: $\beta_1 = 0$ (bottom line).

6. The **Probability** column lists the p-values for the tests of significance for b_0 and b_1.

   a. The p-value for the Y-intercept indicates that the Y-intercept is not significantly different from zero.

   b. The p-value for the slope coefficient provides strong evidence that $\beta_1 \neq 0$, meaning that there is a positive association between concentration and absorbance.

7. This row lists the standard error of the regression (s) and its degrees of freedom.

NOTE I have not yet described some items in this print-out. The information described here (items 1 to 7) is sufficient to determine whether or not there is an association between absorbance and concentration and to define a regression equation that can be used to estimate concentration from absorbance values. The remaining items in this print-out provide information on the strength of the association, and are described in the next section, Analysis of Variance for Regression. ■

Predicted Concentration and Residuals.   In the following table, the Absorbance and Concentration columns contain the X and Y data values used to compute the LSR line. The Predicted Concentration column contains predicted values computed using the equation for the LSR line, Concentration = $[-0.04 + 10.8\ (\text{Absorbance})]$. The Residuals column contains the differences between actual and predicted concentration values.

Absorbance	Concentration	Predicted Concentration	Residuals (Actual − Predicted)
.11	1	1.148	−0.148
.21	2	2.228	−0.228
.24	3	2.552	+0.448
.36	4	3.848	+0.152
.50	5	5.360	−0.360
.55	6	5.900	+0.100
.69	7	7.412	−0.412
.70	8	7.520	+0.480
.81	9	8.708	+0.292
.95	10	10.22	−0.220

Analysis of Variance for Regression

The description of linear regression analysis presented in the previous section is adequate for the simplest applications, where the objective is to predict values for some Y-variable based on values for a single X-predictor variable. If the slope is significantly different from zero, this is a sufficient basis for concluding there is an association between X and Y. However, the slope does not provide an easily interpretable scale for assessing the strength of the association. We must know the strength of the association if we are to assess the precision of estimates for Y computed using the LSR line. Also, many applications of regression involve predicting values for a Y-variable based on data for *multiple* X-variables. For example, health professionals who use a regression approach for estimating a person's percentage of body fat based on simple measurements of body dimensions often use multiple body dimension variables. Men tend to deposit most of their fat around the waist, while many women deposit their fat in the hips, buttocks, and/or thighs. To produce a single regression equation to predict percentage of body fat for a wide diversity of individuals, the equation might include circumference measurements for waist, hips, and thighs as multiple X-variables. A table of coefficients in the regression print-out would include a separate line for each of these multiple X-variables, with a separate t-test for each slope coefficient associated with each X-variable. However, we also need an overall test of significance for the entire regression equation. The Analysis of Variance components of the regression print-out provide both a measure of the strength of the association and an overall test of significance for the regression equation.

When I first introduced the concept of statistical association, I also introduced the terms "pattern" and "noise." Pattern is the systematic variation in the value of the Y-variable that is associated with change in the value of the X-variable. For example, people who are taller tend to weigh more than people who are shorter. Noise is the random variation among individuals that results in deviation from the pattern. For example, a shorter person who has more body fat could weigh more than a taller person who has less body fat. In this example, random variation in the amount of body fat introduces noise to the general pattern that taller people tend to weigh more than shorter people. Analysis of variance for regression partitions the variation among data values for the Y-variable into a *pattern component* due to the association with X and a *noise component* that reflects random deviations from the pattern.

Analysis of variance for regression is based on **sums of squares** that represent total, pattern, and noise variation in the data values for the Y-variable. The calculation of these pattern and noise components of the total variation in the Y data values is described below:

Total	=	Pattern	+	Noise
Total variation of Y	=	Variation explained by regression model	+	Unexplained variation
$\sum_{i=1}^{n} (y_i - \bar{y})^2$	=	$\sum (\hat{y}_i - \bar{y})^2$	+	$\sum (y_i - \hat{y}_i)^2$
Sum of Squares Total	=	Sum of Squares Regression	+	Sum of Squares Residual
SS_{total}	=	$SS_{regression}$	+	$SS_{residual}$

These sums of squares are the numerator terms for the calculation of variances (described in Chapter 2 and Chapter 11). Three "squared deviations" are computed for each y-value in the data used to compute the LSR line, and these are summed across all n y-values in the sample. The Sum of Squares regression ($SS_{regression}$) represents the deviations of the predicted values of Y on the LSR line from a horizontal line with a Y-intercept of $\bar{y}$ (Figure 13.12A). The greater the slope of the LSR line, the larger will be $SS_{regression}$. Some statistics computer programs label this quantity "Model" (i.e., SS_{Model}).

The Sum of Squares Residual ($SS_{residual}$) represents the differences between predicted values of Y and the observed values of Y for each X-value in the data used to compute the LSR line. That is, $SS_{residual}$ represents the deviations of points in a scatterplot away from the LSR line (Figure 13.12B). Some statistics programs give this quantity the label "Error" (i.e., SS_{Error}).

The Sum of Squares Total (SS_{total}) represents the total variation of the Y data values about their mean (Figure 13.12C).

- If there is a strong association between Y and X, the slope of the true LSR line is not zero. The more the slope of the LSR line deviates from horizontal, the larger will be the squared differences between the predicted values of Y on the LSR line and the mean of Y $(\hat{y}_i - \bar{y})^2$. The result is that the $SS_{regression}$ will be a larger value relative to the $SS_{residual}$ that reflects random variation of data points away from the LSR line. Hence, a larger proportion of the total variance of Y will be due to the fact that the mean of Y (μ_y) systematically changes (increases or decreases) as the value of the X-variable changes. (See Figure 13.13A.)

- If there is no association between Y and X, the slope of the true LSR line is 0 and the Y-intercept = μ_y, the true mean of the Y-variable. The computed LSR regression line will be an approximately horizontal line, with all values of $\hat{y}$ close to the mean of Y $(\bar{y})$. This comes from the fact that the LSR line *must* pass through the point $(\bar{x}, \bar{y})$. This indicates that the true mean of the Y-variable (μ_Y) is the same for all values of X. The $SS_{regression} = \Sigma(\hat{y}_i - \bar{y})^2$ reflects only random sampling variation of $\bar{y}$ and b_1, and will be small compared to the $SS_{residual}$. Hence, the pattern variation will be only a small proportion of total variation. (See Figure 13.13B.)

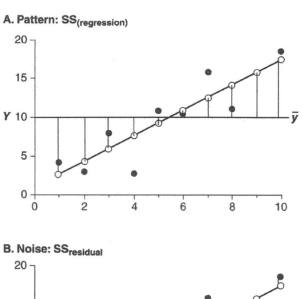

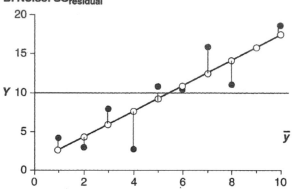

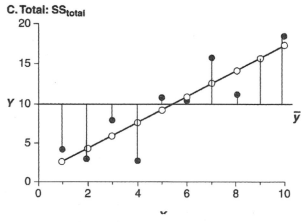

FIGURE 13.12   Analysis of variance for regression.   The three panels above display the "pattern" (regression) and "noise" (residual) components of the variation in values of the Y-variable reported in ANOVA tables for regression. In part A the sum of squared differences $\Sigma(\hat{y}_i - \bar{y})^2$ measures the deviation of the regression line from a horizontal line with a Y-intercept of $\bar{y}$. The filled circles are observed values of the Y-variable, the open circles on the regression line are the predicted values of Y for each observed value x_i, y_i. The vertical lines display the differences $\hat{y}_i - \bar{y}$. In part B the sum of squared differences $\Sigma(y_i - \hat{y}_i)^2$ measures the deviation of points in the scatterplot from the regression line. The vertical lines display the differences $(y_i - \hat{y}_i)$. In part C the sum of squared differences $\Sigma(y_i - \bar{y})^2$ measures the deviations of the individual y-values from their mean. The vertical lines display the differences $(y_i - \bar{y})$. The SS_{total} is the numerator of the equation to compute the variance of Y, as described in Chapter 2.

A. Strong Association

Pattern: $SS_{regression} = \Sigma(\hat{x}_i - \bar{y})^2$

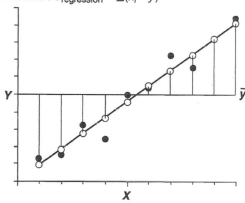

Noise: $SS_{residual} = \Sigma(y_i - y_i)^2$

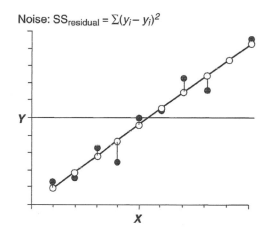

B. Weak Association

Pattern: $SS_{regression} = \Sigma(\hat{y}_i - \bar{y})^2$

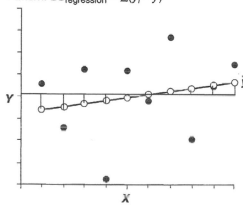

Noise: $SS_{residual} = \Sigma(y_i - \bar{y}_i)^2$

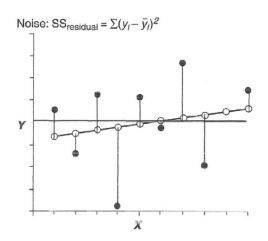

FIGURE 13.13   Regression ANOVA for strong and weak associations.   In part A, a strong association is shown. $SS_{regression}$ is much larger than $SS_{residual}$. In part B, a weak association is shown. $SS_{regression}$ is much smaller than $SS_{residual}$.

The regression ANOVA table presents the partitioning of total variance of Y between regression and residual variance.   This analysis of variance table looks similar to those presented in Chapter 11, but the purpose of a regression ANOVA table is very different. The Chapter 11 analysis of variance is used to test the significance of *differences* among means for three or more groups. The regression ANOVA table is used to test the significance of *associations* between two or more variables.

The ANOVA-based test of significance for the association between X and Y compares the relative sizes of the regression and residual "mean squares" (another term for variance, used in the context of analysis of variance). **Mean squares (MS)** are computed by dividing the sums of squares by their associated degrees of freedom. The degrees of freedom for $SS_{regression}$ is equal to the number of X-variables (and their associated slope coefficients b_1, b_2, b_3, etc.) in the equation for the LSR line. In this chapter I present only the simplest case with one X-variable and $df_{regression} = 1$. Degrees of freedom for $SS_{residual}$ are computed as the sample size n minus the number of estimated coefficients in the equation for the LSR line. For the simple regression line $Y = b_0 + b_1X$, $df_{residual} = n - 2$. The F_{test} statistic is computed as the ratio of $MS_{regression}$ divided by $MS_{residual}$. If there is no association, predicted y-values $\hat{y}_i$ will be approximately equal to $\bar{y}$ and $MS_{regression}$ will be small, as will be the value of F_{test}. Large values for the F_{test} statistic mean that variance due to pattern is much larger than variance due to noise, indicating that there is an association between the X- and Y-variables.

The p-value for the F_{test} statistic is used to test the overall hypothesis that there is an association between the Y-variable and one or more X-variables. If $p \leq 0.05$, you can conclude there is strong evidence that the slope of the LSR line is not equal to zero (i.e., there is an association between X- and Y-variables).

> **NOTE** Both the F-test in the regression ANOVA table and the t-test for the slope can be used to evaluate the evidence for an association. When there is only one X-variable, the results of these two tests are identical. However, when more than one X-variable is included in the regression equation to predict the value of Y, the t-tests are used to test the significance of the slope coefficient for each X-variable separately, whereas the ANOVA F-test evaluates the significance of the overall regression equation. ∎

Other regression statistics based on ANOVA for regression.   There are two other regression statistics shown on the regression print-out of Example 13.3 that are based on the ANOVA table, s and R^2.

1. The **standard error of the regression** (s) is equal to the square root of $MS_{residual}$ in the ANOVA table. This is equivalent to the pooled estimate of the standard deviation used in pooled-variance t-tests (Chapter 9) and in analysis of variance (Chapter 11).

$$s = \sqrt{\frac{\Sigma(y_i - \hat{y})^2}{n - 2}} = \sqrt{\frac{SS_{residual}}{df_{residual}}} = \sqrt{MS_{residual}}$$

2. The **coefficient of determination** (R^2) describes the proportion of the total variation in the response variable Y that is "explained" by the fit of the regression line to the data. The R^2 statistic has values between 0 and 1. This is a measure of the strength of the association, measuring the proportion of total variation that is associated with pattern.

$$R^2 = \frac{\Sigma(\hat{y}_i - \bar{y})^2}{\Sigma(y_i - \bar{y})^2} = \frac{SS_{regression}}{SS_{total}}$$

Example 13.4 contains an interpretation of the ANOVA-related components in the print-out of Example 13.3.

Relationship between correlation and regression analyses.   The coefficient of determination (R^2) from regression analysis is equal to the square of the Pearson's correlation coefficient r for the association between X- and Y-variables.

$$r = \sqrt{R^2}$$

NOTE $\sqrt{R^2}$ cannot tell you the sign of Pearson's correlation coefficient r. The sign of the correlation coefficient will be the same as the sign of the slope coefficient from the regression model. ■

EXAMPLE 13.4

Interpreting ANOVA for regression. This regression print-out is the same one presented in Example 13.3 to assess the association between the color of a solution and the concentration of a chemical. Here I describe the ANOVA-related components.

Dependent variable is: Concentration

$2 \rightarrow$ R-squared $= 98.9\%$ R-squared (adjusted) $= 98.8\%$

$s = 0.3371$ with $10 - 2 = 8$ degrees of freedom

Source	Sums of Squares	df	Mean Square	F-ratio
1 { Regression	81.5909	1	81.5909	718
Residual	0.909134	8	0.113642	

Variable	Coefficient	S.E. of Coeff	t-ratio	Probability
Constant	−0.038578	0.2326	−0.166	0.8724
Absorbance	10.8038	0.4032	26.8	≤0.0001

Explanation of labels in the regression print-out:

1. In the ANOVA table, the variance due to the fact that the mean of Y (concentration) changes as X (absorbance) changes ($MS_{regression}$) is $718\times$ larger than the random variation of Y about the LSR line ($MS_{residual}$), producing an F-value of 718. If there is no association between absorbance and concentration, the expected value of F_{test} is approximately 1.0. Hence, we are quite confident in concluding there is an association between concentration and absorbance. The ANOVA table produced by this statistics computer program does not provide a p-value for F_{test} because this test of significance is the same as the t-test for the slope b_1 (presented in the table of coefficients below the ANOVA table).

2. This print-out displays two R-squared (R^2) values. The R^2 (adjusted) value is the most appropriate for assessing the strength of the association, as this value is adjusted for the number of X-variables included in the LSR equation. This statistics computer program presents R^2 values in % format, as this is the way that most scientists discuss this statistic. However, the actual value for R^2 is 0.988. The interpretation of this R^2 value is that 98.8% of the variance in concentration is explained by the regression on absorbance (strong pattern with very little noise). Because this R^2 value is very close to 1.0 (100%), we can conclude that the association is very strong.

NOTE Based on this R^2 value, we can compute the Pearson's correlation coefficient for this association as $r = \sqrt{0.988} = 0.994$. However, to determine the sign of the correlation, we must look for the sign of the slope coefficient at the bottom of the print-out ($b_1 = +10.8$). Because the slope coefficient is positive, the value for Pearson's $r = +0.994$. ■

Diagnostics to Evaluate Assumptions for Linear Regression

Even though scatterplots can be effective for identifying substantial violations of the assumptions for linear regression, less obvious violations may be missed, especially if there is a fair amount of noise in the association. Most statistics computer programs that perform regression analysis have options for producing graphics that provide a better basis to evaluate whether or not the assumptions are valid. These specialized graphics are described below.

Residuals plots.   Residual plots are used to determine whether the linearity and equal-variance assumptions of regression are valid. The graphic is a scatterplot, with the regression residuals plotted on the Y-axis and the predicted values of Y (or the X-values) plotted on the X-axis. A random distribution of residual points about the horizontal line at zero indicates the relationship is linear. If the spread of the points about the horizontal zero-line appears to be fairly constant from one end of the line to the other, the equal-variance assumption is valid (see Figure 13.14 for examples).

Scatterplot (with LSR line)   Residuals vs. Predicted Plot

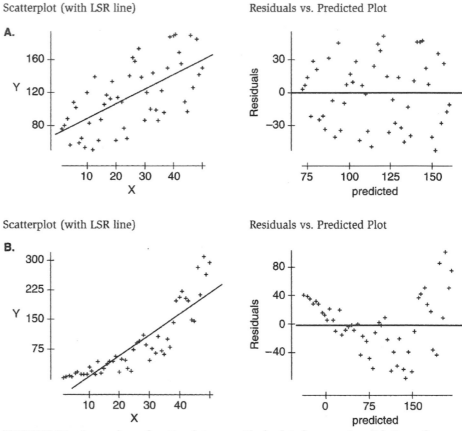

Scatterplot (with LSR line)   Residuals vs. Predicted Plot

FIGURE 13.14   Comparison of scatterplots vs. residuals plots for assessing violations of assumptions for linear regression.   In part A, the pair of plots indicate that the linearity and equal-variance assumptions are valid. In part B, the pair of plots indicate that both the linearity and equal-variance assumptions are violated. These violations are more obvious in the residuals plot.

Normal quantile plot of residuals. The sole purpose of the normal quantile plot of residuals is to determine if the residuals are Normally distributed. If the residuals are Normally distributed, the Normality assumption of regression analysis is fulfilled. Figure 13.15 has two such plots displayed.

> **NOTE** The Normal quantile plot looks very similar to a scatterplot, and students sometimes get the two types of graphic confused. Normal quantile plots provide information about the shape of the residuals distribution only, and tell you *nothing* about the nature of the association between X- and Y-variables. ■

Normal Quantile Plots for Residuals

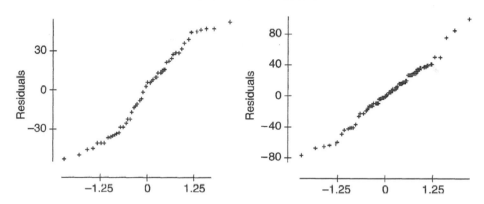

FIGURE 13.15   Determining the shape of the residuals distribution from the Normal quantile plots for residuals. These Normal quantile plots are for the same data used to generate the scatterplots and residuals plots in Figure 13.14. The plot on the left was generated from the data displayed in part A of Figure 13.14, and the plot on the right was generated from the data displayed in part B of that figure. The Normal quantile plot on the left indicates the residuals distribution is "flat" compared to a bell-shaped Normal distribution. This is a minor deviation from Normality and would probably not invalidate the regression analysis. The Normal quantile plot on the right indicates that the residuals are approximately Normally distributed. However, the residuals plot in Figure 13.14 indicated these data violated the linearity and equal-variances assumption.

Final Cautions for Regression Analyses

A number of problems can occur when performing regression analysis if the investigator does not carefully assess whether or not the data meet the assumptions of regression or inappropriately uses the regression equation to predict values for the Y-variable. This section describes the nature and consequences of some of these problems.

Influential observations and outliers.   **Influential observations** are data points that have a disproportionately large influence on the estimated slope and Y-intercept of the LSR line (i.e., the values for b_0 and b_1 change substantially when the points are deleted from the data). Usually, influential observations are outliers along the X-axis (a situation that would violate the Normality assumption). Their large influence on the regression line is due to a "teeter-totter" effect (described later). Just as a small child can lift a larger child on the teeter-totter if they are further from the fulcrum, so too can an outlier (x_i far from $\bar{x}$) exert more "leverage" on the slope of the line than values closer to the mean of X. Figure 13.16 shows an example of an influential observation.

Data values for the Y variable that are **outliers** typically result in data points that are far from the regression line (have large residual values). These outliers generally do not have a major influence on the LSR line, but they inflate the value of $MS_{residual}$, reducing the value of R^2 and increasing the standard error of the regression. The result is that the strength of the association appears to be less than is the case for most of the data values, and the width of confidence intervals are greater than they would be if the outlier was not present (see Figure 13.17).

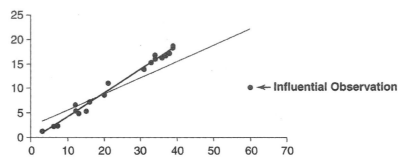

FIGURE 13.16 **The problem of influential observations.** The X-value for one data point is an outlier and therefore this point in the scatterplot is an influential observation. The dark regression line was computed without this data point, whereas the light regression line was computed with the data point included. Note the substantial change in the slope of the line caused by this one observation.

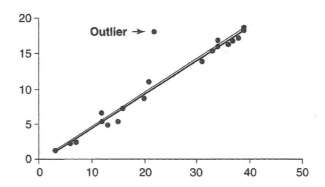

FIGURE 13.17 **The effect of outliers on least-squares regression.** One data point in this scatterplot has an outlier Y-value. The dark regression line was computed with this data point excluded, and the light regression line was computed with the data point included. Because the X-value for this data point was close to the mean of X, this outlier had minimal effect on the slope or Y-intercept of the regression line. However, the R^2 for the regression decreased from 0.99 to 0.90 when the outlier was included, and the standard error of the regression (s) more than doubled from 0.702 to 1.914. The latter increase in s would cause confidence and prediction intervals to be more than twice as wide with the outlier included than if the outlier were not present.

What should you do about outliers and influential observations?   First make sure the data values are correct (no measurement or typographical errors). Then try to obtain additional data for both X and Y from individuals that are similar to the outlier or influential observation. If these additional data suggest that the original "problem" data point is really an oddball, you might consider dropping that data point and recomputing the regression. This will provide estimates for the Y-variable that are closer to the true values for the majority of individuals in the population. However, you should *not* drop outliers and influential data values without careful consideration and a solid justification.

The problem of inappropriate use of the regression equation.   There are three issues to be aware of:

1. *Predicting Y for an x_i value beyond the range of X-values used to compute the LSR line has a high risk of bias and error.* In many situations, associations can be linear over part of the range of X, but grossly nonlinear outside that range. Many biological associations are sigmoid (S-shaped), with a "linear" part for the middle of the range of X, but they are nonlinear at the extremes of X. For example, the association between organism size and age is often sigmoid (Figure 13.10). Other associations are parabolic, such that Y increases more or less linearly as X increases up to some point, but then Y decreases as X increases beyond this point. Whenever you use a LSR line to predict values for some Y-variable, *always* use knowledge and common sense to evaluate whether or not the predicted Y-values make sense.

   For example, Figure 13.18 displays data obtained to study the association between environmental temperature and metabolic rate of a reptile. Like all cold-blooded organisms, metabolic rate of reptiles is strongly influenced by environmental temperature, as indicated by the tight clustering of data points around the LSR line. Based on this regression analysis, a biologist might feel confident about the precision of predicted values for metabolic rate obtained from the regression equation. However, the true relationship between metabolic rate and temperature is distinctly nonlinear across a wider range of temperature values. Using this regression equation to predict metabolic rate of a reptile in boiling water (100°C) would generate a predicted value that makes no sense at all.

2. *Using regression to evaluate time trends or coincident variation over time may violate the independent observations assumption.* Like all other statistical analyses, regression analysis is based on the assumption that the data values represent a randomized, unbiased sample of independent observations from the population of interest. A common violation of this independence assumption is that the x_i- and y_i-values are multiple measurements on the same individual(s) over time. For example, Y could be annual tree-ring width and X summer rainfall. In many situations, there are carry-over effects between one time period and the next. For example, the growth of a tree, as measured by annual ring width, is influenced not only by weather during the time the annual ring is formed, but also by starch reserves, foliage, and roots that were affected by

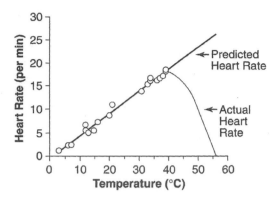

FIGURE 13.18   The problem of predicting values for the *Y*-variable for *X*-values beyond the range of data used to compute the LSR regression line.   This graph shows the data points and LSR line for an association between environmental temperature and metabolic rate of a reptile species. The investigators did not want to kill their study subjects and did not expose the reptiles to lethally high temperatures. Using the LSR line (heavy black line) to predict metabolic rate at temperatures above 40°C grossly overestimates actual heart rate (thin black line).

weather the preceding year. The result is that ring width in one year is *not* independent of ring width for the prior year. In such circumstances, special analyses are required to address the lack of independence in time series data.

3. *Statistically significant associations between two variables documented using regression analysis do NOT provide sufficient evidence to infer the X- and Y-variables have a cause-effect relationship.* The three criteria for evidence to support claims of cause-effect relationship described for correlation analyses are equally applicable when regression analysis is used to describe the statistical association. When there is sufficient evidence to support an inference of cause-effect relationship, this may enhance our confidence in the predictions obtained using the LSR equation. However, it is *not* necessary that a cause-effect relationship be documented before we use a LSR equation to predict values for a *Y*-variable.

Computing the Confidence Interval for $\mu_{Y|X}$, the Mean of *Y* for a Given *X* Value

Predicted values of *Y* obtained from the regression equation ($\hat{y}_i = b_0 + b_1 x_i$) are estimates of the mean of *Y* for a given *X* value ($\mu_{Y|X}$). The vertical line in $Y|X$ is read as "given," so $Y|X$ is read as "the value of *Y*, given the value of *X*." For example, a predicted value $\hat{y}$ for percentage of body fat estimated by entering a waist circumference value of 90 cm into a regression equation, represents an estimate of the mean percentage of body fat for all individuals with a 90-cm waist. In Chapter 7, I explained why simply presenting the value of the sample mean is inadequate, and explained the rationale for confidence intervals. A confidence interval conveys information both for the estimated value of the mean and the precision of this estimate. Just as an indication of precision was required when presenting $\bar{x}$ and $\hat{p}$, so too is an indication of precision required when presenting $\hat{y}$ as an estimate of $\mu_{Y|X}$.

In the computations for the regression line, there are three sample statistics that are independent and can vary from one sample to the next if repeated random samples were taken from the same population: $\bar{x}$, $\bar{y}$, and b_1. (Because the Y-intercept is computed from these sample statistics, it is not independent of these statistics.) Figure 13.11 displayed multiple regression lines computed using data from a simulation of taking repeated, independent samples from the same population. The variation in the positions and slopes of these LSR lines is due solely to random sampling variation.

Because the LSR line *must* past through the point $(\bar{x}, \bar{y})$, random sampling variation in the value of the slope causes the LSR line to behave like a playground teeter-totter. Variation of $\hat{y}$ values on the LSR line for values of X near $\bar{x}$ is constrained to the sampling variation of $\bar{x}$. By forcing the line to pass through the point $(\bar{x}, \bar{y})$, this point acts like the fulcrum of the teeter-totter. Random sampling variation in the value of the slope b_1 will result in more variation among $\hat{y}$ values for X-values out at the edges of the range of X (at the ends of the teeter-totter) than for X-values near $\bar{x}$ (the fulcrum). Hence, $\hat{y}_i$ values for x_i values far from $\bar{x}$ are more variable (less precise). The result is that to maintain the same level of confidence that the interval will include the true mean $\mu_{Y|X}$, a wider confidence interval is required at the ends of the LSR line than at the center.

The $100(1 - \alpha)\%$ confidence interval for the mean of Y for a given value of the X-variable $(\mu_{Y|X})$ is computed as:

$$\text{Confidence Interval} = \hat{y} \pm t_{\alpha/2, \text{df}=n-2} \; \overbrace{s\sqrt{\underbrace{\frac{1}{n}}_{\substack{\text{Standard error}\\\text{of the mean}}} + \underbrace{\frac{(x_i - \bar{x})^2}{(n-1)S_x^2}}_{\substack{\text{"Teeter-totter"}\\\text{effect}}}}}^{\text{Standard error of the estimate } \hat{y}}$$

In this equation, $s = \sqrt{\text{MS}_{\text{residual}}}$ is the standard error of the regression, and S_x^2 is the variance (standard deviation squared) of the data values for variable X used to compute the LSR line.

> **NOTE** Some statistics computer programs do not automatically provide S_x^2 as part of the regression print-out. If not, you should obtain a separate print-out of sample statistics for both X and Y variables (e.g., mean, standard deviation). ■

The standard error of the estimate $\hat{y}$ includes two terms that represent the two different sources of random variation that affect predicted y-values. The standard error of the sampling distribution of $\bar{y}|X$, is computed much the same as the standard error of any mean (note that $S\sqrt{1/n} = S/\sqrt{n}$). The teeter-totter effect on the precision of predicted y-values is quantified by the term $(x_i - \bar{x})^2/(n - 1)S_x^2$. When we compute the confidence interval for $\mu_{Y|X}$ for a specific x_i value far from $\bar{x}$, the value of this teeter-totter term will be large and will increase the standard error of the estimate. This will result in a wider confidence interval, indicating the lower precision of estimates of $\mu_{Y|X}$ for x_i values far from $\bar{x}$. When we compute the confidence interval for $\mu_{Y|X}$ for an x_i value close to $\bar{x}$, the value of this teeter-totter term will be small. This will reduce the standard error of the estimate and reduce the width of the confidence interval, indicating the estimate is more precise.

Interpretation: Because 95% of all intervals computed by this method will include the true value for the mean of the Y-variable for all individuals who are characterized by a particular value of the X-variable, we can be 95% confident that the specific interval we compute for a particular Y-variable and X-value will include the true value for $\mu_{Y|X}$.

Computing the Prediction Interval for an Individual Y, Given X

The concept of a **prediction interval** is new, and so it requires some initial explanation. When we compute a 95% confidence interval for μ, we interpret this interval to mean that there is a probability of 0.95 that the interval will include the population mean μ. However, when we use regression to predict a value $\hat{y}_i$, we are not only interested in predicting the mean value of $Y(\mu_{Y|X})$, but also the specific Y-value for an *individual* who happens to have a particular value for the X-variable $(Y|X)$. For example, when a chemist obtains an absorbance reading for a particular sample and uses a regression equation to estimate concentration of a chemical, the purpose is to estimate concentration for *that individual sample*, not the mean of all samples with the same absorbance. As you have learned in previous chapters, y_i values for individuals will exhibit much more random variation than the values of the mean $\bar{y}$ from repeated sampling from the same population. If we can assume the population distribution of Y is Normal, the Empirical Rule states that approximately 95% of all individual Y-values should fall within the range of $\mu_Y \pm 2\sigma_Y$. However, we don't know μ_Y or σ_Y, and we must estimate these values with random variables $\bar{y}$ and S_Y. To compute a confidence interval for the Y-value of an individual (distinguished from the confidence interval for $\mu_{Y|X}$, by calling the interval for an individual the *prediction interval*), the standard error of the estimate must include random variation among individual y_i values, as well as the sampling variation of the mean $(\bar{y})$. The estimate of the standard deviation among individual y_i values is quantified by the standard error of the regression $s = \sqrt{\text{MS}_{\text{residual}}}$. The estimate of random sampling variation for the mean of Y is the standard error of the mean $S_{\bar{y}} = S/\sqrt{n}$. As was the case for the confidence interval of $\mu_{Y|X}$, additional sampling variation in the values of $\hat{y}_i$ results from the teeter-totter effect described above. Hence, the standard error of $Y|X$ must also include this source of variation. Combining all three sources of random variation that influence the predicted value of Y given X, the standard error of the estimate of an individual's y-value is computed as:

$$\text{Standard Error} = \sqrt{\underbrace{\text{MS}_{\text{residual}}}_{\substack{\text{Variation of} \\ \text{individuals}}} + \underbrace{\frac{\text{MS}_{\text{residual}}}{n}}_{\substack{\text{Variation of} \\ \text{the mean}}} + \underbrace{\frac{\text{MS}_{\text{residual}}\left(x_i - \bar{x}\right)^2}{(n-1)S_x^2}}_{\substack{\text{Teeter-totter} \\ \text{effect}}}}$$

The $\text{MS}_{\text{residual}}$ represents variance of individual Y-values about the regression line. The term $\text{MS}_{\text{residual}}/n$ represents variance of the sampling distribution of $\bar{y}$ for the given X-value. The term $\text{MS}_{\text{residual}}(x_i - \bar{x})^2/(n-1)S_x^2$ represents the teeter-totter effect of sampling variation in the value of b_1, as described above for confidence intervals. The three $\text{MS}_{\text{residual}}$ terms are factored out and the square root taken to obtain the standard error of the regression (s). This simplifies the formula for computing the $100(1-\alpha)\%$ prediction interval for an individual's y-value given $x_i(Y|X)$, as shown:

$$\text{Prediction Interval} = \hat{y}_i \pm t_{\alpha/2,\text{df}=n-2} \; s\sqrt{1 + \frac{1}{n} + \frac{(x_i - \bar{x})^2}{(n-1)S_x^2}}$$

Interpretation: Because 95% of all intervals computed by this method will include the true value for the Y-variable for any one individual who is characterized by a particular value of the X-variable, we can be 95% confident that the specific interval we compute for a particular Y-variable and X-value will include the true value for a specific individual $(Y|X)$.

> **NOTE** Most statistics computer programs provide the option for computing both confidence and prediction intervals, so you will rarely need to actually use this formula. ■

In Example 13.5 we return to the situation described earlier, a colorimetric test. Now we compute the confidence and prediction intervals for an individual Y.

■ **EXAMPLE 13.5**

Computing confidence and prediction intervals for predicted Y-values computed from the LSR equation. In Example 13.3, I described a regression analysis to quantify the association between color intensity from a colorimetric test and concentration of a chemical in a solution. Here I show the calculations for the 95% confidence interval for mean concentration, given an absorbance reading of 0.45 from the colorimetric test, and the 95% prediction interval for the concentration of the chemical in a specific sample.

a. 95% confidence interval for mean concentration when absorbance $x_i = 0.45$

Using the equation for the LSR line, we compute the predicted concentration given Absorbance = 0.45 as follows:

$\hat{y}_i = -0.04 + 10.8(0.45) = 4.82$

The 95% confidence interval is computed as:

$$\hat{y}_i \pm t_{0.025,\text{df}=8} \; s\sqrt{\frac{1}{n} + \frac{(x_i - \bar{x})^2}{(n-1)S_x^2}}$$

$$= 4.82 \pm 2.306\,(0.3371)\sqrt{\frac{1}{10} + \frac{(0.45 - 0.51265)^2}{(10-1)(0.2787)^2}}$$

$$= 4.82 \pm 2.306\,(0.3371)\,\sqrt{0.1 + 0.005615}$$

$$= 4.82 \pm 0.25$$

(S_x^2) is the variance of the X-variable, obtained from the summary statistics printouts for absorbance $(= 0.2787^2)$.

Interpretation: Because 95% of all intervals computed by this method will include the true value for $\mu_{Y|X}$, we can be 95% confident that the interval computed above includes the true mean concentration of the chemical for samples with absorbance of 0.45.

b. 95% prediction interval for concentration of an individual sample with $x_i = 0.45$

$$\hat{y}_i \pm t_{\alpha/2, df = n-2} \; s \; \sqrt{1 + \frac{1}{n} + \frac{(x_i - \bar{x})^2}{(n - 1)S_x^2}}$$

$$= 4.82 \pm 2.306 \, (0.3371) \sqrt{1 + \frac{1}{10} + \frac{(0.45 - 0.51265)^2}{(10 - 1)(0.2787)^2}}$$

$$= 4.82 \pm 2.306 \, (0.3371) \sqrt{1 + 0.1 + 0.005615}$$

$$= 4.82 \pm 0.82$$

Interpretation: Because 95% of all intervals computed by this method will include the true value for the *Y*-variable for a given *X*, we can be 95% confident that the interval computed above includes the true concentration of the chemical in this specific sample with absorbance of 0.45.

NOTE If you compare the second to last step in the calculations for confidence and prediction intervals, you will see that the only difference is that the prediction interval includes a term "1 +" under the square root sign that is not present for the confidence interval. If you must do these calculations by hand, you can save effort (and avoid potential calculation errors), by recording your intermediate results for the confidence interval and use these same numbers plus 1 to obtain the standard error for the prediction interval. ■

Example 13.6 uses the regression concepts presented thus far in a medical study related to heart disease.

■ **EXAMPLE 13.6**

Applying regression to study the association between polyphenol concentration in red wine and suppression of endothelin-1. A number of comparative studies of the prevalence of heart disease in various countries and cultures suggest that consumption of red wine may reduce risk of cardiovascular disease. A laboratory experiment by Corder and others (*Nature* 414: 863–864) documented that alcohol-free extracts of red wine inhibit the production of endothelin-1 (ET-1) in cultured bovine aorta endothelial cells. ET-1 is crucial in the development of coronary atherosclerosis. Further analysis demonstrated that polyphenol components of red wine exerted the strongest inhibitory effect on ET-1 production. To allow for comparison of different wines with regard to their potential for suppressing production of ET-1, these researchers measured the amount of polyphenols in 23 different red wines and the concentration of wine extract necessary to inhibit ET-1 production by 50% (IC-50). The following analyses are used to document the strength of this association and to provide for predicting IC-50 based on measurement of polyphenol content in red wine.

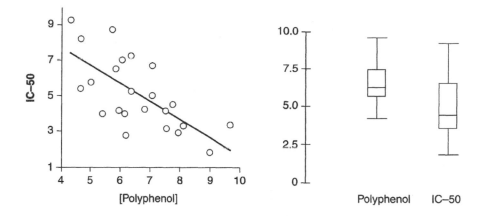

Summary Statistics Variable	Count	Mean	Std Dev	Min	Max
Polyphenol	23	6.57	1.37	4.3	9.6
IC-50	23	5.13	2.01	1.9	9.3

Assessment of Assumptions

a. *Data obtained by randomized unbiased study design.* The wines were not selected by random sampling from all possible red wines. Hence, the extrapolation of these results to all red wines is uncertain. We will assume that the variables were measured by appropriate methods.

b. *The true association is linear.* The data points in the scatterplot above appear to form a more or less linear pattern.

c. *The variance of IC-50 is similar across the range of polyphenol concentration.* Variation of IC-50 decreases somewhat as polyphenol concentration increases. This may violate this assumption and will be further explored using the residuals versus predicted plot.

d. *Both variables are Normally distributed.* Boxplots for the data distributions of both variables are more or less symmetric. Normality assumption fulfilled.

Regression Print-Out

Dependent variable is: IC-50

R-squared = 46.1% R-squared (adjusted) = 43.5%

s = 1.511 with 23 − 2 = 21 degrees of freedom

Source	Sums of Squares	df	Mean Square	F-ratio
Regression	41.0088	1	41.0088	18
Residual	47.9642	21	2.28401	

Variable	Coefficient	S.E. of Coeff	t-ratio	Probability
Constant	11.6856	1.578	7.4	< 0.0001
Polyphenol	−0.997143	0.2353	−4.24	0.0004

Residuals vs Predicted Plot

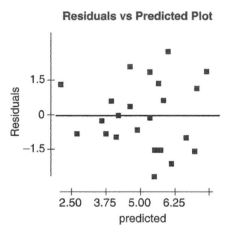

Normal Quantile Plot

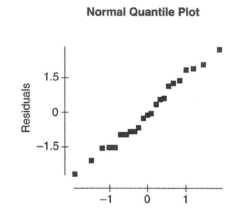

Post-Regression Assessment of Assumptions

1. *Variance of Y is constant across range of X:* In the residuals vs. predicted plot, the spread of points in the vertical direction (variance of residuals) increases as the predicted value for IC-50 increases (and polyphenol content decreases). This pattern indicates that wines with high polyphenol content were similar in their strong inhibitory effect on ET-1 production. In contrast, wines with low polyphenol content vary more in their inhibitory effect on ET-1 production. This violation of the equal variance assumption does not invalidate the LSR line. However, the 95% confidence and prediction intervals computed for this line may be biased, underestimating precision for wines with high polyphenol content and overestimating precision for those with low polyphenol content.

2. *Residuals are Normally distributed:* The Normal quantile plot displays a linear pattern, so this assumption is valid.

Conclusion: The higher the polyphenol content in a red wine, the more effective that wine is for suppressing production of ET-1 [IC-50 $= 11.7 - 0.997$ (polyphenol), $R^2 = 0.435$, $p = 0.0004$). Hence, red wines with high polyphenol content may provide greater benefit for reducing risk of heart disease than wines with lower polyphenol content.

Predicted values, residuals, and 95% confidence/prediction intervals.

Polyphenol (X)	IC-50 (Y)	Predicted ($\hat{y}$)	Residual	Margin of Error 95% C.I.	95% P.I.
4.30	9.30	7.41	1.89	1.29	3.40
4.65	5.40	7.06	−1.66	1.14	3.34
4.65	8.20	7.06	1.14	1.14	3.34
5.00	5.75	6.72	−0.96	1.01	3.30
5.40	4.00	6.32	−2.32	0.87	3.26
5.70	8.75	6.02	2.73	0.78	3.24
5.85	6.50	5.87	0.63	0.74	3.23
5.95	4.20	5.77	−1.57	0.72	3.22
6.05	7.05	5.67	1.38	0.70	3.22
6.15	4.00	5.57	−1.57	0.69	3.22
6.20	2.80	5.52	−2.72	0.68	3.22
6.35	5.25	5.37	−0.12	0.66	3.21
6.35	7.25	5.37	1.88	0.66	3.21
6.80	4.25	4.92	−0.67	0.66	3.21
7.05	5.05	4.67	0.38	0.70	3.22
7.05	6.75	4.67	2.08	0.70	3.22
7.50	4.20	4.22	−0.02	0.80	3.24
7.55	3.20	4.17	−0.97	0.81	3.25
7.75	4.55	3.97	0.58	0.87	3.26
7.95	2.95	3.77	−0.82	0.94	3.28
8.10	3.35	3.62	−0.27	0.99	3.30
9.00	1.90	2.73	−0.83	1.36	3.42
9.70	3.40	2.03	1.37	1.66	3.56

(Source: C.I. = confidence interval; P.I. = prediction interval.)

Example Calculations

a. 95% confidence interval for polyphenol $= 4.30$ ($\hat{y} = 11.7 - 0.997(4.30) = 7.41$)

$$\hat{y}_i \pm t_{0.025, df=21} \, s \sqrt{\frac{1}{n} + \frac{(x_i - \bar{x})^2}{(n-1)S_x^2}}$$

$$= 7.41 \pm 2.08 \,(1.511) \sqrt{\frac{1}{23} + \frac{(4.30 - 6.57)^2}{(23-1)(1.37)^2}}$$

$$= 7.41 \pm 2.08 \,(1.511) \sqrt{0.0435 + 0.1248}$$

$$= \mathbf{7.41 \pm 1.29}$$

Interpretation: Because 95% of all intervals computed by this method will include the true value for $\mu_{Y|X}$ we are 95% confident that the interval computed above includes the true mean IC-50 for wines with polyphenol concentrations of 4.30.

b. 95% prediction interval for polyphenol = 4.30

$$\hat{y}_i \pm t_{0.025, df=21} \, s \sqrt{1 + \frac{1}{n} + \frac{(x_i - \bar{x})^2}{(n-1)S_x^2}}$$

$$= 7.41 \pm 2.08 \, (1.511) \sqrt{1 + \frac{1}{23} + \frac{(4.30 - 6.57)^2}{(23-1)(1.37)^2}}$$

$$= 7.41 \pm 2.08 \, (1.511) \sqrt{1 + 0.0435 + 0.1248}$$

$$= 7.41 \pm \mathbf{3.40}$$

Interpretation: Because 95% of all intervals computed by this method will include the true value of *Y* for an individual characterized by a specific value of *X*, we can be 95% confident that the interval computed above includes the true IC-50 for a specific wine that has a polyphenol content of 4.30.

c. 95% confidence interval for polyphenol = 6.35 ($\hat{y}$ = 11.7 − 0.997(6.35) = 5.37)

$$\hat{y}_i \pm t_{0.025, df=21} \, s \sqrt{\frac{1}{n} + \frac{(x_i - \bar{x})^2}{(n-1)S_x^2}}$$

$$= 5.37 \pm 2.08 \, (1.511) \sqrt{\frac{1}{23} + \frac{(6.35 - 6.57)^2}{(23-1)(1.37)^2}}$$

$$= 5.37 \pm 2.08 \, (1.511) \sqrt{0.0435 + 0.00117}$$

$$= 5.37 \pm \mathbf{0.66}$$

Interpretation: Because 95% of all intervals computed by this method will include the true value for $\mu_{Y|X}$ we can be 95% confident that the interval computed above includes the true mean IC-50 for wines with polyphenol concentrations of 6.35.

d. 95% prediction interval for polyphenol = 6.35

$$\hat{y}_i \pm t_{0.025, df=21} \, s \sqrt{1 + \frac{1}{n} + \frac{(x_i - \bar{x})^2}{(n-1)S_x^2}}$$

$$= 5.37 \pm 2.08 \, (1.511) \sqrt{1 + \frac{1}{23} + \frac{(6.35 - 6.57)^2}{(23-1)(1.37)^2}}$$

$$= 5.37 \pm 2.08 \, (1.511) \sqrt{1 + 0.0435 + 0.00117}$$

$$= 5.37 \pm \mathbf{3.21}$$

Interpretation: Because 95% of all intervals computed by this method will include the true value of *Y* for an individual characterized by a specific value of *X*, we can be 95% confident that the interval computed above includes the true IC-50 for a specific wine that has a polyphenol content of 6.35.

The graph that follows displays the 95% confidence and prediction intervals for all values of *X* (polyphenol content) within the range of the data. In this graph you can see how estimates for IC-50 are more precise for polyphenol values in the middle of the range of data; you can also see that the width of the confidence interval increases for polyphenol values near the upper and lower limits of the range. This is due to the teeter-totter effect described earlier. Also note that the width of the prediction intervals is much wider than that of the confidence intervals. This reflects the fact that individuals exhibit much more random variation than means of groups of individuals. Hence, we can obtain precise estimates of mean IC-50 for all wines with a particular polyphenol content, whereas estimates of IC-50 for any one wine are less likely to be close to the true value.

Graph of predicted IC-50 (heavy solid line) with 95% confidence Intervals (narrow solid lines) and 95% prediction intervals (narrow dashed lines).

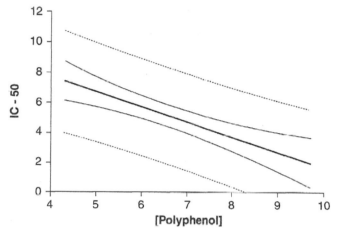

CHAPTER SUMMARY

1. There is an **association** between two variables if the values of one variable tend to change in a systematic manner when the values of the other variable change.

2. **Correlation analysis** computes a statistic that indicates the strength and nature (positive/negative) of linear associations.
   a. **Pearson's correlation coefficient *r*** is used to evaluate sample data as evidence that a *linear* association exists between two Normally distributed quantitative variables.
   b. **Spearman's rank correlation coefficient r_s** is used to evaluate sample data for evidence of an association between two quantitative variables when: (1) there is reason to believe that the distributions for one or both of the two variables are *not* Normal, or (2) a scatterplot indicates the association is nonlinear, but monotonic.

3. **Least-Squares Regression analysis** evaluates the strength and nature of linear association in a manner that allows for using the statistical association between two variables to *predict* values for one variable based on values for the associated variable.
   a. The slope and *Y*-intercept for the LSR line are computed so as to minimize the vertical distance between the *x, y* data points in a scatterplot and the LSR line fit through the middle of the data points.
   b. The difference between an observed value of the *Y*-variable and the predicted value from the LSR line is called a **residual**.
   c. If the slope of the LSR line is significantly different from zero, this provides sufficient evidence to support a conclusion that there is an association between the two variables.

d. The analysis of variance table in standard regression analyses partitions the total variance of the Y-variable into pattern and noise components.

e. The **coefficient of determination (R^2)** represents the proportion of total variation in the Y-variable that is attributable to the pattern of the association; this is a measure of the strength of the association.

f. The **standard error of the regression (s)** represents the random variation of the data points about the pattern of the LSR line.

4. The **100(1 − α)% confidence interval for $\mu_{Y|X}$** (true mean of Y for a given X-value) is computed as

$$\hat{y}_i \pm t_{\alpha/2, df=n-2} \; s \sqrt{\frac{1}{n} + \frac{(x_i - \bar{x})^2}{(n-1)S_x^2}}$$

where $\hat{y}_i$ = predicted value of Y for a given X-value, $t_{\alpha/2, df=n-2}$ is the critical value from the t-distribution, with error rate = α and degrees of freedom = $n - 2$, s = standard error of regression, n = sample size, x_i is the X-value used in the LSR equation to compute $\hat{y}_i$, $\bar{x}$ is the mean of the X data values, and S_x^2 is the variance of the X data values.

5. The **100(1 − α)% prediction interval for $Y|X$** (true value for an individual Y given an X-value) is

$$\hat{y}_i \pm t_{\alpha/2, df=n-2} \; s \sqrt{1 + \frac{1}{n} + \frac{(x_i - \bar{x})^2}{(n-1)S_x^2}}$$

6. For both correlation and regression, simply documenting a statistically significant association between two variables is not sufficient evidence to conclude that the variables are linked in a cause-effect relationship. Additional evidence is required to support such a conclusion including:

a. Similar statistical associations must be documented in multiple locations.

b. The influence of lurking variables on the X- and Y-variables must be eliminated.

c. A plausible, testable cause-effect mechanism must be identified.

14 Tests of Association Between Two Categorical Variables

INTRODUCTION There is an **association** between two variables if the values of one variable tend to change in a systematic manner when the values of the other variable change. For categorical variables, this means that whether or not an individual falls into a specific class of one categorical variable is influenced by its membership in a specific class of a second categorical variable. In an example presented in Chapter 3, I described how whether or not a person is admitted to a hospital may depend on whether or not that person has health insurance. The strongest association or *pattern* in this example would be "No insurance" → "No admittance." However, real-world associations like this are often more subtle; people without health insurance are less likely to be admitted to hospitals. The *noise* of such association relates to other factors that might influence the membership of individuals in classes of the categorical variables. For example, admission of uninsured people to hospitals may depend on whether the hospital is public or private and the urgency of the medical problem.

In this chapter, I present methods for describing and quantifying the association between two categorical variables measured on the same study subjects. I will present a test of significance used to evaluate sample data as evidence to test hypotheses about associations in the real world. The nature of the test for assessing association between categorical variables is quite different from tests of association for quantitative variables. However, the for-

mat of the tests of significance for associations is the same. We will describe the sampling distribution for a test statistic under the Null hypothesis that there is *no* association. We will use this sampling distribution to determine the probability of obtaining the observed value of the sample statistic. If this probability is very small, we will conclude there is strong evidence that an association exists in the real world. As described in the previous chapter, identifying a statistically significant association is only the first step in a more complex process for identifying cause-effect relationships between the two variables. ■

GOALS AND OBJECTIVES

In this chapter you will learn how to assess the evidence provided by sample data to determine if there is an association between categorical variables. You will understand that a statistically significant association between two variables does not necessarily mean that the variables have a cause-effect relationship.

By the end of this chapter you will be able to:

1. Perform and interpret the χ^2 test of Independence to assess the evidence for an association between two categorical variables.
2. Interpret contingency tables, including standardized residuals, to explore the nature of an association between two categorical variables.

14.1
χ^2 Test of Independence

1. **Application:** The χ^2 **test of independence** (also called **contingency table analysis**) is used to evaluate whether or not there is an association between individuals falling into specific classes of one categorical variable and their membership in classes of a second categorical variable. Although the test is called a *test of independence,* the concept of independence is the opposite of that association. Hence, if we reject the Null hypothesis of independence, this is the equivalent of concluding that there is an association.

   This test can also be used to evaluate whether or not membership of individuals in various classes of a categorical variable is independent of their status as measured by a *quantitative* variable. However, the values of the quantitative variable must be transformed to *classes* of a categorical variable. For example, to evaluate whether or not age-at-death is independent of whether or not a person smokes, age-at-death could be expressed in classes such as 0–10, 11–20, 21–30, ..., 71–80, 81–90, ≥ 91. These classes do *not* have to be of equal "width," but you should have a reasonable rationale for your choices of class limits.

   It can sometimes be difficult to distinguish when to use the χ^2 test of independence versus the χ^2 test of homogeneity (described in Chapter 12). The best indicator for which test is appropriate is the nature of the scientific question. If the question focuses on differences between populations, the test of homogeneity is appropriate. If the question pertains to associations between variables or the influence of one variable on another, the test of independence is appropriate.

   Assumptions:

   a. The individuals in the sample represent independent observations chosen by a randomized, unbiased procedure, so the expected values for sample proportions of individuals in all classes of both categorical variables ($\hat{p}_{i,j}$) are equal to the true population proportions ($P_{i,j}$).

   b. The populations from which the individuals were selected are sufficiently large that you can assume true proportions (P values) for the various classes of the categorical variables are constant (population size N is more than 100 times bigger than the sample size n).

   c. The sample size is sufficiently large that the average of expected values across all cells in the contingency table is 5 or more. That is, the number of classes of the row variable multiplied by the number of classes of the column variable divided into the overall sample size is 5 or more. For example, if the row variable had 2 classes, and the column variable had 5 classes, the contingency table would have a total of 10 "cells," defined by all combinations of the classes of these two variables. Hence, the required overall sample size N (such that the average expected value across all cells 5 or more) would be $N \geq 10 \times 5 = 50$.

2. **Statement of Hypotheses**

   - H_0: $P_{r,c} = P_r \times P_c$
   - H_a: $P_{r,c} \neq P_r \times P_c$

   where $P_{r,c}$ is the proportion of the population that has the combination of characteristics corresponding to cell r, c in the contingency table, defined by the combination of class r for the row variable and class c for the column variable. P_r is the proportion of the population that falls into class r of the row variable. P_c is the proportion of the population that falls into class c of the column variable.

NOTE The proportion of individuals in a population that fall into a specified class is equal to the *probability* that a randomly chosen individual from the population will belong to that class. ■

The Null hypothesis for the χ^2 test of independence is based on the assumption that the two variables are *independent*. If the variables are independent, there is no association. The Null hypothesis statement is based on the application of the simple multiplication rule for probabilities, as follows:

a. According to the simple multiplication rule for probabilities, *if two events A and B are independent*, the probability of event A and event B occurring can be computed as: $P[A \text{ and } B] = P[A] \times P[B]$. In this context, event A would be that an individual falls into class r or the row variable and event B would be that the same individual falls into class c of the column variable.

b. If the two variables that define the rows and columns of the contingency table are independent, the probability that a sample unit will fall into a particular cell should be equal to the product of the proportion of all individuals in class r times the proportion of all individuals in class c for that cell. If the row and column variables are *not* independent, the observed proportion of sample units that fall into a particular cell of the contingency table will not equal the expected value computed using the simple multiplication rule for probabilities.

c. Because this test of independence generally involves multiple comparisons between observed and expected proportions, a simultaneous test format is used to control overall, experiment-wise Type I error rate.

3. **The χ^2 Test Statistic**

To calculate the χ^2 test statistic, follow the steps a–c, and then use those results in step d.

a. *Set up the contingency table.* The table is a matrix of cells with the rows in the matrix defined by the class designations of one categorical variable and the columns in the matrix defined by class designations of the other categorical variable. The observed count in each cell represents the number of individuals in the sample that exhibit a combination of specific values for the two categorical variables.

■ **EXAMPLE** The contingency table below was constructed from a random sample of death certificates for 500 residents of Muncie, Indiana, filed during the period 1851 to 1900. Each sample unit (death certificate) was classified as to gender of the person, and the individual's age at time of death. Counts of individuals in each age/gender combination were recorded in the cells of the contingency table. For each row and column, the total number of individuals that were classified into each separate age and gender class were also tallied. (These totals are called the **marginal totals**.) ■

Age-at-Death → Gender ▼	0 to 5	6 to 20	21 to 40	41 to 60	61 to 80	Row Totals
Male	34	11	11	57	112	225
Female	50	14	33	83	95	275
Column Totals	84	25	44	140	207	500

b. *Compute the proportion of the total number of sample units (N) in the table that fall into each row and column.* (These are pooled estimates of $\hat{p}$ for each class of the row and column variables.) In the contingency table below, row and column proportions ($\hat{p}_r$ and $\hat{p}_c$) are listed below the row and column totals. For example, $\hat{p}_{male} = 0.45$, or 45% of the sample was male, and 16.8% of the sample ($\hat{p} = 0.168$) died at age 0 to 5.

Age-at-Death → Gender ▼	0 to 5	6 to 20	21 to 40	41 to 60	61 to 80	Row Totals
Male	34 **37.8**	11 **11.25**	11 **19.8**	57 **63**	112 **93.15**	225 0.45
Female	50 **46.2**	14 **13.75**	33 **24.2**	83 **77**	95 **113.85**	275 0.55
Column Totals	84 0.168	25 0.05	44 0.088	140 0.28	207 0.414	500

c. *Compute the expected number of individuals in each cell of the contingency table.* The row and column proportions $\hat{p}_r$ and $\hat{p}_c$ are estimates of probabilities that a randomly selected individual would fall into a particular row or column class. If the two variables are independent, the simple multiplication rule states the probability that a randomly selected individual would fall into a specific cell in the contingency table, defined by that individual's row *and* column classes, is estimated by $\hat{p}_{r,c} = (\hat{p}_r)(\hat{p}_c)$. The expected *count* of individuals who would fall into that cell if the two variables are independent is computed by multiplying this probability $\hat{p}_{r,c}$ times the total number of individuals in the sample ($N = 500$ in the example).

▓ **EXAMPLE**   The expected value **37.8** in the top-left cell of the contingency table above, corresponding to males that died at ages between 0–5, was computed as the column proportion (0.168) multiplied by the row proportion (0.45) multiplied by the total number of sample units (500). If the two variables are independent, these expected counts should be very close to the observed counts in the table. Compare the observed counts with the expected values in boldface in the contingency table above.   ▓

Important: The sum of expected values across a row and down a column should equal the row and column totals. This is a useful check for calculation errors if the test is done by hand rather than by statistics software.

d. *Compute the χ^2_{test} statistic.* This statistic summarizes the deviations between observed vs. expected counts across all the cells in the table.

$$\chi^2_{test} = \sum \frac{(\text{Observed} - \text{Expected})^2}{\text{Expected}}$$

where the summation is for all cells in the body of the table, *not* including the margins.

Example (from the preceding contingency table):

$$\chi^2_{test} = \frac{(34 - 37.8)^2}{37.8} + \frac{(11 - 11.25)^2}{11.25} + \frac{(11 - 19.8)^2}{19.8} + \frac{(57 - 63)^2}{63} + \frac{(112 - 93.15)^2}{93.15} +$$

$$+ \frac{(50 - 46.2)^2}{46.2} + \frac{(14 - 13.75)^2}{13.75} + \frac{(33 - 24.2)^2}{24.2} + \frac{(83 - 77)^2}{77} + \frac{(95 - 113.85)^2}{113.85}$$

$$\chi^2_{test} = 15.79$$

4. **Sampling Distribution of χ^2_{test}**

   If the Null hypothesis is true, the probability distribution for the χ^2_{test} statistic is a χ^2 distribution with degrees of freedom df = (number of rows − 1) × (number of columns − 1). In the example above, df = $(2 - 1)(5 - 1)$ = 4. See Chapter 12 for a description of the χ^2 distribution.

5. **p-Value**

   a. To obtain an approximate p-value from the χ^2 table (Appendix Table 8), read down the leftmost column to the row that corresponds to the degrees of freedom in your contingency table. Read across this row until you find two values that bracket the χ^2_{test} value computed from the contingency table. The two column headers for these χ^2 table values are p-values that bracket the probability of getting particular χ^2_{test} values due to random variation alone.

   ■ **EXAMPLE**   With χ^2_{test} = 15.79, and df = 4, the p-value for the contingency table relating gender to age-at-death is: $0.0025 < p < 0.005$. Based on these results, we can conclude that there is strong evidence that age-at-death is *not* independent of gender. The χ^2_{test} statistic does not reflect whether observed counts are greater than or less than expected counts, only whether they are *different*. Hence, this is always a two-tailed test of significance. Note: Most spreadsheet software has a function that provides exact p-values, given a χ^2_{test} value and the degrees of freedom (see detailed description in Chapter 12). If statistics software is used to perform this test of significance, an exact p-value is provided. ▨

   b. *Stating conclusions:*

   If $p \leq 0.05$: The probability associated with the random-variation explanation for the observed differences between observed counts of individuals and expected counts is sufficiently low to reject this explanation. If an appropriate study design has been correctly implemented, the only remaining likely explanation is that the two variables are *not* independent. That is, there is an association between the variables.

   ■ **EXAMPLE**   "Age-at-death was not independent of gender ($p < 0.005$)." This statement would be followed by specific descriptions of how the two variables were not independent, as described below. ▨

   NOTE Even when the data provide strong evidence for a statistical association between two categorical variables, this is *not* sufficient to conclude that there is a cause-effect relationship. Additional evidence is required to support such a claim, as described in Chapter 13   ■

If 0.05 < p ≤ 0.10: The probability associated with the random-variation explanation for the differences between observed and expected counts is sufficiently high that we cannot confidently exclude this explanation. However, the probability is sufficiently low that we might suspect that the two variables are not independent. Some scientists would conclude that the results "suggest" the two variables are not independent, but more data are needed to verify this.

■ **EXAMPLE**   "The results suggest that age-at-death is not independent of gender (p = 0.089), but more data are needed to verify this conclusion." In this circumstance, you would generally *not* provide additional description about how the variables are associated.   ■

If p > 0.10: The probability associated with the random-variation explanation for the differences between observed and expected counts is so high that this explanation cannot be excluded. Hence, there is insufficient evidence to claim that there is an association between the two variables.

■ **EXAMPLE**   "The results provided no evidence for an association between age-at-death and gender (p = 0.65)."   ■

14.2
Evaluating the Nature of Associations Between Categorical Variables

If the results of the χ^2 test of independence provide sufficient evidence to conclude that the two variables are not independent, the next and most obvious question is, "What is the nature of the association?" At this point you are in the same situation as occurred when the χ^2 test of homogeneity determined that some sample proportions differed significantly from others, but you didn't yet know which proportions were different. Although there is no multiple comparisons procedure for the test of independence, there are two ways to identify which part(s) of the contingency table are most responsible for rejecting the Null hypothesis. You can:

1. *Determine which cells have the largest difference between observed and expected values.* These cells contribute most to the significant χ^2_{test} value. This assessment is based on **adjusted residuals**, computed as follows:

$$\text{Resid}_{adj} = \frac{O - E}{\sqrt{E(1 - \hat{p}_r)(1 - \hat{p}_c)}}$$

Cells with large positive residuals have many more individuals than would be expected if the row and column variables were independent. Cells with large negative residuals have much fewer individuals than expected.

NOTE Most computer statistics programs will provide adjusted residuals as part of the contingency table print-out.   ■

■ **EXAMPLE**   In this example we compute adjusted residuals for the gender vs. age-at-death contingency table. The computation below for the adjusted residual in the top-left cell in the table (males that died between ages 0 to 5) is:

$$\text{Resid}_{adj} = \frac{34 - 37.8}{\sqrt{37.8(1 - 0.45)(1 - 0.168)}} = -0.914$$

The boldface numbers in the contingency table below are adjusted residuals.

Age-at-Death → Gender ▼	0 to 5	6 to 20	21 to 40	41 to 60	61 to 80	Row Totals
Male	34 37.8 **−0.91**	11 11.25 **−0.10**	11 19.8 **−2.79**	57 63 **−1.2**	112 93.15 **+3.44**	225 0.45
Female	50 46.2 **+0.91**	14 13.75 **+0.10**	33 24.2 **+2.79**	83 77 **+1.2**	95 113.85 **−3.44**	275 0.55
Column Totals	84 0.168	25 0.05	44 0.088	140 0.28	207 0.414	500

Residuals analysis of the contingency table indicates that men had a higher death rate in the 61–80 age class compared to women (as indicated by the large positive residual for men and large negative residual for women). This is consistent with much other data that document generally longer life spans for women than men. Also, more females than expected die at ages 21–40, suggesting that women of childbearing age were more likely to die than men in that age class during the late 1800s. The identification of these differences between the genders was guided by the analysis of the data, rather than based on hypotheses stated before looking at the data. Therefore, you would have to test these specific hypotheses regarding differences between the genders with a new sample before you would have sufficiently strong evidence to make these claims.   ■

2. *Use subset χ^2 analysis.* If one row or column in the table is primarily responsible for the significant χ^2_{test} statistic, and that row or column is deleted, then the χ^2_{test} statistic for the reduced contingency table may no longer be significant. This would provide evidence that the lack of independence occurs in the deleted class of the categorical variable. The most efficient way to proceed with this tedious procedure is to first delete rows or columns with the largest adjusted residual values.

■ **EXAMPLE**   This example will illustrate subset χ^2 analysis for age-at-death vs. gender contingency table analysis.

Column Dropped	χ^2_{test}	df	*p*-Value
None	15.78	4	$p < 0.005$
61 to 80 yr old	4.15	3	$p < 0.25$

Conclusion: Differences in the proportion of males vs. females that died in age class 61−80 account for the significant departure from independence for age-at-death and gender. When this age class is deleted from the contingency table, there is insufficient evidence to reject the Null hypothesis that age-at-death is independent of gender for the remaining age classes.

NOTE This analysis indicates that the large adjusted residual for females of reproductive age is not significant. Again, because the analysis was guided by patterns observed in the data, this conclusion should be verified by an independent analysis using new data. ■

Example 14.1 illustrates the use of the χ^2 test of independence in a study to determine whether the drop-out rate of college students from their exercise program is associated with their exercise regime.

■ EXAMPLE 14.1

Assessing the association between exercise regime and drop-out rate. In Example 11.3 I described a study that compared weight loss by overweight college women among three exercise regimes: 1 × 30 (exercised 1 time per day for 30 minutes), 2 × 15 (exercised 2 times per day for 15 minutes), and 3 × 10 (exercised 3 times per day for 10 minutes). The comparison of mean weight loss by women in these three treatments done in Example 11.3 indicated there was no significant difference. However, the researcher observed that more of the study subjects in the 3 × 10 group dropped-out before the end of the study than in the other groups. These subjects complained about the inconvenience of fitting three exercise periods per day into their schedule. The researcher wants to determine if the drop-out rate of subjects is associated with the exercise regime. *Assessing Assumptions:* As in much human research where completely random sampling is not possible, it is difficult to assess if the sample of study subjects constitutes a representative sample from the population of interest. Also, if study subjects communicated with each other, their decisions to quit or continue the exercise program may not have been independent.

The population of interest for this study, all college-aged, overweight females, can be assumed to number in the tens of thousands. Hence, the assumption that $N > 100n$ is fulfilled.

With two classes of the outcome variable (completed and dropped-out) and three classes of the treatment variable (1 × 30, 2 × 15, and 3 × 10), there are six cells in the contingency table. Hence, 36/6 = 6 fulfills the assumption that the average expected count in each cell is 5 or more.

The results of a χ^2 test of independence are given in the contingency table below. Here Obs = observed value, Exp = expected value, and StResid = standard residual.

Exercise Regime → ↓ Outcome	1x30		2x15	3x10	Row Totals/ Proportions
Completed	12	Obs	10	8	30
	10	Exp	10	10	0.83
	0.63	StResid	0.00	−0.63	
Dropped-Out	0		2	4	6
	2		2	2	0.17
	−1.41		0.00	1.41	
Column Totals/ Proportions	12		12	12	36
	0.33		0.33	0.33	

$$\chi^2_{test} = \frac{(12-10)^2}{10} + \frac{(10-10)^2}{10} + \frac{(8-10)^2}{10} + \frac{(0-2)^2}{2} + \frac{(2-2)^2}{2} + \frac{(4-2)^2}{2}$$

$$= 4.8$$

df $= (3-1)(2-1) = 2$

$p < 0.10$ (from Appendix Table 8)

$p = 0.091$ (exact p-value from computer statistics program)

Conclusion: The results suggest that drop-out rate was related to the exercise regime ($p = 0.091$), but more data are needed to verify this.

> **NOTE** The available data suggest regimes that require multiple bouts of exercise per day have higher drop-out rates. This is indicated by the lower than expected drop-out rate for individuals doing the 1×30 regime (standardized residual -1.41) and the higher than expected drop-out rate for individuals doing the 3×10 regime (standardized residual $+1.41$). ∎

CHAPTER SUMMARY

1. There is an **association** between two categorical variables if membership in a specific class of one categorical variable is influenced by membership in a specific class of a second categorical variable.

2. The χ^2 **test of independence** (also called **contingency table analysis**) is used to evaluate sample data for evidence that there is an association between two categorical variables.

   a. This test is based on the simple multiplication rule that states $P[A \text{ and } B] = P[A] \times P[B]$ if and only if events A and B are independent. In this context, $P[A \text{ and } B]$ is estimated by the proportion of individuals in the sample who are members in class A of categorical variable 1 and also members of class B of categorical variable 2. $P[A]$ is estimated as the proportion of individuals in the sample who are members of class A. $P[B]$ is estimated as the proportion of individuals in the sample who are members of class B.

   b. Although the test is called a test of independence, the concept of independence is the opposite of that of association. Hence, rejecting the Null hypothesis of independence is the equivalent of concluding that there is an association.

3. If the overall Null hypothesis of independence is rejected based on the χ^2 test of independence, standardized residuals can be used to assess the nature of the association between the categorical variables. These residuals provide a measure the size of the deviation between observed sample statistics and expected values if the two variables were independent. Combinations of [A and B] in the contingency table that have large residuals indicate the nature of the association.

Appendix:
Statistical Tables

TABLE 1 Random Numbers.

Line	A	B	C	D	E	F	G	H	I	J	K	L
101	4947	3969	8489	3037	9647	7848	4673	4651	9815	9696	1325	9234
102	2910	3703	4783	4056	2928	0307	2597	3706	5040	5504	1506	6951
103	8929	4010	1386	6838	3648	2597	1965	7534	2586	7902	1762	9949
104	6064	0850	0782	6017	4211	0520	7601	5609	2202	2534	9263	2545
105	8719	2555	2358	5158	5489	5379	4362	9468	5227	3673	4335	4245
106	4551	1119	3068	6804	6233	2762	5297	0244	1509	0499	6120	4778
107	7110	2708	9821	6728	9212	0552	1532	0432	3032	2615	9962	3354
108	6806	3007	9199	8038	9679	1779	9495	5482	9896	0395	2134	7769
109	7363	0545	7662	8246	9328	5440	8612	8376	1763	3304	5030	6474
110	8354	8604	2828	9933	6325	5911	0096	7787	5390	6812	3458	4235
111	1960	3858	3988	3131	9570	9085	4957	7968	6398	6030	0145	9523
112	9569	7343	7451	5049	2794	0841	6848	5969	0863	1683	6088	9233
113	1349	0908	9871	0270	1635	0998	2891	9344	9889	1557	8880	3888
114	9413	9751	0157	2037	4299	3347	3089	4930	7778	8037	9615	3917
115	2699	9177	8106	2767	3942	9912	6458	6861	1903	9865	6528	5174
116	9197	1199	5380	9394	8399	5050	9503	9416	9944	3743	0971	9282
117	8646	4113	1953	6697	0377	2524	0168	7842	7577	8903	8852	1956
118	6322	5653	8647	9209	8746	6164	7868	7766	7264	5268	8183	9917
119	6046	7885	3893	9854	9046	2252	7753	1331	6265	3339	0832	4526
120	8034	9421	0330	0917	8837	9248	4805	5665	9484	1418	4124	7758
121	4722	8128	4377	0498	1023	7984	0912	6838	0293	3309	5417	3647
122	5174	4906	9394	5044	5762	2955	5529	9890	9944	7098	0258	5830
123	6218	6700	0570	6110	9102	2825	8063	3095	3704	3170	2274	9299
124	5924	4352	7146	6961	3954	5781	7598	2061	9327	7053	0393	2005
125	2546	8423	5111	2140	6543	2850	3092	6801	1008	0245	6541	9430
126	9563	3601	0990	0634	8940	9815	4664	2394	9411	9622	4432	2748
127	4936	9841	4854	0756	5827	5850	4105	1373	5937	3577	7574	8710
128	0211	5836	8171	2371	1092	4491	0224	3447	6817	0490	2121	6868
129	2587	7966	9624	9593	7404	0088	2176	5699	0190	2678	7696	7274
130	0583	5334	9529	4574	9374	0337	1368	3873	6759	6433	6800	9266
131	1417	9028	0964	0509	8441	3044	0098	2884	2184	6279	0950	9478
132	5737	1217	3248	9489	5095	2723	2528	2167	6797	9073	9152	4689
133	2814	7226	0886	8326	1576	0168	1133	9002	5807	1078	1848	2647
134	6827	1198	8606	2957	5658	5614	9233	0420	0582	8625	0954	1477
135	2381	3478	0620	9587	3663	3627	7761	1846	9163	0494	4121	5823
136	4848	5337	9722	8160	8308	5320	1854	8065	1611	7647	8829	5314
137	8694	4171	9364	1306	4527	1058	0431	9677	3328	3414	9467	8518
138	4390	9786	9313	9377	3885	4242	9355	5630	6973	4791	7990	2976
139	2046	9910	3038	9711	5710	4317	6312	9977	1094	2943	6305	9462
140	7387	0542	2437	5728	0573	4626	3044	8485	1102	8554	5996	6085
141	3198	4270	3028	9003	9226	8228	8041	8195	5722	5154	8521	5356
142	6865	7193	0245	4928	7649	2312	0326	0595	9526	1026	6500	4855
143	2131	2544	4940	0099	6303	9334	2471	6498	9694	7842	4222	9616
144	0696	6674	3735	7036	9234	0485	8444	8269	2615	1640	2997	1846
145	6794	0493	7347	9386	6142	4646	9429	2523	3459	9267	4480	1129

TABLE 2   Table of Probabilities for the Standard Normal Distribution (Z-Values).

Boldface numbers in the leftmost column represent the first two digits of the Z-value. Boldface numbers in the top row (labeled z) represent the third digit of the Z-value. Tabled values located at the intersection of a specific row and column are the probabilities $P[Z \leq z]$, as shown in the accompanying figure. For example, the probability in boldface in part A of this distribution is the probability $P[Z \leq -1.96]$, and the probability in boldface in part B is the probability $P[Z \leq +1.96]$. To determine probabilities $P[Z \geq z]$, subtract the tabled probability value from 1 $(1 - P[Z \leq z] = P[Z \geq z])$. Part A of this distribution is for negative Z-values, and part B is for positive Z-values.

Part A (for Negative Z-values)

z	0.00	0.01	0.02	0.03	0.04	0.05	0.06	0.07	0.08	0.09
−3.6	0.0002	0.0002	0.0001	0.0001	0.0001	0.0001	0.0001	0.0001	0.0001	0.0001
−3.5	0.0002	0.0002	0.0002	0.0002	0.0002	0.0002	0.0002	0.0002	0.0002	0.0002
−3.4	0.0003	0.0003	0.0003	0.0003	0.0003	0.0003	0.0003	0.0003	0.0003	0.0002
−3.3	0.0005	0.0005	0.0005	0.0004	0.0004	0.0004	0.0004	0.0004	0.0004	0.0003
−3.2	0.0007	0.0007	0.0006	0.0006	0.0006	0.0006	0.0006	0.0005	0.0005	0.0005
−3.1	0.0010	0.0009	0.0009	0.0009	0.0008	0.0008	0.0008	0.0008	0.0007	0.0007
−3.0	0.0013	0.0013	0.0013	0.0012	0.0012	0.0011	0.0011	0.0011	0.0010	0.0010
−2.9	0.0019	0.0018	0.0018	0.0017	0.0016	0.0016	0.0015	0.0015	0.0014	0.0014
−2.8	0.0026	0.0025	0.0024	0.0023	0.0023	0.0022	0.0021	0.0021	0.0020	0.0019
−2.7	0.0035	0.0034	0.0033	0.0032	0.0031	0.0030	0.0029	0.0028	0.0027	0.0026
−2.6	0.0047	0.0045	0.0044	0.0043	0.0041	0.0040	0.0039	0.0038	0.0037	0.0036
−2.5	0.0062	0.0060	0.0059	0.0057	0.0055	0.0054	0.0052	0.0051	0.0049	0.0048
−2.4	0.0082	0.0080	0.0078	0.0075	0.0073	0.0071	0.0069	0.0068	0.0066	0.0064
−2.3	0.0107	0.0104	0.0102	0.0099	0.0096	0.0094	0.0091	0.0089	0.0087	0.0084
−2.2	0.0139	0.0136	0.0132	0.0129	0.0125	0.0122	0.0119	0.0116	0.0113	0.0110
−2.1	0.0179	0.0174	0.0170	0.0166	0.0162	0.0158	0.0154	0.0150	0.0146	0.0143
−2.0	0.0228	0.0222	0.0217	0.0212	0.0207	0.0202	0.0197	0.0192	0.0188	0.0183
−1.9	0.0287	0.0281	0.0274	0.0268	0.0262	0.0256	**0.0250**	0.0244	0.0239	0.0233
−1.8	0.0359	0.0351	0.0344	0.0336	0.0329	0.0322	0.0314	0.0307	0.0301	0.0294
−1.7	0.0446	0.0436	0.0427	0.0418	0.0409	0.0401	0.0392	0.0384	0.0375	0.0367
−1.6	0.0548	0.0537	0.0526	0.0516	0.0505	0.0495	0.0485	0.0475	0.0465	0.0455
−1.5	0.0668	0.0655	0.0643	0.0630	0.0618	0.0606	0.0594	0.0582	0.0571	0.0559
−1.4	0.0808	0.0793	0.0778	0.0764	0.0749	0.0735	0.0721	0.0708	0.0694	0.0681
−1.3	0.0968	0.0951	0.0934	0.0918	0.0901	0.0885	0.0869	0.0853	0.0838	0.0823
−1.2	0.1151	0.1131	0.1112	0.1093	0.1075	0.1056	0.1038	0.1020	0.1003	0.0985
−1.1	0.1357	0.1335	0.1314	0.1292	0.1271	0.1251	0.1230	0.1210	0.1190	0.1170
−1.0	0.1587	0.1562	0.1539	0.1515	0.1492	0.1469	0.1446	0.1423	0.1401	0.1379
−0.9	0.1841	0.1814	0.1788	0.1762	0.1736	0.1711	0.1685	0.1660	0.1635	0.1611
−0.8	0.2119	0.2090	0.2061	0.2033	0.2005	0.1977	0.1949	0.1922	0.1894	0.1867
−0.7	0.2420	0.2389	0.2358	0.2327	-0.2296	0.2266	0.2236	0.2206	0.2177	0.2148
−0.6	0.2743	0.2709	0.2676	0.2643	0.2611	0.2578	0.2546	0.2514	0.2483	0.2451
−0.5	0.3085	0.3050	0.3015	0.2981	0.2946	0.2912	0.2877	0.2843	0.2810	0.2776
−0.4	0.3446	0.3409	0.3372	0.3336	0.3300	0.3264	0.3228	0.3192	0.3156	0.3121
−0.3	0.3821	0.3783	0.3745	0.3707	0.3669	0.3632	0.3594	0.3557	0.3520	0.3483
−0.2	0.4207	0.4168	0.4129	0.4090	0.4052	0.4013	0.3974	0.3936	0.3897	0.3859
−0.1	0.4602	0.4562	0.4522	0.4483	0.4443	0.4404	0.4364	0.4325	0.4286	0.4247
−0.0	0.5000	0.4960	0.4920	0.4880	0.4840	0.4801	0.4761	0.4721	0.4681	0.4641

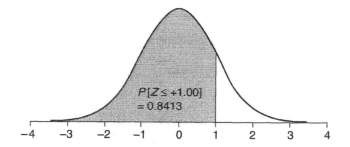

TABLE 2   (Continued)

Part B (for Positive Z-values)

z	0.00	0.01	0.02	0.03	0.04	0.05	0.06	0.07	0.08	0.09
+0.0	0.5000	0.5040	0.5080	0.5120	0.5160	0.5199	0.5239	0.5279	0.5319	0.5359
+0.1	0.5398	0.5438	0.5478	0.5517	0.5557	0.5596	0.5636	0.5675	0.5714	0.5753
+0.2	0.5793	0.5832	0.5871	0.5910	0.5948	0.5987	0.6026	0.6064	0.6103	0.6141
+0.3	0.6179	0.6217	0.6255	0.6293	0.6331	0.6368	0.6406	0.6443	0.6480	0.6517
+0.4	0.6554	0.6591	0.6628	0.6664	0.6700	0.6736	0.6772	0.6808	0.6844	0.6879
+0.5	0.6915	0.6950	0.6985	0.7019	0.7054	0.7088	0.7123	0.7157	0.7190	0.7224
+0.6	0.7257	0.7291	0.7324	0.7357	0.7389	0.7422	0.7454	0.7486	0.7517	0.7549
+0.7	0.7580	0.7611	0.7642	0.7673	0.7704	0.7734	0.7764	0.7794	0.7823	0.7852
+0.8	0.7881	0.7910	0.7939	0.7967	0.7995	0.8023	0.8051	0.8078	0.8106	0.8133
+0.9	0.8159	0.8186	0.8212	0.8238	0.8264	0.8289	0.8315	0.8340	0.8365	0.8389
+1.0	0.8413	0.8438	0.8461	0.8485	0.8508	0.8531	0.8554	0.8577	0.8599	0.8621
+1.1	0.8643	0.8665	0.8686	0.8708	0.8729	0.8749	0.8770	0.8790	0.8810	0.8830
+1.2	0.8849	0.8869	0.8888	0.8907	0.8925	0.8944	0.8962	0.8980	0.8997	0.9015
+1.3	0.9032	0.9049	0.9066	0.9082	0.9099	0.9115	0.9131	0.9147	0.9162	0.9177
+1.4	0.9192	0.9207	0.9222	0.9236	0.9251	0.9265	0.9279	0.9292	0.9306	0.9319
+1.5	0.9332	0.9345	0.9357	0.9370	0.9382	0.9394	0.9406	0.9418	0.9429	0.9441
+1.6	0.9452	0.9463	0.9474	0.9484	0.9495	0.9505	0.9515	0.9525	0.9535	0.9545
+1.7	0.9554	0.9564	0.9573	0.9582	0.9591	0.9599	0.9608	0.9616	0.9625	0.9633
+1.8	0.9641	0.9649	0.9656	0.9664	0.9671	0.9678	0.9686	0.9693	0.9699	0.9706
+1.9	0.9713	0.9719	0.9726	0.9732	0.9738	0.9744	0.9750	0.9756	0.9761	0.9767
+2.0	0.9772	0.9778	0.9783	0.9788	0.9793	0.9798	0.9803	0.9808	0.9812	0.9817
+2.1	0.9821	0.9826	0.9830	0.9834	0.9838	0.9842	0.9846	0.9850	0.9854	0.9857
+2.2	0.9861	0.9864	0.9868	0.9871	0.9875	0.9878	0.9881	0.9884	0.9887	0.9890
+2.3	0.9893	0.9896	0.9898	0.9901	0.9904	0.9906	0.9909	0.9911	0.9913	0.9916
+2.4	0.9918	0.9920	0.9922	0.9925	0.9927	0.9929	0.9931	0.9932	0.9934	0.9936
+2.5	0.9938	0.9940	0.9941	0.9943	0.9945	0.9946	0.9948	0.9949	0.9951	0.9952
+2.6	0.9953	0.9955	0.9956	0.9957	0.9959	0.9960	0.9961	0.9962	0.9963	0.9964
+2.7	0.9965	0.9966	0.9967	0.9968	0.9969	0.9970	0.9971	0.9972	0.9973	0.9974
+2.8	0.9974	0.9975	0.9976	0.9977	0.9977	0.9978	0.9979	0.9979	0.9980	0.9981
+2.9	0.9981	0.9982	0.9982	0.9983	0.9984	0.9984	0.9985	0.9985	0.9986	0.9986
+3.0	0.9987	0.9987	0.9987	0.9988	0.9988	0.9989	0.9989	0.9989	0.9990	0.9990
+3.1	0.9990	0.9991	0.9991	0.9991	0.9992	0.9992	0.9992	0.9992	0.9993	0.9993
+3.2	0.9993	0.9993	0.9994	0.9994	0.9994	0.9994	0.9994	0.9995	0.9995	0.9995
+3.3	0.9995	0.9995	0.9995	0.9996	0.9996	0.9996	0.9996	0.9996	0.9996	0.9997
+3.4	0.9997	0.9997	0.9997	0.9997	0.9997	0.9997	0.9997	0.9997	0.9997	0.9998
+3.5	0.9998	0.9998	0.9998	0.9998	0.9998	0.9998	0.9998	0.9998	0.9998	0.9998
+3.6	0.9998	0.9998	0.9999	0.9999	0.9999	0.9999	0.9999	0.9999	0.9999	0.9999

TABLE 3 Table of Probabilities for Binomial Random Variables.[1]

			Probabilities $P[X]$								
n	*X*	*C*	0.01	0.02	0.03	0.04	0.05	0.06	0.07	0.08	0.09
2	0	1	.9801	.9604	.9409	.9216	.9025	.8836	.8649	.8464	.8281
	1	2	.0198	.0392	.0582	.0768	.0950	.1128	.1302	.1472	.1638
	2	1	.0001	.0004	.0009	.0016	.0025	.0036	.0049	.0064	.0081
3	0	1	.9703	.9412	.9127	.8847	.8574	.8306	.8044	.7787	.7536
	1	3	.0294	.0576	.0847	.1106	.1354	.1590	.1816	.2031	.2236
	2	3	.0003	.0012	.0026	.0046	.0071	.0102	.0137	.0177	.0221
	3	1				.0001	.0001	.0002	.0003	.0005	.0007
4	0	1	.9606	.9224	.8853	.8493	.8145	.7807	.7481	.7164	.6857
	1	4	.0388	.0753	.1095	.1416	.1715	.1993	.2252	.2492	.2713
	2	6	.0006	.0023	.0051	.0088	.0135	.0191	.0254	.0325	.0402
	3	4			.0001	.0002	.0005	.0008	.0013	.0019	.0027
	4	1									.0001
5	0	1	.9510	.9039	.8587	.8154	.7738	.7339	.6957	.6591	.6240
	1	5	.0480	.0922	.1328	.1699	.2036	.2342	.2618	.2866	.3086
	2	10	.0010	.0038	.0082	.0142	.0214	.0299	.0394	.0498	.0610
	3	10		.0001	.0003	.0006	.0011	.0019	.0030	.0043	.0060
	4	5						.0001	.0001	.0002	.0003
	5	1									
6	0	1	.9415	.8858	.8330	.7828	.7351	.6899	.6470	.6064	.5679
	1	6	.0571	.1085	.1546	.1957	.2321	.2642	.2922	.3164	.3370
	2	15	.0014	.0055	.0120	.0204	.0305	.0422	.0550	.0688	.0833
	3	20		.0002	.0005	.0011	.0021	.0036	.0055	.0080	.0110
	4	15					.0001	.0002	.0003	.0005	.0008
	5	6									
	6	1									
7	0	1	.9321	.8681	.8080	.7514	.6983	.6485	.6017	.5578	.5168
	1	7	.0659	.1240	.1749	.2192	.2573	.2897	.3170	.3396	.3578
	2	21	.0020	.0076	.0162	.0274	.0406	.0555	.0716	.0886	.1061
	3	35		.0003	.0008	.0019	.0036	.0059	.0090	.0128	.0175
	4	35				.0001	.0002	.0004	.0007	.0011	.0017
	5	21								.0001	.0001
	6	7									
	7	1									

[1]n = sample size or number of independent observations, X = the number of observations out of n that meet the specified criterion, C = the number of possible combinations of observations that result in X out of n units that meet the specified criterion. Tabled values are probabilities $P[X]$. If $P[X] < 0.0001$, no probability is listed. However, this does *not* mean that $P[X] = 0$, only that it is less than 0.0001.

TABLE 3   (Continued)

		Probabilities $P[X]$								
n	X	0.1	0.15	0.2	0.25	0.3	0.35	0.4	0.45	0.5
2	0	.8100	.7225	.6400	.5625	.4900	.4225	.3600	.3025	.2500
	1	.1800	.2550	.3200	.3750	.4200	.4550	.4800	.4950	.5000
	2	.0100	.0225	.0400	.0625	.0900	.1225	.1600	.2025	.2500
3	0	.7290	.6141	.5120	.4219	.3430	.2746	.2160	.1664	.1250
	1	.2430	.3251	.3840	.4219	.4410	.4436	.4320	.4084	.3750
	2	.0270	.0574	.0960	.1406	.1890	.2389	.2880	.3341	.3750
	3	.0010	.0034	.0080	.0156	.0270	.0429	.0640	.0911	.1250
4	0	.6561	.5220	.4096	.3164	.2401	.1785	.1296	.0915	.0625
	1	.2916	.3685	.4096	.4219	.4116	.3845	.3456	.2995	.2500
	2	.0486	.0975	.1536	.2109	.2646	.3105	.3456	.3675	.3750
	3	.0036	.0115	.0256	.0469	.0756	.1115	.1536	.2005	.2500
	4	.0001	.0005	.0016	.0039	.0081	.0150	.0256	.0410	.0625
5	0	.5905	.4437	.3277	.2373	.1681	.1160	.0778	.0503	.0313
	1	.3281	.3915	.4096	.3955	.3602	.3124	.2592	.2059	.1563
	2	.0729	.1382	.2048	.2637	.3087	.3364	.3456	.3369	.3125
	3	.0081	.0244	.0512	.0879	.1323	.1811	.2304	.2757	.3125
	4	.0005	.0022	.0064	.0146	.0284	.0488	.0768	.1128	.1562
	5		.0001	.0003	.0010	.0024	.0053	.0102	.0185	.0312
6	0	.5314	.3771	.2621	.1780	.1176	.0754	.0467	.0277	.0156
	1	.3543	.3993	.3932	.3560	.3025	.2437	.1866	.1359	.0938
	2	.0984	.1762	.2458	.2966	.3241	.3280	.3110	.2780	.2344
	3	.0146	.0415	.0819	.1318	.1852	.2355	.2765	.3032	.3125
	4	.0012	.0055	.0154	.0330	.0595	.0951	.1382	.1861	.2344
	5	.0001	.0004	.0015	.0044	.0102	.0205	.0369	.0609	.0938
	6			.0001	.0002	.0007	.0018	.0041	.0083	.0156
7	0	.4783	.3206	.2097	.1335	.0824	.0490	.0280	.0152	.0078
	1	.3720	.3960	.3670	.3115	.2471	.1848	.1306	.0872	.0547
	2	.1240	.2097	.2753	.3115	.3177	.2985	.2613	.2140	.1641
	3	.0230	.0617	.1147	.1730	.2269	.2679	.2903	.2918	.2734
	4	.0026	.0109	.0287	.0577	.0972	.1442	.1935	.2388	.2734
	5	.0002	.0012	.0043	.0115	.0250	.0466	.0774	.1172	.1641
	6		.0001	.0004	.0013	.0036	.0084	.0172	.0320	.0547
	7				.0001	.0002	.0006	.0016	.0037	.0078

TABLE 3   (Continued)

			Probabilities P[X]
n	X	C	0.01	0.02	0.03	0.04	0.05	0.06	0.07	0.08	0.09
8	0	1	.9227	.8508	.7837	.7214	.6634	.6096	.5596	.5132	.4703
	1	8	.0746	.1389	.1939	.2405	.2793	.3113	.3370	.3570	.3721
	2	28	.0026	.0099	.0210	.0351	.0515	.0695	.0888	.1087	.1288
	3	56	.0001	.0004	.0013	.0029	.0054	.0089	.0134	.0189	.0255
	4	70			.0001	.0002	.0004	.0007	.0013	.0021	.0031
	5	56							.0001	.0001	.0002
	6	28									
	7	8									
	8	1									
9	0	1	.9135	.8337	.7602	.6925	.6302	.5730	.5204	.4722	.4279
	1	9	.0830	.1531	.2116	.2597	.2985	.3292	.3525	.3695	.3809
	2	36	.0034	.0125	.0262	.0433	.0629	.0840	.1061	.1285	.1507
	3	84	.0001	.0006	.0019	.0042	.0077	.0125	.0186	.0261	.0348
	4	126			.0001	.0003	.0006	.0012	.0021	.0034	.0052
	5	126						.0001	.0002	.0003	.0005
	6	84									
	7	36									
	8	9									
	9	1									
10	0	1	.9044	.8171	.7374	.6648	.5987	.5386	.4840	.4344	.3894
	1	10	.0914	.1667	.2281	.2770	.3151	.3438	.3643	.3777	.3851
	2	45	.0042	.0153	.0317	.0519	.0746	.0988	.1234	.1478	.1714
	3	120	.0001	.0008	.0026	.0058	.0105	.0168	.0248	.0343	.0452
	4	210			.0001	.0004	.0010	.0019	.0033	.0052	.0078
	5	252					.0001	.0001	.0003	.0005	.0009
	6	210									.0001
	7	120									
	8	45									
	9	10									
	10	1									
11	0	1	.8953	.8007	.7153	.6382	.5688	.5063	.4501	.3996	.3544
	1	11	.0995	.1798	.2433	.2925	.3293	.3555	.3727	.3823	.3855
	2	55	.0050	.0183	.0376	.0609	.0867	.1135	.1403	.1662	.1906
	3	165	.0002	.0011	.0035	.0076	.0137	.0217	.0317	.0434	.0566
	4	330			.0002	.0006	.0014	.0028	.0048	.0075	.0112
	5	462					.0001	.0002	.0005	.0009	.0015
	6	462								.0001	.0002
	7	330									
	8	165									
	9	55									
	10	11									
	11	1									

TABLE 3   (Continued)

		Probabilities $P[X]$								
n	X	0.1	0.15	0.2	0.25	0.3	0.35	0.4	0.45	0.5
8	0	.4305	.2725	.1678	.1001	.0576	.0319	.0168	.0084	.0039
	1	.3826	.3847	.3355	.2670	.1977	.1373	.0896	.0548	.0313
	2	.1488	.2376	.2936	.3115	.2965	.2587	.2090	.1569	.1094
	3	.0331	.0839	.1468	.2076	.2541	.2786	.2787	.2568	.2188
	4	.0046	.0185	.0459	.0865	.1361	.1875	.2322	.2627	.2734
	5	.0004	.0026	.0092	.0231	.0467	.0808	.1239	.1719	.2188
	6		.0002	.0011	.0038	.0100	.0217	.0413	.0703	.1094
	7			.0001	.0004	.0012	.0033	.0079	.0164	.0313
	8					.0001	.0002	.0007	.0017	.0039
9	0	.3874	.2316	.1342	.0751	.0404	.0207	.0101	.0046	.0020
	1	.3874	.3679	.3020	.2253	.1556	.1004	.0605	.0339	.0176
	2	.1722	.2597	.3020	.3003	.2668	.2162	.1612	.1110	.0703
	3	.0446	.1069	.1762	.2336	.2668	.2716	.2508	.2119	.1641
	4	.0074	.0283	.0661	.1168	.1715	.2194	.2508	.2600	.2461
	5	.0008	.0050	.0165	.0389	.0735	.1181	.1672	.2128	.2461
	6	.0001	.0006	.0028	.0087	.0210	.0424	.0743	.1160	.1641
	7			.0003	.0012	.0039	.0098	.0212	.0407	.0703
	8				.0001	.0004	.0013	.0035	.0083	.0176
	9						.0001	.0003	.0008	.0020
10	0	.3487	.1969	.1074	.0563	.0282	.0135	.0060	.0025	.0010
	1	.3874	.3474	.2684	.1877	.1211	.0725	.0403	.0207	.0098
	2	.1937	.2759	.3020	.2816	.2335	.1757	.1209	.0763	.0439
	3	.0574	.1298	.2013	.2503	.2668	.2522	.2150	.1665	.1172
	4	.0112	.0401	.0881	.1460	.2001	.2377	.2508	.2384	.2051
	5	.0015	.0085	.0264	.0584	.1029	.1536	.2007	.2340	.2461
	6	.0001	.0012	.0055	.0162	.0368	.0689	.1115	.1596	.2051
	7		.0001	.0008	.0031	.0090	.0212	.0425	.0746	.1172
	8			.0001	.0004	.0014	.0043	.0106	.0229	.0439
	9					.0001	.0005	.0016	.0042	.0098
	10							.0001	.0003	.0010
11	0	.3138	.1673	.0859	.0422	.0198	.0088	.0036	.0014	.0005
	1	.3835	.3248	.2362	.1549	.0932	.0518	.0266	.0125	.0054
	2	.2131	.2866	.2953	.2581	.1998	.1395	.0887	.0513	.0269
	3	.0710	.1517	.2215	.2581	.2568	.2254	.1774	.1259	.0806
	4	.0158	.0536	.1107	.1721	.2201	.2428	.2365	.2060	.1611
	5	.0025	.0132	.0388	.0803	.1321	.1830	.2207	.2360	.2256
	6	.0003	.0023	.0097	.0268	.0566	.0985	.1471	.1931	.2256
	7		.0003	.0017	.0064	.0173	.0379	.0701	.1128	.1611
	8			.0002	.0011	.0037	.0102	.0234	.0462	.0806
	9				.0001	.0005	.0018	.0052	.0126	.0269
	10						.0002	.0007	.0021	.0054
	11								.0002	.0005

TABLE 3   (Continued)

						Probabilities $P[X]$					
n	X	C	0.01	0.02	0.03	0.04	0.05	0.06	0.07	0.08	0.09
12	0	1	.8864	.7847	.6938	.6127	.5404	.4759	.4186	.3677	.3225
	1	12	.1074	.1922	.2575	.3064	.3413	.3645	.3781	.3837	.3827
	2	66	.0060	.0216	.0438	.0702	.0988	.1280	.1565	.1835	.2082
	3	220	.0002	.0015	.0045	.0098	.0173	.0272	.0393	.0532	.0686
	4	495		.0001	.0003	.0009	.0021	.0039	.0067	.0104	.0153
	5	792				.0001	.0002	.0004	.0008	.0014	.0024
	6	924							.0001	.0001	.0003
	7	792									
	8	495									
	9	220									
	10	66									
	11	12									
	12	1									
13	0	1	.8775	.7690	.6730	.5882	.5133	.4474	.3893	.3383	.2935
	1	13	.1152	.2040	.2706	.3186	.3512	.3712	.3809	.3824	.3773
	2	78	.0070	.0250	.0502	.0797	.1109	.1422	.1720	.1995	.2239
	3	286	.0003	.0019	.0057	.0122	.0214	.0333	.0475	.0636	.0812
	4	715		.0001	.0004	.0013	.0028	.0053	.0089	.0138	.0201
	5	1287				.0001	.0003	.0006	.0012	.0022	.0036
	6	1716						.0001	.0001	.0003	.0005
	7	1716									
	8	1287									
	9	715									
	10	286									
	11	78									
	12	13									
	13	1									
14	0	1	.8687	.7536	.6528	.5647	.4877	.4205	.3620	.3112	.2670
	1	14	.1229	.2153	.2827	.3294	.3593	.3758	.3815	.3788	.3698
	2	91	.0081	.0286	.0568	.0892	.1229	.1559	.1867	.2141	.2377
	3	364	.0003	.0023	.0070	.0149	.0259	.0398	.0562	.0745	.0940
	4	1001		.0001	.0006	.0017	.0037	.0070	.0116	.0178	.0256
	5	2002				.0001	.0004	.0009	.0018	.0031	.0051
	6	3003						.0001	.0002	.0004	.0008
	7	3432									.0001
	8	3003									
	9	2002									
	10	1001									
	11	364									
	12	91									
	13	14									
	14	1									

TABLE 3   (Continued)

		Probabilities $P[X]$								
n	X	0.1	0.15	0.2	0.25	0.3	0.35	0.4	0.45	0.5
12	0	.2824	.1422	.0687	.0317	.0138	.0057	.0022	.0008	.0002
	1	.3766	.3012	.2062	.1267	.0712	.0368	.0174	.0075	.0029
	2	.2301	.2924	.2835	.2323	.1678	.1088	.0639	.0339	.0161
	3	.0852	.1720	.2362	.2581	.2397	.1954	.1419	.0923	.0537
	4	.0213	.0683	.1329	.1936	.2311	.2367	.2128	.1700	.1208
	5	.0038	.0193	.0532	.1032	.1585	.2039	.2270	.2225	.1934
	6	.0005	.0040	.0155	.0401	.0792	.1281	.1766	.2124	.2256
	7		.0006	.0033	.0115	.0291	.0591	.1009	.1489	.1934
	8		.0001	.0005	.0024	.0078	.0199	.0420	.0762	.1208
	9			.0001	.0004	.0015	.0048	.0125	.0277	.0537
	10					.0002	.0008	.0025	.0068	.0161
	11						.0001	.0003	.0010	.0029
	12								.0001	.0002
13	0	.2542	.1209	.0550	.0238	.0097	.0037	.0013	.0004	.0001
	1	.3672	.2774	.1787	.1029	.0540	.0259	.0113	.0045	.0016
	2	.2448	.2937	.2680	.2059	.1388	.0836	.0453	.0220	.0095
	3	.0997	.1900	.2457	.2517	.2181	.1651	.1107	.0660	.0349
	4	.0277	.0838	.1535	.2097	.2337	.2222	.1845	.1350	.0873
	5	.0055	.0266	.0691	.1258	.1803	.2154	.2214	.1989	.1571
	6	.0008	.0063	.0230	.0559	.1030	.1546	.1968	.2169	.2095
	7	.0001	.0011	.0058	.0186	.0442	.0833	.1312	.1775	.2095
	8		.0001	.0011	.0047	.0142	.0336	.0656	.1089	.1571
	9			.0001	.0009	.0034	.0101	.0243	.0495	.0873
	10				.0001	.0006	.0022	.0065	.0162	.0349
	11					.0001	.0003	.0012	.0036	.0095
	12							.0001	.0005	.0016
	13									.0001
14	0	.2288	.1028	.0440	.0178	.0068	.0024	.0008	.0002	.0001
	1	.3559	.2539	.1539	.0832	.0407	.0181	.0073	.0027	.0009
	2	.2570	.2912	.2501	.1802	.1134	.0634	.0317	.0141	.0056
	3	.1142	.2056	.2501	.2402	.1943	.1366	.0845	.0462	.0222
	4	.0349	.0998	.1720	.2202	.2290	.2022	.1549	.1040	.0611
	5	.0078	.0352	.0860	.1468	.1963	.2178	.2066	.1701	.1222
	6	.0013	.0093	.0322	.0734	.1262	.1759	.2066	.2088	.1833
	7	.0002	.0019	.0092	.0280	.0618	.1082	.1574	.1952	.2095
	8		.0003	.0020	.0082	.0232	.0510	.0918	.1398	.1833
	9			.0003	.0018	.0066	.0183	.0408	.0762	.1222
	10				.0003	.0014	.0049	.0136	.0312	.0611
	11					.0002	.0010	.0033	.0093	.0222
	12						.0001	.0005	.0019	.0056
	13							.0001	.0002	.0009
	14									.0001

TABLE 3   (Continued)

			Probabilities $P[X]$								
n	X	C	0.01	0.02	0.03	0.04	0.05	0.06	0.07	0.08	0.09
15	0	1	.8601	.7386	.6333	.5421	.4633	.3953	.3367	.2863	.2430
	1	15	.1303	.2261	.2938	.3388	.3658	.3785	.3801	.3734	.3605
	2	105	.0092	.0323	.0636	.0988	.1348	.1691	.2003	.2273	.2496
	3	455	.0004	.0029	.0085	.0178	.0307	.0468	.0653	.0857	.1070
	4	1365		.0002	.0008	.0022	.0049	.0090	.0148	.0223	.0317
	5	3003			.0001	.0002	.0006	.0013	.0024	.0043	.0069
	6	5005						.0001	.0003	.0006	.0011
	7	6435								.0001	.0001
	8	6435									
	9	5005									
	10	3003									
	11	1365									
	12	455									
	13	105									
	14	15									
	15	1									
16	0	1	.8515	.7238	.6143	.5204	.4401	.3716	.3131	.2634	.2211
	1	16	.1376	.2363	.3040	.3469	.3706	.3795	.3771	.3665	.3499
	2	120	.0104	.0362	.0705	.1084	.1463	.1817	.2129	.2390	.2596
	3	560	.0005	.0034	.0102	.0211	.0359	.0541	.0748	.0970	.1198
	4	1820		.0002	.0010	.0029	.0061	.0112	.0183	.0274	.0385
	5	4368			.0001	.0003	.0008	.0017	.0033	.0057	.0091
	6	8008					.0001	.0002	.0005	.0009	.0017
	7	11440								.0001	.0002
	8	12870									
	9	11440									
	10	8008									
	11	4368									
	12	1820									
	13	560									
	14	120									
	15	16									
	16	1									

TABLE 3   (Continued)

n	X	Probabilities $P[X]$								
		0.1	0.15	0.2	0.25	0.3	0.35	0.4	0.45	0.5
15	0	.2059	.0874	.0352	.0134	.0047	.0016	.0005	.0001	.0000
	1	.3432	.2312	.1319	.0668	.0305	.0126	.0047	.0016	.0005
	2	.2669	.2856	.2309	.1559	.0916	.0476	.0219	.0090	.0032
	3	.1285	.2184	.2501	.2252	.1700	.1110	.0634	.0318	.0139
	4	.0428	.1156	.1876	.2252	.2186	.1792	.1268	.0780	.0417
	5	.0105	.0449	.1032	.1651	.2061	.2123	.1859	.1404	.0916
	6	.0019	.0132	.0430	.0917	.1472	.1906	.2066	.1914	.1527
	7	.0003	.0030	.0138	.0393	.0811	.1319	.1771	.2013	.1964
	8		.0005	.0035	.0131	.0348	.0710	.1181	.1647	.1964
	9		.0001	.0007	.0034	.0116	.0298	.0612	.1048	.1527
	10			.0001	.0007	.0030	.0096	.0245	.0515	.0916
	11				.0001	.0006	.0024	.0074	.0191	.0417
	12					.0001	.0004	.0016	.0052	.0139
	13						.0001	.0003	.0010	.0032
	14								.0001	.0005
	15									
16	0	.1853	.0743	.0281	.0100	.0033	.0010	.0003	.0001	.0000
	1	.3294	.2097	.1126	.0535	.0228	.0087	.0030	.0009	.0002
	2	.2745	.2775	.2111	.1336	.0732	.0353	.0150	.0056	.0018
	3	.1423	.2285	.2463	.2079	.1465	.0888	.0468	.0215	.0085
	4	.0514	.1311	.2001	.2252	.2040	.1553	.1014	.0572	.0278
	5	.0137	.0555	.1201	.1802	.2099	.2008	.1623	.1123	.0667
	6	.0028	.0180	.0550	.1101	.1649	.1982	.1983	.1684	.1222
	7	.0004	.0045	.0197	.0524	.1010	.1524	.1889	.1969	.1746
	8	.0001	.0009	.0055	.0197	.0487	.0923	.1417	.1812	.1964
	9		.0001	.0012	.0058	.0185	.0442	.0840	.1318	.1746
	10			.0002	.0014	.0056	.0167	.0392	.0755	.1222
	11				.0002	.0013	.0049	.0142	.0337	.0667
	12					.0002	.0011	.0040	.0115	.0278
	13						.0002	.0008	.0029	.0085
	14							.0001	.0005	.0018
	15								.0001	.0002
	16									

TABLE 3   (Continued)

							Probabilities $P[X]$				
n	X	C	0.01	0.02	0.03	0.04	0.05	0.06	0.07	0.08	0.09
17	0	1	.8429	.7093	.5958	.4996	.4181	.3493	.2912	.2423	.2012
	1	17	.1447	.2461	.3133	.3539	.3741	.3790	.3726	.3582	.3383
	2	136	.0117	.0402	.0775	.1180	.1575	.1935	.2244	.2492	.2677
	3	680	.0006	.0041	.0120	.0246	.0415	.0618	.0844	.1083	.1324
	4	2380		.0003	.0013	.0036	.0076	.0138	.0222	.0330	.0458
	5	6188			.0001	.0004	.0010	.0023	.0044	.0075	.0118
	6	12376					.0001	.0003	.0007	.0013	.0023
	7	19448							.0001	.0002	.0004
	8	24310									
	9	24310									
	10	19448									
	11	12376									
	12	6188									
	13	2380									
	14	680									
	15	136									
	16	17									
	17	1									
18	0	1	.8345	.6951	.5780	.4796	.3972	.3283	.2708	.2229	.1831
	1	18	.1517	.2554	.3217	.3597	.3763	.3772	.3669	.3489	.3260
	2	153	.0130	.0443	.0846	.1274	.1683	.2047	.2348	.2579	.2741
	3	816	.0007	.0048	.0140	.0283	.0473	.0697	.0942	.1196	.1446
	4	3060		.0004	.0016	.0044	.0093	.0167	.0266	.0390	.0536
	5	8568			.0001	.0005	.0014	.0030	.0056	.0095	.0148
	6	18564					.0002	.0004	.0009	.0018	.0032
	7	31824							.0001	.0003	.0005
	8	43758									.0001
	9	48620									
	10	43758									
	11	31824									
	12	18564									
	13	8568									
	14	3060									
	15	816									
	16	153									
	17	18									
	18	1									

TABLE 3   (Continued)

					Probabilities $P[X]$					
n	X	0.1	0.15	0.2	0.25	0.3	0.35	0.4	0.45	0.5
17	0	.1668	.0631	.0225	.0075	.0023	.0007	.0002	.0000	.0000
	1	.3150	.1893	.0957	.0426	.0169	.0060	.0019	.0005	.0001
	2	.2800	.2673	.1914	.1136	.0581	.0260	.0102	.0035	.0010
	3	.1556	.2359	.2393	.1893	.1245	.0701	.0341	.0144	.0052
	4	.0605	.1457	.2093	.2209	.1868	.1320	.0796	.0411	.0182
	5	.0175	.0668	.1361	.1914	.2081	.1849	.1379	.0875	.0472
	6	.0039	.0236	.0680	.1276	.1784	.1991	.1839	.1432	.0944
	7	.0007	.0065	.0267	.0668	.1201	.1685	.1927	.1841	.1484
	8	.0001	.0014	.0084	.0279	.0644	.1134	.1606	.1883	.1855
	9		.0003	.0021	.0093	.0276	.0611	.1070	.1540	.1855
	10			.0004	.0025	.0095	.0263	.0571	.1008	.1484
	11			.0001	.0005	.0026	.0090	.0242	.0525	.0944
	12				.0001	.0006	.0024	.0081	.0215	.0472
	13					.0001	.0005	.0021	.0068	.0182
	14						.0001	.0004	.0016	.0052
	15							.0001	.0003	.0010
	16									.0001
	17									
18	0	.1501	.0536	.0180	.0056	.0016	.0004	.0001	.0000	.0000
	1	.3002	.1704	.0811	.0338	.0126	.0042	.0012	.0003	.0001
	2	.2835	.2556	.1723	.0958	.0458	.0190	.0069	.0022	.0006
	3	.1680	.2406	.2297	.1704	.1046	.0547	.0246	.0095	.0031
	4	.0700	.1592	.2153	.2130	.1681	.1104	.0614	.0291	.0117
	5	.0218	.0787	.1507	.1988	.2017	.1664	.1146	.0666	.0327
	6	.0052	.0301	.0816	.1436	.1873	.1941	.1655	.1181	.0708
	7	.0010	.0091	.0350	.0820	.1376	.1792	.1892	.1657	.1214
	8	.0002	.0022	.0120	.0376	.0811	.1327	.1734	.1864	.1669
	9		.0004	.0033	.0139	.0386	.0794	.1284	.1694	.1855
	10		.0001	.0008	.0042	.0149	.0385	.0771	.1248	.1669
	11			.0001	.0010	.0046	.0151	.0374	.0742	.1214
	12				.0002	.0012	.0047	.0145	.0354	.0708
	13					.0002	.0012	.0045	.0134	.0327
	14						.0002	.0011	.0039	.0117
	15							.0002	.0009	.0031
	16								.0001	.0006
	17									.0001
	18									

TABLE 3   (Continued)

						Probabilities $P[X]$					
n	X	C	0.01	0.02	0.03	0.04	0.05	0.06	0.07	0.08	0.09
19	0	1	.8262	.6812	.5606	.4604	.3774	.3086	.2519	.2051	.1666
	1	19	.1586	.2642	.3294	.3645	.3774	.3743	.3602	.3389	.3131
	2	171	.0144	.0485	.0917	.1367	.1787	.2150	.2440	.2652	.2787
	3	969	.0008	.0056	.0161	.0323	.0533	.0778	.1041	.1307	.1562
	4	3876		.0005	.0020	.0054	.0112	.0199	.0313	.0455	.0618
	5	11628			.0002	.0007	.0018	.0038	.0071	.0119	.0183
	6	27132				.0001	.0002	.0006	.0012	.0024	.0042
	7	50388						.0001	.0002	.0004	.0008
	8	75582								.0001	.0001
	9	92378									
	10	92378									
	11	75582									
	12	50388									
	13	27132									
	14	11628									
	15	3876									
	16	969									
	17	171									
	18	19									
	19	1									
20	0	1	.8179	.6676	.5438	.4420	.3585	.2901	.2342	.1887	.1516
	1	20	.1652	.2725	.3364	.3683	.3774	.3703	.3526	.3282	.3000
	2	190	.0159	.0528	.0988	.1458	.1887	.2246	.2521	.2711	.2818
	3	1140	.0010	.0065	.0183	.0364	.0596	.0860	.1139	.1414	.1672
	4	4845		.0006	.0024	.0065	.0133	.0233	.0364	.0523	.0703
	5	15504			.0002	.0009	.0022	.0048	.0088	.0145	.0222
	6	38760				.0001	.0003	.0008	.0017	.0032	.0055
	7	77520						.0001	.0002	.0005	.0011
	8	125970							.0001	.0002	
	9	167960									
	10	184756									
	11	167960									
	12	125970									
	13	77520									
	14	38760									
	15	15504									
	16	4845									
	17	1140									
	18	190									
	19	20									
	20	1									

TABLE 3 (Continued)

		Probabilities $P[X]$								
n	X	0.1	0.15	0.2	0.25	0.3	0.35	0.4	0.45	0.5
19	0	.1351	.0456	.0144	.0042	.0011	.0003	.0001	.0000	.0000
	1	.2852	.1529	.0685	.0268	.0093	.0029	.0008	.0002	.0000
	2	.2852	.2428	.1540	.0803	.0358	.0138	.0046	.0013	.0003
	3	.1796	.2428	.2182	.1517	.0869	.0422	.0175	.0062	.0018
	4	.0798	.1714	.2182	.2023	.1491	.0909	.0467	.0203	.0074
	5	.0266	.0907	.1636	.2023	.1916	.1468	.0933	.0497	.0222
	6	.0069	.0374	.0955	.1574	.1916	.1844	.1451	.0949	.0518
	7	.0014	.0122	.0443	.0974	.1525	.1844	.1797	.1443	.0961
	8	.0002	.0032	.0166	.0487	.0981	.1489	.1797	.1771	.1442
	9		.0007	.0051	.0198	.0514	.0980	.1464	.1771	.1762
	10		.0001	.0013	.0066	.0220	.0528	.0976	.1449	.1762
	11			.0003	.0018	.0077	.0233	.0532	.0970	.1442
	12				.0004	.0022	.0083	.0237	.0529	.0961
	13				.0001	.0005	.0024	.0085	.0233	.0518
	14					.0001	.0006	.0024	.0082	.0222
	15						.0001	.0005	.0022	.0074
	16							.0001	.0005	.0018
	17								.0001	.0003
	18									
	19									
20	0	.1216	.0388	.0115	.0032	.0008	.0002	.0000	.0000	.0000
	1	.2702	.1368	.0576	.0211	.0068	.0020	.0005	.0001	.0000
	2	.2852	.2293	.1369	.0669	.0278	.0100	.0031	.0008	.0002
	3	.1901	.2428	.2054	.1339	.0716	.0323	.0123	.0040	.0011
	4	.0898	.1821	.2182	.1897	.1304	.0738	.0350	.0139	.0046
	5	.0319	.1028	.1746	.2023	.1789	.1272	.0746	.0365	.0148
	6	.0089	.0454	.1091	.1686	.1916	.1712	.1244	.0746	.0370
	7	.0020	.0160	.0545	.1124	.1643	.1844	.1659	.1221	.0739
	8	.0004	.0046	.0222	.0609	.1144	.1614	.1797	.1623	.1201
	9	.0001	.0011	.0074	.0271	.0654	.1158	.1597	.1771	.1602
	10		.0002	.0020	.0099	.0308	.0686	.1171	.1593	.1762
	11			.0005	.0030	.0120	.0336	.0710	.1185	.1602
	12			.0001	.0008	.0039	.0136	.0355	.0727	.1201
	13				.0002	.0010	.0045	.0146	.0366	.0739
	14					.0002	.0012	.0049	.0150	.0370
	15						.0003	.0013	.0049	.0148
	16							.0003	.0013	.0046
	17								.0002	.0011
	18									.0002
	19									
	20									

TABLE 4   Probabilities for the Student's t-Distribution.
Tabled values are $t_{critical}$ values for **df** $= n - 1$, that correspond to the right-tail probabilities (p-values) listed in boldface at the top of each column; $P[t \geq t_{critical}] = p$-value. Only positive t-values are listed because the t-distribution is symmetric, so $P[t \leq -t_{critical}] = [t \geq +t_{critical}]$.

Example: with df $= 30$, $P[t \geq 1.055] = P[t \leq -1.055] = 0.15$.

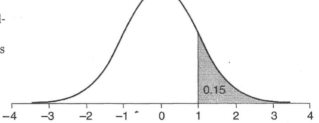

					Right—tail Probability (p-value)							
df	0.25	0.20	0.15	0.10	0.05	0.025	0.02	0.01	0.005	0.0025	0.001	0.0005
1	1.000	1.376	1.963	3.078	6.314	12.71	15.89	31.82	63.66	127.3	318.3	636.6
2	0.816	1.061	1.386	1.886	2.920	4.303	4.849	6.965	9.925	14.09	22.33	31.60
3	0.765	0.978	1.250	1.638	2.353	3.182	3.482	4.541	5.841	7.453	10.21	12.92
4	0.741	0.941	1.190	1.533	2.132	2.776	2.999	3.747	4.604	5.598	7.173	8.610
5	0.727	0.920	1.156	1.476	2.015	2.571	2.757	3.365	4.032	4.773	5.894	6.869
6	0.718	0.906	1.134	1.440	1.943	2.447	2.612	3.143	3.707	4.317	5.208	5.959
7	0.711	0.896	1.119	1.415	1.895	2.365	2.517	2.998	3.499	4.029	4.785	5.408
8	0.706	0.889	1.108	1.397	1.860	2.306	2.449	2.896	3.355	3.833	4.501	5.041
9	0.703	0.883	1.100	1.383	1.833	2.262	2.398	2.821	3.250	3.690	4.297	4.781
10	0.700	0.879	1.093	1.372	1.812	2.228	2.359	2.764	3.169	3.581	4.144	4.587
11	0.697	0.876	1.088	1.363	1.796	2.201	2.328	2.718	3.106	3.497	4.025	4.437
12	0.695	0.873	1.083	1.356	1.782	2.179	2.303	2.681	3.055	3.428	3.930	4.318
13	0.694	0.870	1.079	1.350	1.771	2.160	2.282	2.650	3.012	3.372	3.852	4.221
14	0.692	0.868	1.076	1.345	1.761	2.145	2.264	2.624	2.977	3.326	3.787	4.140
15	0.691	0.866	1.074	1.341	1.753	2.131	2.249	2.602	2.947	3.286	3.733	4.073
16	0.690	0.865	1.071	1.337	1.746	2.120	2.235	2.583	2.921	3.252	3.686	4.015
17	0.689	0.863	1.069	1.333	1.740	2.110	2.224	2.567	2.898	3.222	3.646	3.965
18	0.688	0.862	1.067	1.330	1.734	2.101	2.214	2.552	2.878	3.197	3.610	3.922
19	0.688	0.861	1.066	1.328	1.729	2.093	2.205	2.539	2.861	3.174	3.579	3.883
20	0.687	0.860	1.064	1.325	1.725	2.086	2.197	2.528	2.845	3.153	3.552	3.850
21	0.686	0.859	1.063	1.323	1.721	2.080	2.189	2.518	2.831	3.135	3.527	3.819
22	0.686	0.858	1.061	1.321	1.717	2.074	2.183	2.508	2.819	3.119	3.505	3.792
23	0.685	0.858	1.060	1.319	1.714	2.069	2.177	2.500	2.807	3.104	3.485	3.768
24	0.685	0.857	1.059	1.318	1.711	2.064	2.172	2.492	2.797	3.091	3.467	3.745
25	0.684	0.856	1.058	1.316	1.708	2.060	2.167	2.485	2.787	3.078	3.450	3.725
26	0.684	0.856	1.058	1.315	1.706	2.056	2.162	2.479	2.779	3.067	3.435	3.707
27	0.684	0.855	1.057	1.314	1.703	2.052	2.158	2.473	2.771	3.057	3.421	3.689
28	0.683	0.855	1.056	1.313	1.701	2.048	2.154	2.467	2.763	3.047	3.408	3.674
29	0.683	0.854	1.055	1.311	1.699	2.045	2.150	2.462	2.756	3.038	3.396	3.660
30	0.683	0.854	1.055	1.310	1.697	2.042	2.147	2.457	2.750	3.030	3.385	3.646
40	0.681	0.851	1.050	1.303	1.684	2.021	2.123	2.423	2.704	2.971	3.307	3.551
50	0.679	0.849	1.047	1.299	1.676	2.009	2.109	2.403	2.678	2.937	3.261	3.496
60	0.679	0.848	1.045	1.296	1.671	2.000	2.099	2.390	2.660	2.915	3.232	3.460
80	0.678	0.846	1.043	1.292	1.664	1.990	2.088	2.374	2.639	2.887	3.195	3.416
100	0.677	0.845	1.042	1.290	1.660	1.984	2.081	2.364	2.626	2.871	3.174	3.390

TABLE 5   Critical Values for the F-Statistic for $\alpha = 0.05$ (one-tailed test).[1]

					DFN					
DFD	**1**	**2**	**3**	**4**	**5**	**6**	**7**	**8**	**9**	**10**
1	161.45	199.50	215.71	224.58	230.16	233.99	236.77	238.88	240.54	241.88
2	18.51	19.00	19.16	19.25	19.30	19.33	19.35	19.37	19.38	19.40
3	10.13	9.55	9.28	9.12	9.01	8.94	8.89	8.85	8.81	8.79
4	7.71	6.94	6.59	6.39	6.26	6.16	6.09	6.04	6.00	5.96
5	6.61	5.79	5.41	5.19	5.05	4.95	4.88	4.82	4.77	4.74
6	5.99	5.14	4.76	4.53	4.39	4.28	4.21	4.15	4.10	4.06
7	5.59	4.74	4.35	4.12	3.97	3.87	3.79	3.73	3.68	3.64
8	5.32	4.46	4.07	3.84	3.69	3.58	3.50	3.44	3.39	3.35
9	5.12	4.26	3.86	3.63	3.48	3.37	3.29	3.23	3.18	3.14
10	4.96	4.10	3.71	3.48	3.33	3.22	3.14	3.07	3.02	2.98
11	4.84	3.98	3.59	3.36	3.20	3.09	3.01	2.95	2.90	2.85
12	4.75	3.89	3.49	3.26	3.11	3.00	2.91	2.85	2.80	2.75
13	4.67	3.81	3.41	3.18	3.03	2.92	2.83	2.77	2.71	2.67
14	4.60	3.74	3.34	3.11	2.96	2.85	2.76	2.70	2.65	2.60
15	4.54	3.68	3.29	3.06	2.90	2.79	2.71	2.64	2.59	2.54
16	4.49	3.63	3.24	3.01	2.85	2.74	2.66	2.59	2.54	2.49
17	4.45	3.59	3.20	2.96	2.81	2.70	2.61	2.55	2.49	2.45
18	4.41	3.55	3.16	2.93	2.77	2.66	2.58	2.51	2.46	2.41
19	4.38	3.52	3.13	2.90	2.74	2.63	2.54	2.48	2.42	2.38
20	4.35	3.49	3.10	2.87	2.71	2.60	2.51	2.45	2.39	2.35
21	4.32	3.47	3.07	2.84	2.68	2.57	2.49	2.42	2.37	2.32
22	4.30	3.44	3.05	2.82	2.66	2.55	2.46	2.40	2.34	2.30
23	4.28	3.42	3.03	2.80	2.64	2.53	2.44	2.37	2.32	2.27
24	4.26	3.40	3.01	2.78	2.62	2.51	2.42	2.36	2.30	2.25
25	4.24	3.39	2.99	2.76	2.60	2.49	2.40	2.34	2.28	2.24
26	4.23	3.37	2.98	2.74	2.59	2.47	2.39	2.32	2.27	2.22
27	4.21	3.35	2.96	2.73	2.57	2.46	2.37	2.31	2.25	2.20
28	4.20	3.34	2.95	2.71	2.56	2.45	2.36	2.29	2.24	2.19
29	4.18	3.33	2.93	2.70	2.55	2.43	2.35	2.28	2.22	2.18
30	4.17	3.32	2.92	2.69	2.53	2.42	2.33	2.27	2.21	2.16
40	4.08	3.23	2.84	2.61	2.45	2.34	2.25	2.18	2.12	2.08
50	4.03	3.18	2.79	2.56	2.40	2.29	2.20	2.13	2.07	2.03
60	4.00	3.15	2.76	2.53	2.37	2.25	2.17	2.10	2.04	1.99
70	3.98	3.13	2.74	2.50	2.35	2.23	2.14	2.07	2.02	1.97
80	3.96	3.11	2.72	2.49	2.33	2.21	2.13	2.06	2.00	1.95
90	3.95	3.10	2.71	2.47	2.32	2.20	2.11	2.04	1.99	1.94
100	3.94	3.09	2.70	2.46	2.31	2.19	2.10	2.03	1.97	1.93
200	3.89	3.04	2.65	2.42	2.26	2.14	2.06	1.98	1.93	1.88

[1] DFD = degrees of freedom for the denominator variance (mean square) and DFN = degrees of freedom for the numerator variance (mean square). Tabled values are $F_{critical}$ with a right-tail area under the probability distribution = **0.05**. If the computed F_{test} value is greater than or equal to the tabled $F_{critical}$ value, the numerator variance is significantly different than the denominator variance, based on a **one-tailed test** with $\alpha = 0.05$.

TABLE 5 (Continued) α = 0.05 (one-tailed test).

					DFN					
DFD	11	12	13	14	15	16	17	18	19	20
1	242.98	243.91	244.69	245.36	245.95	246.46	246.92	247.32	247.69	248.01
2	19.40	19.41	19.42	19.42	19.43	19.43	19.44	19.44	19.44	19.45
3	8.76	8.74	8.73	8.71	8.70	8.69	8.68	8.67	8.67	8.66
4	5.94	5.91	5.89	5.87	5.86	5.84	5.83	5.82	5.81	5.80
5	4.70	4.68	4.66	4.64	4.62	4.60	4.59	4.58	4.57	4.56
6	4.03	4.00	3.98	3.96	3.94	3.92	3.91	3.90	3.88	3.87
7	3.60	3.57	3.55	3.53	3.51	3.49	3.48	3.47	3.46	3.44
8	3.31	3.28	3.26	3.24	3.22	3.20	3.19	3.17	3.16	3.15
9	3.10	3.07	3.05	3.03	3.01	2.99	2.97	2.96	2.95	2.94
10	2.94	2.91	2.89	2.86	2.85	2.83	2.81	2.80	2.79	2.77
11	2.82	2.79	2.76	2.74	2.72	2.70	2.69	2.67	2.66	2.65
12	2.72	2.69	2.66	2.64	2.62	2.60	2.58	2.57	2.56	2.54
13	2.63	2.60	2.58	2.55	2.53	2.51	2.50	2.48	2.47	2.46
14	2.57	2.53	2.51	2.48	2.46	2.44	2.43	2.41	2.40	2.39
15	2.51	2.48	2.45	2.42	2.40	2.38	2.37	2.35	2.34	2.33
16	2.46	2.42	2.40	2.37	2.35	2.33	2.32	2.30	2.29	2.28
17	2.41	2.38	2.35	2.33	2.31	2.29	2.27	2.26	2.24	2.23
18	2.37	2.34	2.31	2.29	2.27	2.25	2.23	2.22	2.20	2.19
19	2.34	2.31	2.28	2.26	2.23	2.21	2.20	2.18	2.17	2.16
20	2.31	2.28	2.25	2.22	2.20	2.18	2.17	2.15	2.14	2.12
21	2.28	2.25	2.22	2.20	2.18	2.16	2.14	2.12	2.11	2.10
22	2.26	2.23	2.20	2.17	2.15	2.13	2.11	2.10	2.08	2.07
23	2.24	2.20	2.18	2.15	2.13	2.11	2.09	2.08	2.06	2.05
24	2.22	2.18	2.15	2.13	2.11	2.09	2.07	2.05	2.04	2.03
25	2.20	2.16	2.14	2.11	2.09	2.07	2.05	2.04	2.02	2.01
26	2.18	2.15	2.12	2.09	2.07	2.05	2.03	2.02	2.00	1.99
27	2.17	2.13	2.10	2.08	2.06	2.04	2.02	2.00	1.99	1.97
28	2.15	2.12	2.09	2.06	2.04	2.02	2.00	1.99	1.97	1.96
29	2.14	2.10	2.08	2.05	2.03	2.01	1.99	1.97	1.96	1.94
30	2.13	2.09	2.06	2.04	2.01	1.99	1.98	1.96	1.95	1.93
40	2.04	2.00	1.97	1.95	1.92	1.90	1.89	1.87	1.85	1.84
50	1.99	1.95	1.92	1.89	1.87	1.85	1.83	1.81	1.80	1.78
60	1.95	1.92	1.89	1.86	1.84	1.82	1.80	1.78	1.76	1.75
70	1.93	1.89	1.86	1.84	1.81	1.79	1.77	1.75	1.74	1.72
80	1.91	1.88	1.84	1.82	1.79	1.77	1.75	1.73	1.72	1.70
90	1.90	1.86	1.83	1.80	1.78	1.76	1.74	1.72	1.70	1.69
100	1.89	1.85	1.82	1.79	1.77	1.75	1.73	1.71	1.69	1.68
200	1.84	1.80	1.77	1.74	1.72	1.69	1.67	1.66	1.64	1.62

TABLE 5   (Continued) $\alpha = 0.05$ (one-tailed test).

DFD	30	40	50	60	70	80	90	100	200
					DFN				
1	250.10	251.14	251.77	252.20	252.50	252.72	252.90	253.04	253.68
2	19.46	19.47	19.48	19.48	19.48	19.48	19.48	19.49	19.49
3	8.62	8.59	8.58	8.57	8.57	8.56	8.56	8.55	8.54
4	5.75	5.72	5.70	5.69	5.68	5.67	5.67	5.66	5.65
5	4.50	4.46	4.44	4.43	4.42	4.41	4.41	4.41	4.39
6	3.81	3.77	3.75	3.74	3.73	3.72	3.72	3.71	3.69
7	3.38	3.34	3.32	3.30	3.29	3.29	3.28	3.27	3.25
8	3.08	3.04	3.02	3.01	2.99	2.99	2.98	2.97	2.95
9	2.86	2.83	2.80	2.79	2.78	2.77	2.76	2.76	2.73
10	2.70	2.66	2.64	2.62	2.61	2.60	2.59	2.59	2.56
11	2.57	2.53	2.51	2.49	2.48	2.47	2.46	2.46	2.43
12	2.47	2.43	2.40	2.38	2.37	2.36	2.36	2.35	2.32
13	2.38	2.34	2.31	2.30	2.28	2.27	2.27	2.26	2.23
14	2.31	2.27	2.24	2.22	2.21	2.20	2.19	2.19	2.16
15	2.25	2.20	2.18	2.16	2.15	2.14	2.13	2.12	2.10
16	2.19	2.15	2.12	2.11	2.09	2.08	2.07	2.07	2.04
17	2.15	2.10	2.08	2.06	2.05	2.03	2.03	2.02	1.99
18	2.11	2.06	2.04	2.02	2.00	1.99	1.98	1.98	1.95
19	2.07	2.03	2.00	1.98	1.97	1.96	1.95	1.94	1.91
20	2.04	1.99	1.97	1.95	1.93	1.92	1.91	1.91	1.88
21	2.01	1.96	1.94	1.92	1.90	1.89	1.88	1.88	1.84
22	1.98	1.94	1.91	1.89	1.88	1.86	1.86	1.85	1.82
23	1.96	1.91	1.88	1.86	1.85	1.84	1.83	1.82	1.79
24	1.94	1.89	1.86	1.84	1.83	1.82	1.81	1.80	1.77
25	1.92	1.87	1.84	1.82	1.81	1.80	1.79	1.78	1.75
26	1.90	1.85	1.82	1.80	1.79	1.78	1.77	1.76	1.73
27	1.88	1.84	1.81	1.79	1.77	1.76	1.75	1.74	1.71
28	1.87	1.82	1.79	1.77	1.75	1.74	1.73	1.73	1.69
29	1.85	1.81	1.77	1.75	1.74	1.73	1.72	1.71	1.67
30	1.84	1.79	1.76	1.74	1.72	1.71	1.70	1.70	1.66
40	1.74	1.69	1.66	1.64	1.62	1.61	1.60	1.59	1.55
50	1.69	1.63	1.60	1.58	1.56	1.54	1.53	1.52	1.48
60	1.65	1.59	1.56	1.53	1.52	1.50	1.49	1.48	1.44
70	1.62	1.57	1.53	1.50	1.49	1.47	1.46	1.45	1.40
80	1.60	1.54	1.51	1.48	1.46	1.45	1.44	1.43	1.38
90	1.59	1.53	1.49	1.46	1.44	1.43	1.42	1.41	1.36
100	1.57	1.52	1.48	1.45	1.43	1.41	1.40	1.39	1.34
200	1.52	1.46	1.41	1.39	1.36	1.35	1.33	1.32	1.26

TABLE 5   Critical Values for the *F*-Statistic for α = 0.05 (two-tailed test).[1]

DFD	DFN									
	1	2	3	4	5	6	7	8	9	10
1	647.79	799.50	864.16	899.58	921.85	937.11	948.22	956.66	963.28	968.63
2	38.51	39.00	39.17	39.25	39.30	39.33	39.36	39.37	39.39	39.40
3	17.44	16.04	15.44	15.10	14.88	14.73	14.62	14.54	14.47	14.42
4	12.22	10.65	9.98	9.60	9.36	9.20	9.07	8.98	8.90	8.84
5	10.01	8.43	7.76	7.39	7.15	6.98	6.85	6.76	6.68	6.62
6	8.81	7.26	6.60	6.23	5.99	5.82	5.70	5.60	5.52	5.46
7	8.07	6.54	5.89	5.52	5.29	5.12	4.99	4.90	4.82	4.76
8	7.57	6.06	5.42	5.05	4.82	4.65	4.53	4.43	4.36	4.30
9	7.21	5.71	5.08	4.72	4.48	4.32	4.20	4.10	4.03	3.96
10	6.94	5.46	4.83	4.47	4.24	4.07	3.95	3.85	3.78	3.72
11	6.72	5.26	4.63	4.28	4.04	3.88	3.76	3.66	3.59	3.53
12	6.55	5.10	4.47	4.12	3.89	3.73	3.61	3.51	3.44	3.37
13	6.41	4.97	4.35	4.00	3.77	3.60	3.48	3.39	3.31	3.25
14	6.30	4.86	4.24	3.89	3.66	3.50	3.38	3.29	3.21	3.15
15	6.20	4.77	4.15	3.80	3.58	3.41	3.29	3.20	3.12	3.06
16	6.12	4.69	4.08	3.73	3.50	3.34	3.22	3.12	3.05	2.99
17	6.04	4.62	4.01	3.66	3.44	3.28	3.16	3.06	2.98	2.92
18	5.98	4.56	3.95	3.61	3.38	3.22	3.10	3.01	2.93	2.87
19	5.92	4.51	3.90	3.56	3.33	3.17	3.05	2.96	2.88	2.82
20	5.87	4.46	3.86	3.51	3.29	3.13	3.01	2.91	2.84	2.77
21	5.83	4.42	3.82	3.48	3.25	3.09	2.97	2.87	2.80	2.73
22	5.79	4.38	3.78	3.44	3.22	3.05	2.93	2.84	2.76	2.70
23	5.75	4.35	3.75	3.41	3.18	3.02	2.90	2.81	2.73	2.67
24	5.72	4.32	3.72	3.38	3.15	2.99	2.87	2.78	2.70	2.64
25	5.69	4.29	3.69	3.35	3.13	2.97	2.85	2.75	2.68	2.61
26	5.66	4.27	3.67	3.33	3.10	2.94	2.82	2.73	2.65	2.59
27	5.63	4.24	3.65	3.31	3.08	2.92	2.80	2.71	2.63	2.57
28	5.61	4.22	3.63	3.29	3.06	2.90	2.78	2.69	2.61	2.55
29	5.59	4.20	3.61	3.27	3.04	2.88	2.76	2.67	2.59	2.53
30	5.57	4.18	3.59	3.25	3.03	2.87	2.75	2.65	2.57	2.51
35	5.48	4.11	3.52	3.18	2.96	2.80	2.68	2.58	2.50	2.44
40	5.42	4.05	3.46	3.13	2.90	2.74	2.62	2.53	2.45	2.39
45	5.38	4.01	3.42	3.09	2.86	2.70	2.58	2.49	2.41	2.35
50	5.34	3.97	3.39	3.05	2.83	2.67	2.55	2.46	2.38	2.32
60	5.29	3.93	3.34	3.01	2.79	2.63	2.51	2.41	2.33	2.27
70	5.25	3.89	3.31	2.97	2.75	2.59	2.47	2.38	2.30	2.24
80	5.22	3.86	3.28	2.95	2.73	2.57	2.45	2.35	2.28	2.21
90	5.20	3.84	3.26	2.93	2.71	2.55	2.43	2.34	2.26	2.19
100	5.18	3.83	3.25	2.92	2.70	2.54	2.42	2.32	2.24	2.18
200	5.10	3.76	3.18	2.85	2.63	2.47	2.35	2.26	2.18	2.11

[1] DFD = degrees of freedom for the denominator variance (mean square) and DFN = degrees of freedom for the numerator variance (mean square). Tabled values are $F_{critical}$ with a right-tail area under the probability distribution = **0.025**. If the computed F_{test} value is greater than or equal to the tabled $F_{critical}$ value, the numerator variance is significantly different than the denominator variance, based on a **two-tailed test** with α = 0.05.

TABLE 5   (Continued) $\alpha = 0.05$ (two-tailed test).

DFD	11	12	13	14	15	16	17	18	19	20
1	973.03	976.71	979.84	982.53	984.87	986.92	988.73	990.35	991.80	993.10
2	39.41	39.41	39.42	39.43	39.43	39.44	39.44	39.44	39.45	39.45
3	14.37	14.34	14.30	14.28	14.25	14.23	14.21	14.20	14.18	14.17
4	8.79	8.75	8.71	8.68	8.66	8.63	8.61	8.59	8.58	8.56
5	6.57	6.52	6.49	6.46	6.43	6.40	6.38	6.36	6.34	6.33
6	5.41	5.37	5.33	5.30	5.27	5.24	5.22	5.20	5.18	5.17
7	4.71	4.67	4.63	4.60	4.57	4.54	4.52	4.50	4.48	4.47
8	4.24	4.20	4.16	4.13	4.10	4.08	4.05	4.03	4.02	4.00
9	3.91	3.87	3.83	3.80	3.77	3.74	3.72	3.70	3.68	3.67
10	3.66	3.62	3.58	3.55	3.52	3.50	3.47	3.45	3.44	3.42
11	3.47	3.43	3.39	3.36	3.33	3.30	3.28	3.26	3.24	3.23
12	3.32	3.28	3.24	3.21	3.18	3.15	3.13	3.11	3.09	3.07
13	3.20	3.15	3.12	3.08	3.05	3.03	3.00	2.98	2.96	2.95
14	3.09	3.05	3.01	2.98	2.95	2.92	2.90	2.88	2.86	2.84
15	3.01	2.96	2.92	2.89	2.86	2.84	2.81	2.79	2.77	2.76
16	2.93	2.89	2.85	2.82	2.79	2.76	2.74	2.72	2.70	2.68
17	2.87	2.82	2.79	2.75	2.72	2.70	2.67	2.65	2.63	2.62
18	2.81	2.77	2.73	2.70	2.67	2.64	2.62	2.60	2.58	2.56
19	2.76	2.72	2.68	2.65	2.62	2.59	2.57	2.55	2.53	2.51
20	2.72	2.68	2.64	2.60	2.57	2.55	2.52	2.50	2.48	2.46
21	2.68	2.64	2.60	2.56	2.53	2.51	2.48	2.46	2.44	2.42
22	2.65	2.60	2.56	2.53	2.50	2.47	2.45	2.43	2.41	2.39
23	2.62	2.57	2.53	2.50	2.47	2.44	2.42	2.39	2.37	2.36
24	2.59	2.54	2.50	2.47	2.44	2.41	2.39	2.36	2.35	2.33
25	2.56	2.51	2.48	2.44	2.41	2.38	2.36	2.34	2.32	2.30
26	2.54	2.49	2.45	2.42	2.39	2.36	2.34	2.31	2.29	2.28
27	2.51	2.47	2.43	2.39	2.36	2.34	2.31	2.29	2.27	2.25
28	2.49	2.45	2.41	2.37	2.34	2.32	2.29	2.27	2.25	2.23
29	2.48	2.43	2.39	2.36	2.32	2.30	2.27	2.25	2.23	2.21
30	2.46	2.41	2.37	2.34	2.31	2.28	2.26	2.23	2.21	2.20
35	2.39	2.34	2.30	2.27	2.23	2.21	2.18	2.16	2.14	2.12
40	2.33	2.29	2.25	2.21	2.18	2.15	2.13	2.11	2.09	2.07
45	2.29	2.25	2.21	2.17	2.14	2.11	2.09	2.07	2.04	2.03
50	2.26	2.22	2.18	2.14	2.11	2.08	2.06	2.03	2.01	1.99
60	2.22	2.17	2.13	2.09	2.06	2.03	2.01	1.98	1.96	1.94
70	2.18	2.14	2.10	2.06	2.03	2.00	1.97	1.95	1.93	1.91
80	2.16	2.11	2.07	2.03	2.00	1.97	1.95	1.92	1.90	1.88
90	2.14	2.09	2.05	2.02	1.98	1.95	1.93	1.91	1.88	1.86
100	2.12	2.08	2.04	2.00	1.97	1.94	1.91	1.89	1.87	1.85
200	2.06	2.01	1.97	1.93	1.90	1.87	1.84	1.82	1.80	1.78

TABLE 5   (Continued) $\alpha = 0.05$ (two-tailed test).

DFD	DFN 30	40	50	60	70	80	90	100	200
1	1001.41	1005.60	1008.12	1009.80	1011.00	1011.91	1012.61	1013.17	1015.71
2	39.46	39.47	39.48	39.48	39.48	39.49	39.49	39.49	39.49
3	14.08	14.04	14.01	13.99	13.98	13.97	13.96	13.96	13.93
4	8.46	8.41	8.38	8.36	8.35	8.33	8.33	8.32	8.29
5	6.23	6.18	6.14	6.12	6.11	6.10	6.09	6.08	6.05
6	5.07	5.01	4.98	4.96	4.94	4.93	4.92	4.92	4.88
7	4.36	4.31	4.28	4.25	4.24	4.23	4.22	4.21	4.18
8	3.89	3.84	3.81	3.78	3.77	3.76	3.75	3.74	3.70
9	3.56	3.51	3.47	3.45	3.43	3.42	3.41	3.40	3.37
10	3.31	3.26	3.22	3.20	3.18	3.17	3.16	3.15	3.12
11	3.12	3.06	3.03	3.00	2.99	2.97	2.96	2.96	2.92
12	2.96	2.91	2.87	2.85	2.83	2.82	2.81	2.80	2.76
13	2.84	2.78	2.74	2.72	2.70	2.69	2.68	2.67	2.63
14	2.73	2.67	2.64	2.61	2.60	2.58	2.57	2.56	2.53
15	2.64	2.59	2.55	2.52	2.51	2.49	2.48	2.47	2.44
16	2.57	2.51	2.47	2.45	2.43	2.42	2.40	2.40	2.36
17	2.50	2.44	2.41	2.38	2.36	2.35	2.34	2.33	2.29
18	2.44	2.38	2.35	2.32	2.30	2.29	2.28	2.27	2.23
19	2.39	2.33	2.30	2.27	2.25	2.24	2.23	2.22	2.18
20	2.35	2.29	2.25	2.22	2.20	2.19	2.18	2.17	2.13
21	2.31	2.25	2.21	2.18	2.16	2.15	2.14	2.13	2.09
22	2.27	2.21	2.17	2.14	2.13	2.11	2.10	2.09	2.05
23	2.24	2.18	2.14	2.11	2.09	2.08	2.07	2.06	2.01
24	2.21	2.15	2.11	2.08	2.06	2.05	2.03	2.02	1.98
25	2.18	2.12	2.08	2.05	2.03	2.02	2.01	2.00	1.95
26	2.16	2.09	2.05	2.03	2.01	1.99	1.98	1.97	1.92
27	2.13	2.07	2.03	2.00	1.98	1.97	1.95	1.94	1.90
28	2.11	2.05	2.01	1.98	1.96	1.94	1.93	1.92	1.88
29	2.09	2.03	1.99	1.96	1.94	1.92	1.91	1.90	1.86
30	2.07	2.01	1.97	1.94	1.92	1.90	1.89	1.88	1.84
35	2.00	1.93	1.89	1.86	1.84	1.82	1.81	1.80	1.75
40	1.94	1.88	1.83	1.80	1.78	1.76	1.75	1.74	1.69
45	1.90	1.83	1.79	1.76	1.74	1.72	1.70	1.69	1.64
50	1.87	1.80	1.75	1.72	1.70	1.68	1.67	1.66	1.60
60	1.82	1.74	1.70	1.67	1.64	1.63	1.61	1.60	1.54
70	1.78	1.71	1.66	1.63	1.60	1.59	1.57	1.56	1.50
80	1.75	1.68	1.63	1.60	1.57	1.55	1.54	1.53	1.47
90	1.73	1.66	1.61	1.58	1.55	1.53	1.52	1.50	1.44
100	1.71	1.64	1.59	1.56	1.53	1.51	1.50	1.48	1.42
200	1.64	1.56	1.51	1.47	1.45	1.42	1.41	1.39	1.32

TABLE 6   Critical Values for the Studentized Range Q-Statistic ($\alpha = 0.05$)

Direction: The Studentized range statistic is used in unplanned multiple comparisons tests, where $k =$ the number of groups being compared. Read down the column with the appropriate k-value at the top of the column to the row with the appropriate degrees of freedom (df). For Tukey's multiple comparisons tests for sample means (HSD and Tukey-Kramer test for unequal sample sizes), degrees of freedom equal df for the Mean Square Error (MSE) term in the analysis of variance table. If the df value falls between two tabled df values, round *down* to the smaller tabled df value. For multiple comparisons tests for three or more sample proportions, df $= \infty$.

df	k = 2	k = 3	k = 4	k = 5	k = 6	k = 7	k = 8	k = 9	k = 10
1	17.97	26.98	32.82	37.08	40.41	43.12	45.4	47.36	49.07
2	6.085	8.331	9.798	10.880	11.750	12.440	13.030	13.540	13.990
3	4.501	5.910	6.825	7.502	8.037	8.478	8.853	9.177	9.462
4	3.927	5.040	5.757	6.287	6.707	7.053	7.347	7.602	7.826
5	3.635	4.602	5.218	5.673	6.033	6.330	6.582	6.802	6.995
6	3.461	4.339	4.896	5.305	5.628	5.895	6.122	6.319	6.493
7	3.344	4.165	4.681	5.060	5.359	5.606	5.815	5.998	6.158
8	3.261	4.041	4.529	4.886	5.167	5.399	5.597	5.767	5.918
9	3.199	3.949	4.415	4.756	5.024	5.244	5.432	5.595	5.739
10	3.151	3.877	4.327	4.654	4.912	5.124	5.305	5.461	5.599
11	3.113	3.820	4.256	4.574	4.823	5.028	5.202	5.353	5.487
12	3.082	3.773	4.199	4.508	4.751	4.950	5.119	5.265	5.395
13	3.055	3.735	4.151	4.453	4.690	4.885	5.049	5.192	5.318
14	3.033	3.702	4.111	4.407	4.639	4.829	4.990	5.131	5.254
15	3.014	3.674	4.076	4.367	4.595	4.782	4.940	5.077	5.198
16	2.998	3.649	4.046	4.333	4.557	4.741	4.897	5.031	5.150
17	2.984	3.628	4.020	4.303	4.524	4.705	4.858	4.991	5.108
18	2.971	3.609	3.997	4.277	4.495	4.673	4.824	4.956	5.071
19	2.960	3.593	3.977	4.253	4.469	4.645	4.794	4.924	5.038
20	2.950	3.578	3.958	4.232	4.445	4.620	4.768	4.896	5.008
24	2.919	3.532	3.901	4.166	4.373	4.541	4.684	4.807	4.915
30	2.888	3.486	3.845	4.102	4.302	4.464	4.602	4.720	4.824
40	2.858	3.442	3.791	4.039	4.232	4.389	4.521	4.635	4.735
60	2.829	3.399	3.737	3.977	4.163	4.314	4.441	4.550	4.646
120	2.800	3.356	3.685	3.917	4.096	4.241	4.363	4.468	4.560
∞	2.772	3.314	3.633	3.858	4.030	4.170	4.286	4.387	4.474

TABLE 7 Critical Values for F_{max} Statistic to Test for Equal Variances.

Directions: Compute the sample variances of the sample or experimental groups and compute $F_{max} = S^2_{largest}/S^2_{smallest}$. Find the column that corresponds to k = the number of experimental or sample groups in the study. Read down that column to the row that corresponds to df = $n - 1$, where n is the smaller of the two sample sizes used to compute $S_{largest}$ and $S_{smallest}$. If your df value falls between two tabled df values, round the df *down* to the smaller tabled df value. If the F_{max} value you computed is larger than the value in the table, there is sufficient evidence ($p \leq 0.05$) to conclude that the population variances are *not* equal.

df	$k = 2$	$k = 3$	$k = 4$	$k = 5$	$k = 6$	$k = 7$	$k = 8$	$k = 9$	$k = 10$
2	39.0	87.5	142	202	266	333	403	475	550
3	15.4	27.8	39.2	50.7	62.0	72.9	83.5	93.9	104
4	9.60	15.5	20.6	25.2	29.5	33.6	37.5	41.1	44.6
5	7.15	10.8	13.7	16.3	18.7	20.8	22.9	24.7	26.5
6	5.82	8.38	10.4	12.1	13.7	15.0	16.3	17.5	18.6
7	4.99	6.94	8.44	9.70	10.8	11.8	12.7	13.5	14.3
8	4.43	6.00	7.18	8.12	9.03	9.78	10.5	11.1	11.7
9	4.03	5.34	6.31	7.11	7.80	8.41	8.95	9.45	9.91
10	3.72	4.85	5.67	6.34	6.92	7.42	7.87	8.28	8.66
12	3.28	4.16	4.79	5.30	5.72	6.09	6.42	6.72	7.00
15	2.86	3.54	4.01	4.37	4.68	4.95	5.19	5.40	5.59
20	2.46	2.95	3.29	3.54	3.76	3.94	4.10	4.24	4.37
30	2.07	2.40	2.61	2.78	2.91	3.02	3.12	3.21	3.29
60	1.67	1.85	1.96	2.04	2.11	2.17	2.22	2.26	2.30

TABLE 8 Table of Probabilities for the χ^2 Distribution.

Tabled values are $\chi^2_{critical}$ values for specific degrees of freedom df that correspond to the right-tail probabilities listed in boldface at the top of each column; $P[\chi^2 \geq \chi^2_{critical}] = p\text{-value}$.

	Right-tail Probability (p-value)											
df	0.25	0.20	0.15	0.10	0.05	0.025	0.02	0.01	0.005	0.0025	0.001	0.0005
1	1.32	1.64	2.07	2.71	3.84	5.02	5.41	6.63	7.88	9.14	10.83	12.12
2	2.77	3.22	3.79	4.61	5.99	7.38	7.82	9.21	10.60	11.98	13.82	15.20
3	4.11	4.64	5.32	6.25	7.81	9.35	9.84	11.34	12.84	14.32	16.27	17.73
4	5.39	5.99	6.74	7.78	9.49	11.14	11.67	13.28	14.86	16.42	18.47	20.00
5	6.63	7.29	8.12	9.24	11.07	12.83	13.39	15.09	16.75	18.39	20.51	22.11
6	7.84	8.56	9.45	10.64	12.59	14.45	15.03	16.81	18.55	20.25	22.46	24.10
7	9.04	9.80	10.75	12.02	14.07	16.01	16.62	18.48	20.28	22.04	24.32	26.02
8	10.22	11.03	12.03	13.36	15.51	17.53	18.17	20.09	21.95	23.77	26.12	27.87
9	11.39	12.24	13.29	14.68	16.92	19.02	19.68	21.67	23.59	25.46	27.88	29.67
10	12.55	13.44	14.53	15.99	18.31	20.48	21.16	23.21	25.19	27.11	29.59	31.42
11	13.70	14.63	15.77	17.28	19.68	21.92	22.62	24.73	26.76	28.73	31.26	33.14
12	14.85	15.81	16.99	18.55	21.03	23.34	24.05	26.22	28.30	30.32	32.91	34.82
13	15.98	16.98	18.20	19.81	22.36	24.74	25.47	27.69	29.82	31.88	34.53	36.48
14	17.12	18.15	19.41	21.06	23.68	26.12	26.87	29.14	31.32	33.43	36.12	38.11
15	18.25	19.31	20.60	22.31	25.00	27.49	28.26	30.58	32.80	34.95	37.70	39.72
16	19.37	20.47	21.79	23.54	26.30	28.85	29.63	32.00	34.27	36.46	39.25	41.31
17	20.49	21.61	22.98	24.77	27.59	30.19	31.00	33.41	35.72	37.95	40.79	42.88
18	21.60	22.76	24.16	25.99	28.87	31.53	32.35	34.81	37.16	39.42	42.31	44.43
19	22.72	23.90	25.33	27.20	30.14	32.85	33.69	36.19	38.58	40.88	43.82	45.97
20	23.83	25.04	26.50	28.41	31.41	34.17	35.02	37.57	40.00	42.34	45.31	47.50
21	24.93	26.17	27.66	29.62	32.67	35.48	36.34	38.93	41.40	43.77	46.80	49.01
22	26.04	27.30	28.82	30.81	33.92	36.78	37.66	40.29	42.80	45.20	48.27	50.51
23	27.14	28.43	29.98	32.01	35.17	38.08	38.97	41.64	44.18	46.62	49.73	52.00
24	28.24	29.55	31.13	33.20	36.42	39.36	40.27	42.98	45.56	48.03	51.18	53.48
25	29.34	30.68	32.28	34.38	37.65	40.65	41.57	44.31	46.93	49.44	52.62	54.95
26	30.43	31.79	33.43	35.56	38.89	41.92	42.86	45.64	48.29	50.83	54.05	56.41
27	31.53	32.91	34.57	36.74	40.11	43.19	44.14	46.96	49.65	52.22	55.48	57.86
28	32.62	34.03	35.71	37.92	41.34	44.46	45.42	48.28	50.99	53.59	56.89	59.30
29	33.71	35.14	36.85	39.09	42.56	45.72	46.69	49.59	52.34	54.97	58.30	60.73
30	34.80	36.25	37.99	40.26	43.77	46.98	47.96	50.89	53.67	56.33	59.70	62.16
40	45.62	47.27	49.24	51.81	55.76	59.34	60.44	63.69	66.77	69.70	73.40	76.10
50	56.33	58.16	60.35	63.17	67.50	71.42	72.61	76.15	79.49	82.66	86.66	89.56
60	66.98	68.97	71.34	74.40	79.08	83.30	84.58	88.38	91.95	95.34	99.61	102.7
70	77.58	79.71	82.26	85.53	90.53	95.02	96.39	100.4	104.2	107.8	112.3	115.6
80	88.13	90.41	93.11	96.58	101.9	106.6	108.1	112.3	116.3	120.1	124.8	128.3
90	98.65	101.1	103.9	107.6	113.2	118.1	119.7	124.1	128.3	132.3	137.2	140.8
100	109.1	111.7	114.7	118.50	124.3	129.6	131.1	135.8	140.2	144.3	149.5	153.2

Index